DATE DUE

DEMCO 38-296

THE INTERNATIONAL SERIES OF MONOGRAPHS ON CHEMISTRY

THE INTERNATIONAL SERIES OF
MONOGRAPHS ON CHEMISTRY

Physical Adsorption: Forces and Phenomena

L. W. Bruch

Professor of Physics
University of Wisconsin

Milton W. Cole

Professor of Physics
Pennsylvania State University

Eugene Zaremba

Professor of Physics
Queen's University, Kingston, Ontario

CLARENDON PRESS · OXFORD

1997

Clarendon Street, Oxford OX2 6DP
New York
Bogota Bombay Buenos Aires
ılaam Delhi Florence Hong Kong
pur Madras Madrid Melbourne
Mexico City Nairobi Paris Singapore Taipei Tokyo Toronto
and associated companies in
Berlin Ibadan

Oxford is a trade mark of Oxford University Press

Published in the United States
by Oxford University Press Inc., New York

© L. W. Bruch, Milton W. Cole, and Eugene Zaremba, 1997

A catalogue record for this book is available from the British Library

Library of Congress Cataloging in Publication Data
(Data available)
ISBN 0 19 855638 1

Typeset by the authors using LaTeX
Printed in Great Britain by
Biddles Ltd, Guildford and King's Lynn

PREFACE

About twenty years ago, as a rapid expansion of surface physics was under-way, Greg Dash and Bill Steele published a pair of influential monographs on physical adsorption defining the language and identifying a family of issues that are common to most work in the field. In the intervening years, many work-ers have contributed to the growth of understanding of physical adsorption. At present, we appear to be in a period of consolidation and it seemed appropriate to us to attempt a monograph that would incorporate the progress that has been made. This book is the result of our efforts.

We focus our attention on the phenomena that occur within the first atomic layer adsorbed on clean surfaces, that is, we are emphasizing monolayer physics. We develop the relation between monolayer physics and 2D physics, but do not go very far in describing the role of the monolayer as a template for multilayer growth. Also, we emphasize atomic adsorption over molecular adsorption; quan-titative understanding of the latter is only now developing.

Even with these restrictions, we have touched upon a great range of concepts, theoretical methods, and experiments, as evidenced by the bibliography included in the book. It has been rewarding and enjoyable to revisit work of almost twenty years ago and to see how well the overall picture is now fitting together. In prodding our colleagues for information and advice, we have encountered reactions of the type most succinctly stated by one who responded to us that '...[it] brought back happy memories of some of the excitement I felt when actively working on these topics.'

We wish to thank, specifically, M. Chan, R. Diehl, S. C. Fain, C. Girardet, R. B. Griffiths, F. Y. Hansen, A. D. Novaco, W. A. Steele, H. Taub, and M. B. Webb for their careful reviews of portions of the text. Also, we thank our students and collaborators who did much to develop our knowledge of the field. We are indebted to the National Science Foundation, the National Research Council of Canada, and the Petroleum Research Fund of the American Chemical Society for support of our research. We have made many impositions on our librarians and thank them for their help.*

Our goals in the literature citations have been to recognize the landmark papers and to give leading references with which developments from the present (1995) may be traced. Readers may recognize imbalances in treatments, as we have sometimes succumbed to the temptation to write about what we know. There is another reason worth noting. We have tried to present the general

*The final writing occurred during a sabbatical leave of one of us (L.W.B.) and it is a pleasure to acknowledge the cooperation of librarians at the Max Planck Institut für Strömungforschung, Göttingen, and the Danish Technical University, Copenhagen, in retriev-ing bibliographic materials.

understanding of monolayer physics that has been achieved, but there remain puzzles that do not fit easily into the picture and that show the limits of our knowledge. Examples include the interactions responsible for adsorption on transition metals, e.g., Xe/Pt(111), molecular adsorption, and adsorption on ionic crystals. We expect the puzzles to be resolved without drastic changes in the concepts used to describe physical adsorption, but a case can be made for each puzzle that it concerns the uncertainties associated with the rather softly determined limits of our understanding. Thus, discussion of the most recent activities becomes extended at places.

Finally, we recognize that we may have made erroneous statements or evaluations in the text. We accept the responsibility and invite correction.

Madison, WI   L. W. B.
University Park, PA   M. W. C.
Kingston, Ontario   E. Z.
October, 1996

CONTENTS

To Nancy, Pamela, and Vanda

1

MONOLAYER PHYSICS

'In the structure of matter there can be no fundamental distinction between chemical and physical forces: it has been customary to call a force <u>chemical</u>, when it is more familiar to chemists, and to call the same force <u>physical</u>, when the physicist discovers an explanation of it.' I. Langmuir, *Phenomena, Atoms, and Molecules* (Philosophical Library, NY, 1950) p.60

' a [chemical] bond does not exist at all: it is a most convenient fiction...both to theoretical and experimental chemists.' C. Coulson, quoted on p.261 of *From Chemical Philosophy to Theoretical Chemistry*, Mary Jo Nye (U. California, Berkeley, 1994)

In setting out to write a volume on physical adsorption and monolayer physics we have had to decide what we mean and understand by these terms and thereby to set some limits on our coverage. Physical adsorption is not an independent basic subject nor is it one with sharp demarcations from the regime of chemical adsorption. The theoretical development has been part of the implementation of quantitative statistical mechanics of dense systems and of methods of the theory of phase transitions. The experimental development has been strongly influenced by advances in capabilities for high vacuum work and by progress with scattering probes of atomically thin layers. There has been for twenty years a strong, stimulating, and fruitful interplay of experiment and theory. We have attempted to capture that flavor in our presentation.

In this chapter we present the setting for the rest of the book. Section 1.1 contains introductory comments. Section 1.2 presents a qualitative discussion of van der Waals forces and Section 1.3 presents an empirical definition of the monolayer regime. Section 1.4 contains a brief history of the subject and Section 1.5 explains the scope of the book. We present a survey of the experimental techniques applied to monolayers in Section 1.6 and conclude in Section 1.7 with a bibliography of books and reviews for areas that overlap our coverage. For general reference, we provide tabulations of various properties of the most commonly studied gases and their condensed phases in Appendix A. Appendix B presents a summary of our surface thermodynamic formalism and notations.

1.1 Introduction

Consider an ideal, atomically flat surface that is exposed to a foreign vapor, held at pressure P and temperature T. In equilibrium with this vapor is a film on the surface; the coverage depends on P, T, and the relevant interaction potentials. When P (but not T) is low, the number of atoms per unit area on the surface is small. These atoms tend to be confined to a narrow domain near the surface, where the attraction provided by the substrate is greatest. This film represents a two-dimensional (2D) phase of matter. Its properties, in general, are qualitatively distinct from those of any 3D phase of the same material. The ability to create such a new system lies at a frontier of modern physics and chemistry. What makes the subject particularly interesting is the wide variety of behavior that may occur as P and T are varied. Included in the possible behaviors are some phases that are similar to 3D phases, e.g., gases, liquids, and solids. More intriguing are those states which are totally new, lacking a 3D analogue. Among these possibilities are commensurate phases, in which the overlayer has its positional order imposed by the substrate, and cases of orientational epitaxy, in which the overlayer has a distinct lattice spacing but has an orientation determined by, but not identical to, the substrate.

The physics of 3D matter is simplified conceptually and computationally by the common topologies of the phase diagrams of quite different materials, as exemplified by the law of corresponding states and universal values of the critical exponents of phase transitions. Similar 'universality' is found for 2D matter; there exist, however, more classes of behavior in 2D because of the numerous possible combinations of adsorbate and substrate. Predicting and systematizing this richness of behavior is one of the challenges to the theory, requiring an accurate assessment of the relevant force laws, a careful analysis of the delicate energetic balance of competing structures, and often the imagination to conceive of the various possible states of 2D matter.

This book is concerned with the problem of physical adsorption. Physical adsorption, sometimes called *physisorption*, refers to the weak binding of atoms or molecules to surfaces. Contrast is made with the case of *chemisorption*, which refers to stronger binding. The distinction in names originates from the presumed role of 'chemical' forces, in the latter case, while only physical interactions, or van der Waals forces, are present in the former case. This dichotomy is oversimplified and even misleading because there is actually a continuum of interaction strengths. In practice, the dividing line between the two descriptions is both subjective and nonuniform. Here we shall adopt the arbitrary value of 0.3 eV as the upper limit for the binding energy of a single adatom in the case of physical adsorption.* In other systems of units, this amounts to about 7 kcal (30 kilojoules) per mole or 3500 K per molecule. As Table 1.1 indicates, physical

---

*At the other extreme, the smallest known binding energy is the case of a hydrogen atom on liquid ^{4}He, for which the energy is 1 ± 0.05 K or about 10^{-4} eV (Walraven 1992; Yu *et al.* 1993).

Table 1.1   *One-atom adsorption energies.*[†]

Adsorbate	^{4}He	Ne	Ar	Xe	H_2
surface	(7.14)	(232)	(930)	(1907)	(89.8)
graphite	193	378	1114	1880	600
MgO	87	271	840	1404	557
Ag	70	159	835	2450	367
Cs	4.4	11	84.5	–	34

adsorption is commonly found for inert gases and saturated molecules. It is not coincidental that these same gases experience weak mutual interactions, as reflected in their low cohesive energies, also appearing in Table 1.1. Note in this table that the holding potential has an unusual dependence on substrate: the most reactive surfaces are the least strongly binding. This trend is explained in Chapter 2.

Several consequences of weak binding make physical adsorption particularly convenient for theoretical and experimental study. The phenomena often reveal conceptual simplicity, making the subject ideal for fundamental research, and have attracted scientists from a wide range of backgrounds. In the rest of this introduction, we present a brief overview of the concepts and phenomena that will be elaborated in the rest of the book. We provide as well a brief history of the field, a survey of experimental techniques, and a summary of the relevant books, conferences, and review articles that give overviews of the field and also show the ties to other regimes of adsorption and of theory.

1.2   Physical interactions

The relevant physical interactions include interatomic interactions, atom–surface interactions, and the effects of the surface on interatomic interactions. Their theoretical characterization involves an adequate knowledge of the electron densities, or wave functions, and their response to external fields. Beginning with the studies of London (1930a,b) and Lennard-Jones (1932a,b) in the 1930s and by the Lifshitz group and McLachlan in the 1950s and 1960s (Dzyaloshinskii *et al.* 1961; McLachlan 1964), we have come to a reliable theoretical understanding of the origin of long-range, attractive interactions. These so-called *dispersion forces* arise purely from quantum mechanics through the coupled (zero-point) motion of the electrons present in the interacting constituents.

[†]Estimated or measured well depths for the adsorption of various gases on various surfaces. In parentheses are the ground state cohesive energy of the bulk adsorbate (from Appendix A). All energies are expressed in K; the conversion is 11.6 K = 1 meV. Values for the graphite, MgO, and Ag surfaces are from Vidali *et al.* (1991). For Cs the sources are: He, Vidali *et al.* (1991); Ne, E. Cheng *et al.* (1994); Ar, A. Chizmeshya (unpublished); H_2, E. Cheng *et al.* (1993b).

Consider, for example, a nonpolar atom at distance z above a solid surface bounded by the x–y plane. Suppose that the atomic electrons' instantaneous configuration corresponds to a dipole moment p in the $+z$ direction. The electrostatic screening by charges of the solid is represented by an 'image dipole', a parallel dipole of strength proportional to p. From first order perturbation theory, the dipole-image-dipole interaction is then attractive, has a strength proportional to p^2, and varies as the inverse cube of the distance. Thus we expect that the adsorption potential varies as

$$V \simeq -k\langle p^2\rangle/z^3, \tag{1.1}$$

where k is a dimensionless constant* that depends on the strength of the substrate's dipole moment (relative to p) and the brackets refer to a quantum-mechanical average. Since $\langle p^2\rangle$ is proportional to the adatom's polarizability (Fano and Cooper 1968), highly polarizable atoms are more strongly attracted than weakly polarizable atoms. As discussed in Chapter 2, qualitatively similar conclusions can be derived, alternatively, from the addition of pair interactions (varying as r^{-6}) between the adatom and the many atoms of the substrate. The distance dependence and its exact prefactor will also be derived there under quite general assumptions.

Equation (1.1) is reliable, however, only at distances z that are sufficiently large that the electrons of the atom do not significantly overlap those of the substrate, because then quantum mechanical exchange and deviation from the point dipole approximation are negligible.† At small separation, a more sophisticated theory is needed, and indeed much effort has been expended in recent years to address this problem. A variety of alternative methods has yielded significant progress toward a quantitative theory of physisorption potentials.

Much of the stimulus for exploring these potentials is the desire to explain the thermodynamic and structural properties of monolayer films. Other sources of relevant data include kinetic measurements, such as atomic scattering, desorption, and diffusion on surfaces. Finally, the fact that these potentials are sensitive to the electronic properties of the surface, especially the inhomogeneous charge density, provides further motivation for their investigation. This last property exemplifies the very broad ramifications of the study of physical adsorption.

1.3   What is a monolayer?

Adsorption experiments are carried out often, but not always, in conditions of thermodynamic equilibrium between film and vapor. One of the many advantages of equilibrium is that this state's properties are independent of the route

---

*We use esu units for all formulae involving electromagnetism, following the prevalent usage in the research literature. In Appendix H, we present a set of conversions for the mixed units that arise in the literature.

†The distance should be large enough that the substrate appears to be a continuum, but not so large that effects of relativistic retardation modify the potential. See Chapter 2.

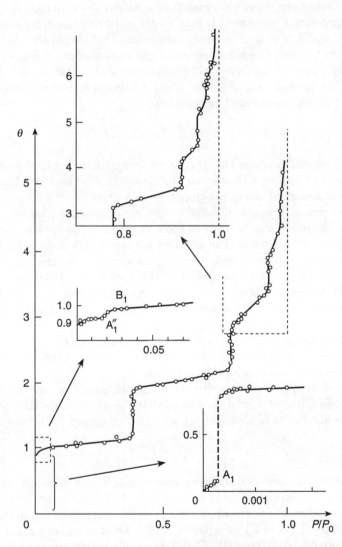

FIG. 1.1. Krypton multilayer adsorption isotherms. Adsorption of Kr multilayer
films on exfoliated graphite at 77.3 K, from Thomy *et al.* (1981) and based
on data from Thomy and Duval (1970). Coverage (in layers) as a function
of pressure relative to the saturated vapor pressure. Insets are blow-ups to
emphasize submonolayer and multilayer transitions at P/P_0 less than 0.001
and greater than 0.8, respectively.

followed to achieve it. Another is the certainty that the chemical potential of the film equals that of the vapor. We may then consider the evolution of film structure as the pressure P increases. It may be observed directly using experimental probes such as diffraction. Thermodynamic data provide another indication of the monolayer through the phenomenon of gas pressure saturation.

It is easy to prove, and physically plausible, that a low density film coexists with a low pressure vapor.* The linear relationship between coverage and pressure in this regime is called Henry's law:

$$P = k_H(T)N/A \quad (\text{small } N/A), \tag{1.2}$$

where $k_H(T)$ depends only on the adsorption potential; an explicit expression for k_H is derived in Chapter 4. Note that the term 'density' here is a 2D quantity, the number of adsorbed atoms per unit area of surface.

At higher pressure, many alternative film structures are possible. The case of physical adsorption usually involves the formation of one or more well-defined layers of film as P increases. The detailed nature of this deposition process is part of the broad problem known generally as *wetting*. For the most part, our concern in this book is the properties of the first layer. Often there is little experimental ambiguity about this concept. As seen in Fig. 1.1, the beginning of the second layer at $P/P_0 \simeq 0.4$ is characterized by an abrupt rise in the coverage as a function of P. Figure 1.2 shows that the dynamical responses of the monolayer, bilayer, and trilayer films measured by inelastic atomic scattering are quite distinct. More than a half-century of theory has been directed toward a quantitative understanding of this behavior.

A classic description is embodied in a simple theory of Langmuir (1918). As discussed in Chapter 4, one version of that theory assumes that adsorption occurs at identical sites of a surface lattice. The Langmuir isotherm,

$$\theta = P/(P + P_L), \tag{1.3}$$

results if interactions between adsorbed atoms on different sites are omitted and the presence of adsorption sites other than the first layer is ignored; here θ is the fraction of sites which are occupied, P_L is the pressure (depending on the adsorption potential and T) at which $\theta = \frac{1}{2}$. An alternative simple model of Brunauer, Emmett, and Teller (BET, 1938) permits higher layers to be occupied and includes 'vertical', but not 'lateral', interactions (i.e., interactions of a first layer site with sites above it, but not with other sites in the same plane). The BET model has often been used to describe multilayer adsorption, in spite of its implausible model assumptions. Over the course of time, much more sophisticated theory has been developed, yielding increasingly realistic descriptions of adsorption. Often the lattice description of adsorption sites is used, which is

---

*Note that the converse is not necessarily true! For example, at sufficiently low temperature the film is condensed into a high density phase while the coexisting vapor pressure is low. See Section 4.1.

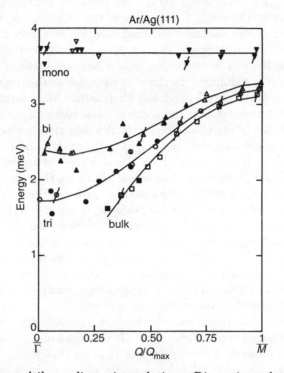

FIG. 1.2. Argon multilayer dispersion relations. Dispersion relation for Ar films of varying thickness adsorbed on Ag(111), as measured by inelastic He atom scattering and comparison with the bulk spectrum, from Gibson *et al.* (1985). These are branches of the spectra that have polarization primarily perpendicular to the substrate surface. The branches shown are for 1-, 2-, and 3-layer films, and the wavenumber is given as a fraction of its value at the Brillouin zone edge.

convenient but precludes any reliable conclusions about the relative stability of alternative structures. Recently, the technique of computer simulation has been applied extensively, providing the possibility of more accurate modelling.

Simple solvable models, such as the BET, often provide insight into issues pertinent to the monolayer concept. One example involves the vertical rise in coverage corresponding to a rapid increase in second layer occupation. We should bear in mind that, at finite T, there is always an incomplete first layer, e.g., vacancies, at the same time that second layer sites are increasing in occupancy. This is one of several sources of ambiguity in the definition of *monolayer completion*. Another is the fact that the discretization assumed in lattice models becomes particularly problematic at this point, since the atoms are becoming increasingly free then to move vertically and because systems exhibit varying degrees of 2D compressibility.* In other words, just at this point of special interest lattice models become less reliable than usual. More fundamental, really, is the attendant breakdown in the concept of a layer. This conclusion is of more than academic interest because physical adsorption is often regarded as a valuable tool for measuring surface area, of both smooth and irregular surfaces.

We return briefly to the regime of low film coverage for further qualitative discussion of monolayer phenomena. Individual atoms are then expected to migrate along the surface, mostly confined to a narrow spread in their z coordinates that depends on the temperature.† This is evidently a quasi-two-dimensional problem. Consider next the case of a condensed solid; a similar description applies. As will be seen in Chapter 4, a quasi-two-dimensional phonon analysis describes the adlayer's dynamics. Finally, when we turn to the problem of phase transitions, we find that correlation lengths become large in only two of the three dimensions. Thus in spite of many complications, much of monolayer physics really involves physics in reduced dimensionality.

Having reached this conclusion, which might be self-evident, we may consider some of the interesting implications of the 2D character of these films. One of the most fascinating issues is a consequence of the fact that fluctuations are larger in 2D than in 3D. For a 2D solid, for example, it was realized in the 1930s that mean-square vibrations about lattice sites diverge at finite temperature. While at first sight this might be deemed to rule out the existence of such solids, more careful analysis yielded the conclusion that solids, but not true crystals,‡ exist in 2D. The 2D solids exhibit qualitatively different (from 3D) positional order and melting behaviors; examples of the 2D diffraction peaks are shown in Fig. 1.3.

---

*An extreme form of behavior is 'de-wetting' of the monolayer, i.e., a decrease of first layer number as the total adsorption increases.

†If the mass is small, quantum effects can contribute a significant spread even at low temperature. For helium and hydrogen, the root-mean-square vibration amplitude in the z-direction is typically 0.2 to 0.5 Å.

‡Here we are adopting 'traditional' solid state physics definitions of a crystal, requiring infinitely long-range positional order and delta function Bragg spots. As seen in Chapter 4, monolayer solids' diffraction patterns exhibit well-defined, narrow peaks; often these are difficult, or practically impossible, to distinguish from the ideal crystal behavior.

This situation has attracted remarkably intense experimental and theoretical attention to these materials.

Our discussion has been intended to illustrate both the complexity and the simplicity of diverse monolayer film phenomena. Approximations are made in both qualitative and quantitative analyses. Often these are implicit; sometimes they are explicit. Often the inadequacy of assumptions has been made manifest by experiments.

1.4 Brief history

Because the experimental study of physical adsorption frequently involves temperatures $T < 100$ K (corresponding to the energy scales), the evolution of the field required the advent of low temperature capability in the laboratory. By itself, however, this was far from sufficient, because meaningful experimental data required adequate surface quality (Ross and Olivier 1964). The data of Halsey and coworkers in the 1950s on graphitized carbon black (Constabaris and Halsey 1957; Sams et al. 1960) provided the first useful contact with the theory of thermodynamic properties of monolayer films. Lander and Morrison detected ordered monolayer structures using low energy electron diffraction in the 1960s. A genuine explosion of the field began around 1970, with the addition of many new techniques and surface preparations. Essential to these measurements was adequate compatibility between two conflicting requirements: sufficient total surface area (for signal-to-noise reasons) and sufficiently extended, flat surfaces (for relevance to theoretical models). This period coincided with the vast improvement and expansion of surface science and vacuum capabilities, eventually permitting the study of well-characterized and clean single surfaces, primarily by scattering methods. It is perhaps instructive to note that much of the progress involved the use of forms of graphite that were available commercially as lubricants.*

Around the time when experiments were beginning to yield plentiful data concerning a wide variety of systems, there occurred significant developments in the theories needed to make reliable predictions. Particularly important was the advent of scaling and renormalization group ideas in the theory of phase transitions and the arrival of computational capability appropriate to these systems and problems.

1.5 Scope of this book

'The literature pertaining to the adsorption of gases by solids is now so vast that it is impossible for any, except those who are specialists in the experimental technique, rightly to appraise the work, which has been done, or to understand the main theoretical programs which require elucidation.' Lennard-Jones (1932a) *The adsorption of gases by solid surfaces*, (Faraday Soc., London), p.333.

---

*Similarly, the pioneering studies of Langmuir focussed on tungsten surfaces, stimulated by the problems and opportunities presented by tungsten filaments and high vacuum capabilities at the General Electric Laboratory.

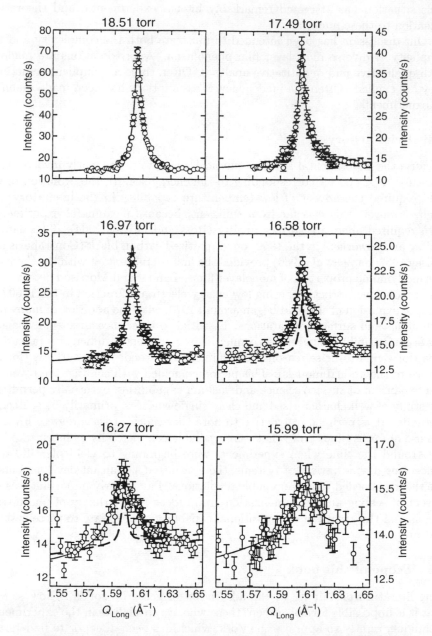

FIG. 1.3. Shapes of diffraction peaks for 2D solid xenon. Synchrotron x-ray scattered intensity from monolayer Xe on a single facet of graphite, from Nuttall *et al.* (1994). As the pressure P is reduced at $T = 140$ K, the pattern evolves from that characteristic of a 2D solid to that of a correlated 2D liquid, aligned orientationally with the substrate. The dashed curves at low P correspond to the shape that would be present if the solid phase were present.

The growth of the field has made the writing of a truly comprehensive one-volume text virtually impossible; see Section 1.7 for a bibliography. Thus the desire actually to complete this monograph has necessitated a stringent set of constraints on the subject matter. For the most part, our discussion will focus on atoms. The fact that many properties of simple, nearly spherical molecules (H_2, CH_4, ...) are adequately described by both the concepts and mathematics is a fortunate bonus; on occasion, reference will be made to such molecular systems either to point out their similarity to atomic systems, where appropriate, or to distinguish differences that arise.

A second restriction in the subject is that we focus on phenomena occurring within a monolayer. Except for occasional comments pertinent to the growth of the second layer and its exchange of atoms with the first layer, thicker films will be essentially ignored. A third restriction is to regular surfaces. The real surfaces found in nature are not regular, either compositionally or geometrically. While the effects of such heterogeneity are present to some extent in any experiment, we believe that the study of the simplest, best-understood systems is a promising route toward an eventual broad understanding of adsorption phenomena.

Finally, and most significantly, the book's contents are influenced by the authors' backgrounds as theoretical physicists. This book centers on the theory and its relation to the experimental results. However, the field has always involved a detailed interplay between theory and experiment. We refer to the experimental techniques at many points in this book and now discuss them in a summary way before proceeding with the main text. We first emphasize an obvious fact: the brevity of the discussion reflects nothing more than our lack of competence to do better.

1.6　Experimental techniques

Thermodynamic measurements such as adsorption isotherms, specific heats, and heats of adsorption mapped out major areas of monolayer phenomena. Diffraction measurements then confirmed the thermodynamic identifications in many instances and provided the first identifications of subtle structural transitions such as that between the incommensurate aligned and incommensurate rotated lattices. For homogeneous phases, the experimental information is packing geometry and energy and length scales. Determining the length scales to *ca.* $\frac{1}{2}$ % and energy scales to 10% leads to tests of assumptions about the effective interactions within the adsorbate. Parameters of first-order phase transitions, such as the thermodynamic discontinuities, can also be part of such tests. Precisely determined critical exponents of a second-order phase transition may help to identify the relevant universality class.

We now enumerate experimental techniques that have been applied to the study of physically adsorbed layers. The accounts of each technique are very brief, but our goals are: (1) to provide an overview that enables the reader to see the way in which the methods complement each other, and (2) to provide leading references where the reader may learn about the details of implementing the methods.

1.6.1 Thermodynamic measurements

The thermodynamic methods usually involve the measurement of a total property of adsorbate and substrate and then a subtraction of the contribution of the substrate. This gives information on the total amount adsorbed and on the energy of the adsorbate. While such experiments usually require materials with large values for the surface-to-volume ratio, there are exceptions that we note.

1.6.1.1 *Volumetric measurements* The classic experiment is the measurement of an adsorption isotherm, the amount adsorbed as a function of 3D gas pressure at constant temperature. The data often display clear signatures of phase transitions in the monolayer regime. Thomy, Duval and Régnier (1981) review such experiments, particularly for adsorption on preparations of graphite. Larher (1992) reviews the technique and the preparation of appropriate samples of lamellar halides; other surveys are: Grillet and Rouquerol (1979) for graphite; Coulomb and Vilches (1984) for MgO; and Shrestha *et al.* (1994) for BN.

There are variants that do not require powders or exfoliated samples. Auger electron spectroscopy (AES) has been used to measure the amount of adsorbate on a single crystal surface (Suzanne *et al.* 1974); small coverages are measured with good sensitivity after a calibration of the peak-to-peak height of an Auger transition of the adsorbate. Ellipsometry has the capability of measuring the thickness of very thin films on a single crystal surface, by exploiting the sensitivity of polarized light with a null geometry (Quentel *et al.* 1975; Hess 1991; Knorr 1992). Small frequency shifts of a resonant oscillator circuit have also been used to measure small increments in adsorbed mass, using a quartz microbalance (Krim 1991) and using a vibrating carbon fiber (Bruschi and Torzo 1991). Another variant is the use of positron annihilation (Hozhabri *et al.* 1989) to probe nonuniform density, such as defects in the layer.

A traditional method of coverage determination is gravimetric; the mass increase upon adsorption is substantial for a highly porous substrate. One of the oldest techniques used to study physical adsorption is the measurement of dielectric properties (McIntosh 1992). For signal-to-noise reasons, this has been most commonly applied to the case of porous media, but it is also applicable to the case of flat surfaces using high sensitivity capacitive methods. One of the characteristic features of the data is a *kink* in the dielectric response at the completion of each layer. Because the method is most sensitive to reactive adsorbates and substrates, it has not been extensively applied to physisorption.

Another signal for monitoring coverage involves the shift of surface plasmons of a metal due to the presence of a film. This is probed by the technique of attenuated total reflection of laser light impinging from a glass prism upon which the metallic substrate is adsorbed (Raether 1988). The surface plasmon resonance is observed to shift in angle by an amount proportional to the film coverage. This method has been shown to be useful for both equilibrium and surface diffusion studies (Herminghaus *et al.* 1994; Qian and Bretz 1988).

1.6.1.2   *Calorimetry*   The classic experiment is the measurement of the specific heat and heats of transformation. Historically, a very important step in establishing the utility of grafoil substrates was the demonstration that the specific heat of a low-density helium monolayer was approximately 1 k_B/atom. The experiments are usually performed on samples with a large surface-to-volume ratio (Bretz *et al.* 1973; Dash 1975; Marx 1985; Ma *et al.* 1988) and in a mode corresponding to a closed cell (fixed volume) experiment in 3D. Signals at low temperatures may be enhanced by the fact that the substrate is 'stiffer' than the adsorbed layer, so that its Debye temperature is much larger than that of the adlayer. Clearly, metals pose special problems in this respect, but *a.c.* specific heat measurements on films adsorbed on single crystal facets have been shown to be feasible for selected adsorbate/substrate pairs (Campbell *et al.* 1980; Campbell and Bretz 1985; Kenny and Richards 1990). Differential calorimeters have been developed for directly measuring isosteric heats of adsorption (Rouquerol *et al.* 1977).

1.6.2   *Structural experiments*

The lattice type and lattice constant of an adsorbed layer can be derived from the measurement of the Bragg reflections in a diffraction experiment. Some of the probes (atoms and electrons) are surface sensitive and do not penetrate significantly into the substrate; they were the first to be used for adsorbates on single crystal surfaces, but have the complication that they are also so strongly coupled in the scattering that quantitative analysis of reflected intensities is nontrivial. Probes that are not intrinsically surface sensitive, such as neutrons and X-rays, ordinarily require large surface-to-volume ratios unless the phenomenon of total external reflection (for X-rays) is exploited.

1.6.2.1   *Atoms and electrons*   Diffraction of electrons and atoms has been used to determine the structures for a wide variety of adsorbate/substrate combinations. Thermal energy atoms and low energy electrons have comparable ability to penetrate the 3D vapor in coexistence with an adsorbate in an equilibrium experiment. For both, the scattering cross-section with adsorbate atoms is of the order of tens of Å^2. Both have very limited ability to penetrate a dense phase, and the detailed analysis of the diffracted intensity is difficult.

Low energy electron diffraction (LEED) uses electrons of a few hundred electron volts (eV), an energy which penetrates an adsorbed layer, and the diffraction experiment shows peaks corresponding to the substrate surface and to the adsorbed layer (Lander and Morrison 1967). Precision in the lattice constant of 0.01 to 0.02 Å has been obtained and the adlayer-substrate separation has been inferred with an accuracy of ~ 0.1 Å. Webb and Lagally (1973) reviewed the use of LEED for surface analysis and Fain (1982) reviewed the use of LEED for a physisorbed layer. A variant with higher precision in the length measurements is

based on techniques of electron microscopy; it is termed THEED, transmission
high energy electron diffraction (Price and Venables 1976; Hamichi *et al.* 1991).
Since the electron energy is much larger than the excitation energy of electronic
states in the adsorbate, effects of electron beam desorption sometimes require
special techniques (Moog *et al.* 1983; Cui and Fain 1989). Techniques of photoelectron spin-polarized spectroscopies have also been applied (Schönhense 1986;
Potthof *et al.* 1995).

Intense and nearly monochromatic thermal energy atomic beams of helium
are created in nozzle jet expansions; length resolutions of ~ 0.01 Å are achieved
in atomic diffraction. This is thus a sensitive probe of surface structure (Boato *et al.* 1978). The atomic beam energies are on the scale of tens of meV, so that there
are important applications in the measurement of adlayer dynamics (Skofronick
and Toennies 1992). Analysis of the He diffraction pattern gives a sensitive
direct measure of the He interaction with the surface (Hoinkes 1980). Scattering
patterns for other gases require careful modeling to yield similar information
(Barker and Auerbach 1985).

1.6.2.2 *Neutrons and X-rays* Both thermal energy neutrons and X-rays have
weak coupling in the scattering; therefore, quantitative analysis of the diffracted
intensity is quite direct. For both, measures are available to enhance their effec-
tiveness in probing adsorbed layers.

Because the thermal neutrons penetrate quite readily, all experiments have
used samples of large surface-to-volume ratio. However, the contribution from
the scattering by the layer is enhanced by choice of adsorbate isotopes (e.g.,
for argon and nitrogen) with large scattering cross-sections (Kjems *et al.* 1974;
Taub *et al.* 1977; Sinha 1987; Godfrin and Lauter 1995). The replacement of
atomic hydrogen by atomic deuterium in an adsorbate molecule has also been
exploited to have both the coherent and incoherent scatterings for nearly the
same adsorbate (Taub 1980, 1988; Freimuth *et al.* 1990).

X-ray scattering (Brady *et al.* 1977; Stephens *et al.* 1979; Zabel and Robinson 1992) is greater for atoms with larger nuclear charge, so the experiments tend
to be done on the heavier inert gases. An exception is the series of experiments
on ethylene, C_2H_4 (Mochrie *et al.* 1984). As more intense sources of X-rays become available, experiments on smaller atoms and molecules will become feasible.
With high intensity and well-collimated sources of X-rays, precisions in lattice
constant on the scale of 0.001 Å are achieved. The analysis of the experiments
has rapidly developed into careful consideration of the line-shapes in diffraction
patterns, to derive information on correlations in monolayer solids and fluids.
After a series of experiments on large surface area graphite preparations, there
now are experiments for grazing incidence reflection from layers on single crystal
surfaces (Nuttall *et al.* 1994; Specht *et al.* 1987). The competing effects of the
substrate corrugation and adatom–adatom interactions in the equilibrium structure of the adlayer are reflected in modulation satellites accessible to diffraction
experiments.

1.6.3 *Dynamics*

Measuring adlayer dynamics, e.g., using inelastic neutron and helium atom scattering for excitation energies and nuclear magnetic resonance for relaxation times and mobility, gives additional confirmation of the character of surface phases deduced from thermodynamic and structural measurements. The change in adatom mobility upon melting of the 2D solid is observed in a magnetic resonance experiment (Richards 1980). The vibrational frequencies of one, two, and three layer solids provide a basis for distinguishing ordered layers whose intraplanar spacings are very similar. The excitation experiments involve measuring the energy change in the inelastic scattering of a probe particle. Because the energies of excitations in the adlayer are on the scale of a few Kelvin to a few tens of Kelvins (0.1 to 4 meV), the probes usually are neutrons and helium atoms with thermal energies. Infrared spectroscopy has also been used to characterize the vibrations of adsorbed molecules (Steele 1993).

There is a strong similarity to the use of thermal energy neutrons in mapping out the dispersion curves of normal modes of 3D solids. The remarks made in the previous section on the applicability of neutron scattering for structural determination apply here (Taub *et al.* 1977; Nielsen *et al.* 1977, 1980a; Lauter 1990). For some preparations of graphite with a considerable alignment of the adsorbate planes, the scattering geometry can be varied and polarization criteria can be used to infer the character of the excitations.

Inelastic thermal energy atomic scattering can be done for adsorbates on a single crystal surface, but the experimental geometries have tended to be such that the dominant momentum transfer is perpendicular to the plane of the adsorbate (Skofronick and Toennies 1992). Then the atom couples primarily to modes with polarization perpendicular to the surface plane. For the monolayer there is only a weak dispersion for that branch of the excitation spectrum, but there is already a marked dispersion observed in scattering from a bilayer. The difference has been used as a discriminant for the monolayer, bilayer, and trilayer (See Fig. 1.2; Gibson and Sibener 1988b; Kern and Comsa 1989).*

A rather different regime is the quasielastic scattering of thermal energy neutrons and atoms, in which diffusion in the adlayer is manifested by a broadening around the elastic scattering peak. This has been applied for adsorbed layers and multilayers of CH_4, exploiting a strongly incoherent neutron scattering (Coulomb *et al.* 1981; Bienfait *et al.* 1987; Bienfait and Gay 1991). It has recently been demonstrated for thermal helium atom scattering from various systems (Frenken and Hinch 1992).

Even with the relatively high resolution attained in electron spectroscopies (Ibach and Mills 1982; Ibach 1991), the energy scale of the excitations in physically adsorbed layers is usually too small for a major application. If the resolution

---

*These statements are based on experience and modeling for monatomic adsorbates. Strong dispersion in a monolayer mode polarized mostly perpendicular to the substrate has been observed for a molecular adsorbate, OCS/NaCl(001) (Glebov *et al.* 1996).

in electron energy loss spectroscopy (EELS) can be brought below 1 meV the situation will change. The only application in the physically adsorbed layers has been to use the EELS signal as a measure of coverage of a scattering molecule (Elliott *et al.* 1993).

Three methods which have been applied to the measurement of helium coverage are worthy of mention. One is third sound, the analogue of a capillary-gravity wave in a pool of liquid, with the substitution of the van der Waals force for gravity. This requires the presence of superfluidity because normal films experience viscous damping within the viscous penetration depth; superfluid films have an inviscid component that experiences no such viscous drag. The most common interpretation of third sound velocities involves a hydrodynamic derivation of the relation

$$c_3^2 = [n_s d\mu/dn]/m, \qquad (1.4)$$

where n_s is the number of superfluid atoms per unit area. Thus the technique is related to the derivative of an adsorption isotherm (Goodstein 1969; E. Cheng *et al.* 1992).

The second method involves another dynamical probe of the fluid. An electron weakly bound to a helium film experiences a drag force that is related to the coverage-dependent compressibility of the film. By measuring the mobility of the electron parallel to the surface, one learns about the oscillatory dependence of this response function of the film (Mugele *et al.* 1994).

A third dynamical probe of helium films also exploits its superfluid properties. The torsional oscillator method is a direct probe of the amount of superfluid that flows without resistance. It has proved capable of revealing the layer-dependent properties of superfluid films on a regular substrate such as graphite (Crowell and Reppy 1993).

Fourier-Transform-Infrared (FTIR) spectroscopy has been developed as a probe of the dynamics of molecular adsorbates on single-crystal insulating surfaces. It has a sensitivity to small perturbations of the admolecule by the adsorption interactions and, with polarization analyses, it is informative on the ordering and orientation of the molecular axes (Chabal 1988; Quattrocci and Ewing 1992; Heidberg *et al.* 1995a).

Other methods that have been applied are nuclear magnetic resonance NMR to probe the mobility in adsorbed layers (Richards 1980) and Mössbauer spectroscopy for a large "tagged" molecule (Shechter *et al.* 1976, 1982; Wang *et al.* 1983).

The kinetics of adsorption, desorption and diffusion have been widely studied in physisorption systems (Menzel 1982; Barker and Auerbach 1985; Tully 1981, 1994). An interesting application that yields information on the adsorption energy is thermal desorption spectroscopy (labelled as TDS, TPD, and TDMS). The basis of the method is the observation by Redhead (1962) that the energy may be inferred from the temperature at which the desorption rate is maximum. One uses a heating rate of a few kelvin per second and, instead of a measurement of an equilibrium vapor pressure as a function of temperature, re-lies on a measurement of the remaining coverage or of the amount desorbed. The analysis requires assumptions or knowledge about the order of magnitude and coverage dependence of the prefactor of the Arrenhius-like desorption rate, but the method may be applied under nonequilibrium conditions ('dosed' layer) or for a rapid survey of the system energetics. While it is most frequently applied to chemisorbed systems (e.g., Thiel *et al.* 1981), there also are applications to physisorbed systems (Wang and Gomer 1980; Wandelt and Hulse 1984; Behm *et al.* 1986). For the physisorbed layers, the temperature at which the desorption rate is maximum is typically 1/30 of the adsorption energy (in kelvin).

1.6.4 *Electronic structure*

The change in the work function of the substrate is a macroscopic measurement indicating that the adatom–substrate complex is distorted in the process of ad-sorption; it is interpreted in terms of the formation of electrostatic dipoles on the adatoms (Palmberg 1971; Aruga and Murata 1989). There is little systematic information on this process for physically adsorbed species, but phenomena of classical electrostatics occur, such as depolarization by the fields of the other ad-sorption dipoles. The interpretation of such phenomena requires an extension of the theory of electrostatic response applicable to nonuniformity at atomic scale distances (Barton 1979; Feibelman 1982).

Although the energy scale of electronic excitations is much larger than that of most of the shifts that occur in physical adsorption, Auger electron spectroscopy and photoemission spectroscopy have contributed some information. Auger spec-troscopy is used for the identification of adsorbed species and as a diagnostic of the purity of the surface (Suzanne *et al.* 1974). Effects on the electronic band structure of the adlayer are observed with angle-resolved photoemission (Horn *et al.* 1978). Final state interactions in Auger excitations are exploited for the determination of overlayer spacings (Kaindl *et al.* 1980; Behm *et al.* 1986).

1.6.5 *Imaging*

The observation of the adsorption site may be a decisive step in assigning the bonding mechanism for an adsorbed species. The case of xenon adsorbed on metal surfaces may be the closest to having this done, using scanning tunneling microscopes (STM) (Eigler and Schweitzer 1990; Eigler *et al.* 1991; Stroscio and Eigler 1991; Zeppenfeld *et al.* 1994a). As indicated in Fig. 1.4 one can directly image physically adsorbed atoms. Because such atoms are weakly bound, however, the probe is strongly perturbing. Thus its utility is restricted to the heavier, more strongly bound, gases.

Field emission microscopy (Gomer 1990) has been used to explore adsorbate equilibrium and transport properties. Its limitation in the physisorption case is that the necessary heating (to equilibrate) tends to induce desorption.

In contrast to the field emission technique, field ion microscopy is not or-dinarily capable of imaging physically adsorbed gases. The reason is that the

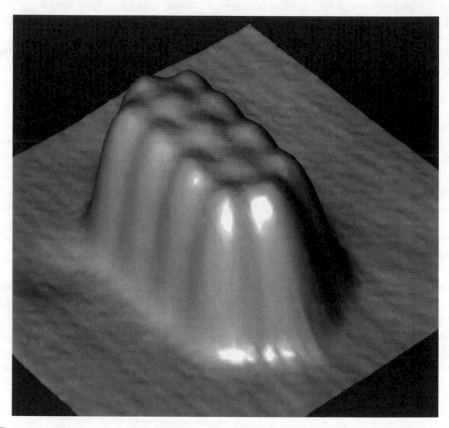

FIG. 1.4. STM image of Xe/Pt(111). An STM image of 13 Xe atoms on a Pt(111) surface, from Weiss and Eigler (1992). The cluster shown is rotated by $\sim 30°$ from the close-packed direction of the substrate. The apparent interatomic spacings are 10–15% smaller than in bulk Xe, while He diffraction data for a full monolayer found a spacing close to the bulk value (Kern *et al.* 1988).

latter method requires electric fields at the tip ($\sim$ 1 volt per angstrom) that are an order of magnitude higher than field emission values, and these very quickly field desorb weakly bound atoms (Tsong 1983, 1990).

1.7 Bibliography

While it is a cliché to repeat that we are 'standing on the shoulders of giants' when writing this monograph, it is important to note that numerous existing reviews and books address aspects of virtually every subject that we discuss. Many of these provide much more detailed information than we do. Therefore, we present, at the outset, a bibliography of these principal sources of information concerning physical adsorption, including major references that show the overlap with other fields.

1.7.1 Monographs: physical adsorption

- Dash, J. G. (1975). *Films on solid surfaces*. Academic, New York.

- De Boer, J. H. (1968). *The dynamical character of adsorption*. Clarendon Press, Oxford.

- Jaroniec, M. and Madey, R. (1988). *Physical adsorption on heterogeneous solids*. Academic, New York.

- Kreuzer, H. J. and Gortel, Z. W. (1986). *Physisorption kinetics*. Springer, Berlin.

- Nicholson, D. and Parsonage, N. G. (1982). *Computer simulation and the statistical mechanics of adsorption*. Academic, London.

- Ross, S. and Olivier, J. P. (1964). *On physical adsorption*. Wiley Interscience, New York.

- Rudzinski, W. and Everett, D. H. (1992). *Adsorption of gases on heterogeneous surfaces*. Academic, New York.

- Steele, W. A. (1974). *The interaction of gases with solid surfaces*. Pergamon, Oxford.

- Young, D. M. and Crowell, A. D. (1962). *Physical adsorption of gases*. Butterworths, London.

1.7.2 Conference proceedings/Advanced Study Institutes

Only about the last twenty years' publications are listed here. The reader will note that this list has an average interval of about two years and there is no indication that this series will ever stop!

- Ricca, F. (ed.) (1972). *Adsorption–desorption phenomena*. Academic, New York.

- Daunt, J. G. and Lerner, E. (ed.) (1973). *Monolayer and submonolayer helium films.* Plenum, New York.

- Bienfait, M. and Suzanne, J. (ed.) (1977). *Journal de Physique,* Vol.**38** Colloque C-4 (1977) Marseille.

- Sinha, S. K. (ed.) (1980). *Ordering in two dimensions.* North Holland, New York.

- Dash, J. G. and Ruvalds, J. (ed.) (1980). *Phase transitions in surface films.* Plenum, New York.

- Benedek, G. and Valbusa, U. (ed.) (1982). *Dynamics of gas–surface interaction.* Springer series in Chemical Physics, Vol.**21**.

- Mutaftschiev, B. (ed.) (1982). *Interfacial aspects of phase transformations.* Reidel, Dordrecht.

- Cole, M. W., Toigo, F., and Tosatti, E. (ed.) (1983). *Statistical mechanics of adsorption. Surface Science,* Vol.**125**.

- Benedek, G., Celli, V., Cole, M. W., Toigo, F., and Weare, J. (ed.) (1984). *Gas–surface interactions and physical adsorption. Surface Science,* Vol.**148**.

- Pullman, B., Jortner, J., Nitzan, A., and Gerber, R. B. (ed.) (1984). *Dynamics on surfaces.* Reidel, Dordrecht.

- *Physical interactions and energy exchange at the gas–solid interface.* (1985). *Faraday Discussions of the Chemical Society,* No.**80**, Royal Society of Chemistry, London.

- *Phase transitions in adsorbed layers.* (1986). *Faraday Symposium,* **20**, *Journal of the Chemical Society Faraday Transactions II,* **82**, 1569–1871.

- *Symposium on adsorption on solid surfaces.* (1989). *Langmuir,* **5**, 550–886.

- Taub, H., Torzo, G., Lauter, H. J., and Fain, S. C. Jr. (ed.) (1991). *Phase transitions in surface films 2.* Plenum, New York.

- *Symposium on quantum fluids and solids.* (1992). *Journal of Low Temperature Physics,* Vol.**89**, No. 1/2 and 3/4.

- Strandburg, K. J. (ed.) (1992). *Bond orientational order in condensed matter systems.* Springer, Berlin.

- Zabel, H. and Robinson, I. K. (ed.) (1992). *Surface X-ray and neutron scattering.* Springer, Berlin.

- Mills, D. L. and Burstein, E. (ed.) (1994). *Scattering from surfaces,* in *Surface Review and Letters,* **1**, 47–210. World Scientific, Singapore.

1.7.3   Related books and chapters

- Avgul, N. N. and Kiselev, A. V. (1970). In *Chemistry and Physics of Carbon*, Vol.**6**, (ed. P. Walker), pp. 1–124. Dekker, New York.

- Benedek, G. (ed.) (1992). *Surface properties of layered materials*. Kluwer, Dordrecht.

- Bortolani, V., March, N. H., and Tosi, M. P. (ed.) (1990). *Interactions of atoms and molecules with solid surfaces*. Plenum, New York.

- Christmann, K. (1991). *Introduction to surface physical chemistry*. Steinkopff Verlag, Darmstadt.

- Davis, II. T. (1996). *Statistical mechanics of phases, interfaces, and thin films*. VCH Publishers, New York.

- Duke, C. B. (ed.) (1994). *Surface science; the first thirty years, Surface Science*, **299/300**.

- Goodman, F. O. and Wachman, H. Y. (1976). *Dynamics of gas–surface scattering*. Academic, New York.

- Höhler, G. and Niekisch, E. A. (ed.) (1982). *Structural studies of surfaces with atomic and molecular beam diffraction*. Springer, Berlin.

- Holloway, S. and Nørskov, J. K. (1991). *Bonding at surfaces*. Liverpool University Press, Liverpool.

- Hulpke, E. (ed.) (1992). *Helium atom scattering from surfaces*. Springer, Berlin.

- Israelachvili, J. N. (1992). *Intermolecular and surface forces*. Academic, London.

- Langbein, D. (1974). *Van der Waals attraction*, Springer Tracts in Modern Physics, Vol.**72**.

- Lennard-Jones, J. E. (1932). *The adsorption of gases by solid surfaces*. Faraday Society, London.

- Mahanty, J. and Ninham, B. W. (1976). *Dispersion forces*. Academic, London.

- March, N. H. (1986). *Chemical bonds outside solid surfaces*. Plenum, New York.

- Nelson, D. R. (1980). In *Fundamental problems in statistical mechanics V*, (ed. E. G. D. Cohen), pp.53–108. North Holland, New York.

- Nelson, D. R. (1983). In *Phase transitions and critical phenomena*, Vol. 7, (ed. C. Domb and J. L. Lebowitz), pp.1–99. Academic, New York.

- O'Connor, D. J., Sexton, B. A., and Smart, R. St. C. (ed). (1992). *Surface analysis methods in materials science*, Springer Series in Surface Science, Vol.**23**. Springer, New York.

- Safran, S.A. (1994). *Statistical thermodynamics of surfaces, interfaces and membranes*. Addison-Wesley, New York.

- Shrimpton, N. D., Cole, M. W., Steele, W. A., and Chan, M. H. W. (1992). In *Surface properties of layered structures*, (ed. G. Benedek), pp.219–60. Kluwer, Dordrecht.

- Somorjai, G. A. (1981). *Chemistry in two dimensions: surfaces*. Cornell, Ithaca.

- Steele, W. A. (1977). In *Chemistry and physics of solid surfaces*, (ed. R. Vanselow and S. Y. Tong), pp.139–53. CRC Press, Cleveland.

- Unertl, W. N. (1997). *Physical structure of solid surfaces*, Vol.1 in the Handbook of Surface Science, Series editors S. Holloway and N. V. Richardson.

- Woodruff, D. P. and Delchar, T. A. (1986). *Modern techniques of surface science*. Cambridge, New York.

- Zangwill, A. (1988). *Physics at surfaces*. Cambridge, New York.

1.7.4 Other major reviews

1.7.4.1 Methodology

- Feibelman, P. J. (1982). *Progress in Surface Science*, **12**, 287–402.

- McTague, J. P., Nielsen, M., and Passell, L. (1978). *Critical Reviews in Solid State and Materials Science*, **8**, 135–55.

- Scoles, G. (ed.) (1988). *Atomic and molecular beam methods*. Oxford, New York.

- Sinha, S. K. (1987). In *Methods of experimental physics*, Vol.23-part B, (ed. D. L. Price and K. Sköld), pp.1–84. Academic, San Diego.

- Tabony, J. (1980). *Progress in Nuclear Magnetic Resonance Spectroscopy*, **14**, 1–26.

- Thomas, R. K. (1982). *Progress in Solid State Chemistry*, **14**, 1–93.

1.7.4.2 Interactions with and atomic scattering from surfaces

- Barker, J. A. and Auerbach, D. J. (1985). *Surface Science Reports*, **4**, 1–99.

- Boato, G. and Cantini, P. (1983). *Advances in Electronics and Electron Physics*, **60**, 95–160.

- Bortolani, V. and Levi, A. C. (1986). *La Rivista del Nuovo Cimento*, **9**, #11. 1–77.

- Cardillo, M. J. (1981). *Annual Review of Physical Chemistry*, **32**, 331–57.

- Cole, M. W., Frankl, D. R., and Goodstein, D. L. (1981). *Reviews of Modern Physics*, **53**, 199–210.

- Frankl, D. R. (1983). *Progress in Surface Science*, **13**, 285–356.

- Hoinkes, H. (1980). *Reviews of Modern Physics*, **52**, 933–70.

- Kern, K. and Comsa, G. (1989). *Advances in Chemical Physics*, **76**, 211–80.

- Schmeits, M. and Lucas, A. A. (1983). *Progress in Surface Science*, **14**, 1–52.

- Steele, W. A. (1993). *Chemical Reviews*, **93**, 2355–78.

- Vidali, G., Ihm, G., Kim, H.-Y., and Cole, M. W. (1991). *Surface Science Reports*, **12**, 133–81.

1.7.4.3 *Monolayer phases*

- Binder, K. (1992). *Annual Review of Physical Chemistry*, **43**, 33–59.

- Binder, K. and Landau, D. P. (1989). *Advances in Chemical Physics*, **76**, 91–152.

- Dash, J. G. and Schick, M. (1978). In *The physics of liquid and solid helium*, Part. II. (ed. K. H. Bennemann and J. B. Ketterson), pp.497–571. Wiley, New York.

- Dash, J. G. (1978). *Physics Reports*, **38C**, 177–226.

- Fain, S. C. Jr. (1982). In *Chemistry and physics of solid surfaces IV*, (ed. R. Vanselow and R. Howe), pp.203–18. Springer, New York.

- Godfrin, H. and Lauter, H. J. (1995). In *Progress in Low Temperature Physics*, Vol. XIV, (ed. W. P. Halperin), Chap. 4, pp.213–320. North Holland, Amsterdam.

- Knorr, G. (1992). *Physics Reports*, **214**, 113–57.

- Marx, R. (1985). *Physics Reports*, **125**, 1–67.

- Ohtani, H., Kao, C.-T., Van Hove, M. A., and Somarjai, G. A. (1986). *Progress in Surface Science*, **23**, 155–316.

- Schick, M. (1981). *Progress in Surface Science*, **11**, 245–92.

- Steele, W. A. (1996). *Langmuir*, **12**, 145–53.

- Strandburg, K. J. (1988). *Reviews of Modern Physics*, **60**, 161–207.

- Suzanne, J. and Gay, J. M. (1997). In *Physical structure of solid surfaces*, Vol. 1 (W. N. Unertl, ed.)

- Vilches, O. E. (1980). *Annual Review of Physical Chemistry*, **31**, 463–90.

- Webb, M. B. and Lagally, M. G. (1973). In *Solid State Physics*, **28** (ed. H. Ehrenreich, F. Seitz, and D. Turnbull), pp.301–405. Academic, New York.

- Weinberg, W. H. (1983). *Annual Review of Physical Chemistry*, **34**, 217–43.

1.7.4.4 *Computer simulation methods*

- Abraham, F. F. (1986). *Advances in Physics*, **35**, 1–111.

- Allen, M. P. and Tildesley, D. J. (1987). *Computer simulation of liquids.* Clarendon Press, Oxford.

- Binder, K. and Heermann, D. W. (1992). *Monte Carlo simulation in statistical physics. An introduction*, (2nd edn). Springer, Berlin.

- Binder, K. (ed.) (1992). *The Monte Carlo method in condensed matter physics*, Springer Topics in Applied Physics, Vol.**71**. Springer, Berlin.

- Frenkel, D. and McTague, J. P. (1980). *Annual Review of Physical Chemistry*, **31**, 491–521.

- Hoover, W. G. (1986). *Molecular dynamics.* Springer, New York.

- Kalos, M. H. and Whitlock, P. A. (1986). *Monte Carlo methods.* Wiley, New York.

- Landau, D. P., Mon, K. K., and Schüttler, H.-B. (ed.) (1993). *Computer simulation in condensed matter physics V.* Springer, Berlin.

- Mouritsen, O. G. (1984). *Computer studies of phase transitions and critical phenomena.* Springer, Berlin.

2

INTERACTIONS

The properties of adsorbed layers are ultimately determined by the interactions among the constituents. These interactions are fundamental to a broad range of adsorption phenomena, including the equilibrium structures of the adsorbate, the phases that exist as a function of temperature and density, and surface kinetics. In this chapter, we develop the concepts that are central to understanding the forces acting in physical adsorption and review the progress that has been made in determining these forces from *a priori* theories or from semi-empirical and empirical constructions using experimental data. Monographs on intermolecular forces include the books of Margenau and Kestner (1971), Mahanty and Ninham (1976), Israelachvili (1992), Stone (1996), and of Hirschfelder, Curtiss, and Bird (1964). Reviews of the forces in physical adsorption include Nicholson and Parsonage (1982), Bruch (1983a), Vidali *et al.* (1991), and Steele (1993).

One of the simplifying aspects of physically, as opposed to chemically, adsorbed layers is the fact the electronic structures of both the adsorbate and the surface are only weakly perturbed in the adsorption process. This is most evident for the large-distance regime where the asymptotic behavior of the adatom–substrate attraction is directly related to properties of the separated atom and substrate. Likewise, the interaction between adsorbates is, to a good first approximation, determined by the interactions between the atoms in the gas phase. Yet, paradoxically, the weakness of the interaction introduces one of the principal complications in that the attractive interaction is strongly dependent on electronic correlation energies, so that a level of approximation sufficient to provide a semiquantitative account of the formation of chemical bonds is insufficient to yield the van der Waals well of the physical adsorption system. Chemical considerations begin to play a role only at the small separations where wave function overlap gives rise to the repulsive part of the interaction. One of our main objectives is to explain how these competing processes of overlap repulsion and van der Waals attraction can be combined to provide a qualitative, and in some cases even quantitative, understanding of the interactions in physically adsorbed layers.

This chapter is organized in four sections. In Section 2.1 we give a broad overview of the theory of intermolecular forces and the major modifications that occur in the context of adsorption. In Section 2.2 we review the theory of the van der Waals attraction and the short-range repulsion in the adatom–substrate potential. In Section 2.3, we describe semiempirical constructions of the potential, especially of its corrugation. Finally, in Section 2.4, we present several examples to illustrate the considerations of the preceding sections. Some of the

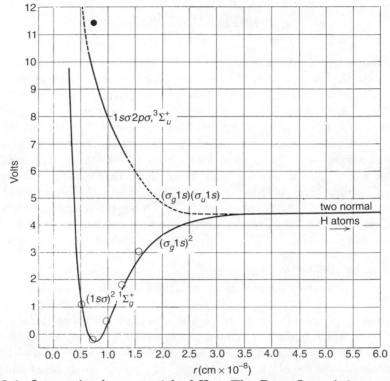

FIG. 2.1. Intramolecular potential of H_2. The Born–Oppenheimer potential energies of the ground state and first excited state of the H–H interaction, adapted from Mulliken (1932). The open and filled circles denote modern results for the ground and excited states, respectively, from Kołos and Wolniewicz (1965) and Cencek *et al.* (1995). The minimum Born–Oppenheimer energy is 0.174475931(1) Hartrees, or about 4.75 eV, lower than the energy of two isolated hydrogen atoms. The first excited state, the $^3\Sigma_u^+$ state shown here, has a minimum not visible on the scale of the figure, with depth about 20 microHartrees at 4 Å (Kleinekathöfer *et al.* 1996).

material has been deferred to Appendices. In Appendix C, we fill in details of the derivation of the van der Waals attraction and in Appendix D, we complete the derivation of the *fragment formula* for the overlap repulsion (Zaremba and Kohn 1977). In Appendix E, we present simple and relatively accurate approximations for the dynamic response of the adatom and substrate and the associated dispersion energy coefficients (Vidali and Cole 1981; Rauber *et al.* 1982).

2.1 Theory of intermolecular forces

The interaction of atoms with solid surfaces has many parallels to the interaction between an isolated pair of atoms. Further, for many physically adsorbed layers, the interactions between the atoms in the adsorbed layer can be represented, to a good approximation, by the sum of the interactions of pairs of atoms. We therefore begin our discussion by considering the simpler situation of interatomic potentials (Slater 1963; Margenau and Kestner 1971).

2.1.1 *Interatomic interactions*

As noted in Chapter 1, it is common in adsorption problems to distinguish between physical and chemical interactions. This distinction is already present for interatomic interactions, which exhibit qualitative differences according to the atomic species being considered. Thus the interaction of two hydrogen atoms to form an H_2 molecule in the singlet ground state is the paradigm of the chemical bond. On the other hand, the interaction of two helium atoms, each consisting of a closed-shell electronic configuration, is of a different type and the interatomic attraction has little to do with the formation of chemical bonds. Fundamentally, though, these two situations are based on the same Coulombic electron–electron and electron–nucleus interactions, and there is no obvious sharp distinction between chemical and physical interactions. We must look more deeply into the problem in order to appreciate where the differences arise.

2.1.1.1 *Definition of the potential* From a quantum chemistry point of view, the interaction between two atoms is simply defined: for a given separation, one must calculate the total electronic ground state energy. Perhaps surprisingly, to achieve a thermodynamic level of accuracy in such a calculation is so formidable a task that it has only been completed for the helium dimer within the last few years (Anderson *et al.* 1993).

To see what is involved, consider the interaction between two atoms, A and B, with nuclear charges Z_A and Z_B located at the positions $\mathbf{R}_A$ and $\mathbf{R}_B$, respectively, and $N = Z_A + Z_B$ electrons. The total electronic Hamiltonian of the system is

$$H = \sum_{i=1}^{N} \frac{\mathbf{p}_i^2}{2m} - \sum_{i=1}^{N} \left[\frac{e^2 Z_A}{|\mathbf{r}_i - \mathbf{R}_A|} + \frac{e^2 Z_B}{|\mathbf{r}_i - \mathbf{R}_B|} \right] + \frac{1}{2} \sum_{i \neq j} \frac{e^2}{|\mathbf{r}_i - \mathbf{r}_j|} + \frac{e^2 Z_A Z_B}{|\mathbf{R}_A - \mathbf{R}_B|} .$$

(2.1)

The total ground state energy of the system is a function of the nuclear separation $R = |\mathbf{R}_A - \mathbf{R}_B|$ and is given by

$$E(R) = \langle \Psi | H | \Psi \rangle ,$$

(2.2)

where the total N-electron wave function $|\Psi\rangle$ must be antisymmetric under particle permutations.

If at large separations the system consists of two neutral atoms in their atomic ground states with a total energy $E_0^A + E_0^B$, the interatomic potential is defined to be

$$V(R) = E(R) - E_0^A - E_0^B .$$ (2.3)

The general shape of the intermolecular potential is that $V(R)$ approaches zero as $R \to \infty$, has an attractive well at interatomic separations typical of a condensed phase (say, a few Å) and then becomes strongly repulsive at small separations. The depth and position of this well depends on the atomic species and the nature of the bonding that occurs.

The determination of the electronic energy in eqn (2.3) is a very complex problem even in the limit of separated atoms. No analytical solutions are available and some form of approximation must be invoked. Which is the most appropriate depends on the situation and is intimately associated with the chemical or physical nature of the interaction. For example, at small separation, a self-consistent-field approximation such as Hartree–Fock (HF) theory is often a useful first approximation for chemically reactive species (Slater 1963; Hurley 1976). The electronic wave function then is approximated as a Slater determinant of molecular orbitals that are optimized to minimize the electronic energy. For two open-shell atoms, the HF theory accounts for the formation of covalent bonds with energies on the scale of electron volts (eV). Further, for an arbitrary pair of monomers, this approach gives the expected short-range repulsive part of the interaction, which is variously described as 'exchange' or 'overlap'.*

Born–Oppenheimer approximation The potential $V(R)$ is termed the adiabatic potential and its application to a dynamical situation is called the Born–Oppenheimer approximation. The internal degrees of freedom of the atoms are assumed to remain in their ground state as a function of the instantaneous nuclear positions. Departures from this condition may arise when the kinetic energy of relative motion of the atoms is large, but they are not significant at the thermal energies typically encountered in physical adsorption. Another complication arises when so-called *curve-crossings* occur in the interactions of excited state atoms; while we do not treat these further in our discussion, they enter in the desorption of atoms by electronic excitation.

Molecular energy levels We may illustrate the approach of molecular orbitals with the H_2 and He_2 examples. The interaction potentials between two hydrogen atoms and two helium atoms are illustrated in Figs. 2.1 and 2.2. The H_2 molecule is strongly bound with a HF potential well[†] 3.635 eV deep (Kołos and Roothaan 1960) compared to the exact value of 4.747 eV. On the other hand, the He–He potential is strictly repulsive in the HF approximation.[‡] A discussion

---

*We prefer to use the term 'overlap' inasmuch as exchange is but one specific contribution to the Hartree–Fock energy.

[†]The HF energy is referenced to the exact energy of the separated hydrogen atoms in order to avoid a pathology of the HF-dissociation limit for H_2.

[‡]The *attractive* HF interaction sometimes obtained at large interatomic separations is believed to be an artifact of the limited basis states used in the calculations.

in terms of molecular orbitals gives insight into the origin of these differences. In both cases, the $1s$ atomic orbitals centered on the two sites hybridize into bonding and antibonding molecular orbitals which may each accommodate two electrons. For H_2, only the bonding orbitals are occupied, and a definite energy lowering with respect to the isolated atom energies is brought about by formation of the molecule. For He_2, with a total of four electrons, both the bonding and antibonding orbitals are doubly occupied, and there is no obvious energy advantage to dimer formation. When the total HF energy is evaluated, it is found that this simple picture is essentially correct: a strong chemical bond is formed in the case of H_2, whereas a net repulsive interaction arises for He_2. To resolve the shallow potential well that actually occurs in He_2 requires a calculation that goes beyond the HF approximation.

The combination of a closed and open shell atomic pair is more complex, and the character of the interaction ranges between chemical and physical bonding, depending on the atomic species. For example, the latter situation is realized when the valence orbital energy of the open-shell atom is well above the valence orbital energy of the closed-shell partner. Then the closed-shell atom is only weakly perturbed by its neighbor, while the relatively shallow open-shell orbital is strongly perturbed. The analogous situation in physisorption is the interaction between a metal (open-shell) and an inert gas (closed-shell) adsorbate.*

To go beyond HF theory requires including the interelectron correlations induced by the repulsive electron–electron interaction, in addition to the quantum statistics effects built into the Slater determinant. Although these improvements can be achieved in a variety of ways, the basic strategy is to improve the ground state wave function by admitting greater variational freedom. One commonly used method is that of configuration interaction, which, in principle, is capable of achieving high accuracy. However, its implementation requires large computational resources, and calculations of thermodynamic accuracy for intermolecular potentials between closed-shell molecules are feasible only for systems with relatively few electrons.

Well-separated atoms   The approaches discussed so far have emphasized the sharing of electrons between the two nuclei. That is, electrons are continually exchanged between the atoms, and a particular subset of electrons cannot be associated with a given nucleus, even though on average each nucleus retains its complement of atomic electrons. At large separations, however, where the interaction between the atoms is weak and the overlap is small, a more natural starting point is to assume that one subset of electrons ($i = 1, \ldots, Z_A$) belongs to atom A and a second ($j = 1, \ldots, Z_B$) to atom B. This idea is formalized by partitioning the N-electron Hamiltonian as

$$H = H_A + H_B + V_{AB} , \qquad (2.4)$$

*The weakness of the interactions may be extreme. In the case of alkali atoms and surfaces interacting with inert gases, the well depths are even smaller than those of homonuclear inert gas pairs (see Cole *et al.* 1994).

where

$$H_A = \sum_{i=1}^{Z_A} \frac{\mathbf{p}_i^2}{2m} - \sum_{i=1}^{Z_A} \frac{e^2 Z_A}{|\mathbf{r}_i - \mathbf{R}_A|} + \frac{1}{2} \sum_{i \neq j}^{Z_A} \frac{e^2}{|\mathbf{r}_i - \mathbf{r}_j|} , \tag{2.5}$$

with a similar expression for H_B, and

$$V_{AB} = - \sum_{i=1}^{Z_A} \frac{e^2 Z_B}{|\mathbf{r}_i - \mathbf{R}_B|} - \sum_{j=1}^{Z_B} \frac{e^2 Z_A}{|\mathbf{r}_j - \mathbf{R}_A|} + \sum_{i=1}^{Z_A} \sum_{j=1}^{Z_B} \frac{e^2}{|\mathbf{r}_i - \mathbf{r}_j|} + \frac{e^2 Z_A Z_B}{|\mathbf{R}_A - \mathbf{R}_B|} . \tag{2.6}$$

V_{AB} represents the Coulomb interaction between the two fragments. If V_{AB} is neglected, the product states $|\Psi_m^A \Psi_n^B\rangle = |\Psi_m^A\rangle|\Psi_n^B\rangle$ are eigenstates of H, and the ground state is $|\Psi_0^A \Psi_0^B\rangle$ with energy $E_0^A + E_0^B$, that is, the value of the total energy at infinite separation.

Because the product state distinguishes between subsets of electrons on the two atoms, it violates the requirement that the full wave function be totally antisymmetric with respect to the interchanges of any pair of electrons. However, it can be shown that exchange between the subsets of two atoms has a negligible effect on the ground state energy, provided that the separation R is large compared to the spatial extent of the atoms. Only when $|\Psi_m^A\rangle$ and $|\Psi_n^B\rangle$ have a significant spatial overlap does the exchange symmetry lead to appreciable corrections to the results obtained on the basis of distinct atomic Hilbert spaces. The size of the energy shift and whether the overlap ultimately leads to repulsion or attraction are indeed in the realm of 'chemistry'.

Van der Waals forces   At large interatomic separations, the Coulomb interaction V_{AB} can be treated as a perturbation to the isolated atomic Hamiltonians. Besides the direct electrostatic interactions arising from permanent multipole moments, V_{AB} induces correlations between electronic motions on the different atoms. The result is an attractive force that arises from the *change* in the total correlation energy of the system. Such an interaction between atoms with no permanent electrostatic moments was hypothesized by van der Waals and is referred to as the *dispersion* or *van der Waals* force.* The leading term in the interaction at large R arises from correlations of the fluctuating dipole moments on the atoms (London 1930a,b) and has the form (Margenau and Kestner 1971):

$$V_{vdW}(R) = -\frac{C_6}{R^6} , \tag{2.7}$$

with

$$C_6 = \frac{3\hbar}{\pi} \int_0^\infty du\, \alpha_a(iu) \alpha_b(iu) , \tag{2.8}$$

---

*The adjective *dispersion* arises because the interaction may be expressed in terms of the oscillator strengths that appear in the index of refraction for light (Hirschfelder *et al.* 1964).

where $\alpha_a(\omega)$ and $\alpha_b(\omega)$ are the dynamic dipole polarizabilities of the two atoms, continued to imaginary frequencies $\omega = \imath u$.[†]

Higher inverse powers R^{-n} are obtained by including the interaction of higher multipole moments. Technically, this is an asymptotic series (Brooks 1952; Dalgarno and Lewis 1956) that fails at small values of R due to an inappropriate use of multipole expansions. In the region where the atomic charge densities are finite, the full form of the Coulomb interaction should be retained in order to avoid the small-R divergences. However, such a calculation still omits effects of the overlap of the atomic wave functions which become appreciable in the region of most physical interest, the vicinity of the potential minimum. Then it is no longer possible to sustain the fiction of atoms with their complements of distinct and distinguishable electrons. The full antisymmetry of the electronic wave function must be retained; exchange and correlation must be treated on an equal footing. Both aspects, a more rigorous treatment of the Coulomb interaction and inclusion of wave function overlap, modify the perturbation theory result for the van der Waals energy and, in principle, require ambitious *ab initio* calculations.

2.1.1.2 *Semiempirical construction* One approach to the construction of the total interaction is to adopt one of the semiempirical methods available in the chemical literature (Hirschfelder *et al.* 1964; Buckingham *et al.* 1988; Meath and Koulis 1991). The key idea is to represent the total interaction as a sum of repulsive and attractive parts:

$$V(R) = V_{rep}(R) + V_{att}(R) \ . \tag{2.9}$$

There is an ambiguity in this definition because only the sum $V(R)$ has a definite physical meaning. Nevertheless, we know that in the limit of large separations R only $V_{att}(R)$ survives, and has the form of the van der Waals interaction eqn (2.7).[*] Since V_{vdW} is, in fact, a result of electron correlation, we are led to rewrite eqn (2.9) as

$$V(R) = V_{HF}(R) + V_{corr}(R) \ , \tag{2.10}$$

where $V_{HF}(R) = E_{HF} - E_{HF}^A - E_{HF}^B$ is the change in energy at the HF level and $V_{corr}(R) = E_{corr} - E_{corr}^A - E_{corr}^B$ is the change in correlation energy. This definition conforms to the usual definition of the correlation energy as the correction to the total energy as obtained in the HF approximation. Although $V_{corr}(R)$ reduces to the van der Waals interaction at large separations, it is extremely difficult to calculate it accurately as a correction to the HF energy since the total correlation energy is dominated by intraatomic correlations. At

---

[†]At distance scales R larger than about $c\hbar/$(ionization energy) ~ 300 Å, the finite speed of light c causes retardation of the interaction and changes the asymptotic dependence to R^{-7}, see the discussion at eqns (2.22) and (2.23).

[*]For illustrative purposes we suppose that only the dipolar dispersion force is relevant. The higher multipole contributions can be treated in a similar way.

large R, a much more direct procedure is to treat the interatomic interactions as a perturbation to the separate atoms in their respective ground states, as described above. However, extensions of the perturbation method to include wave function overlap become exceedingly cumbersome and are ill-suited to obtaining those contributions which HF theory provides so naturally. Progress in combining wave function overlap and a perturbation theory formulation to obtain terms such as the exchange-dispersion energy has come only recently (Tang *et al.* 1995).

Damped dispersion forces Some progress in combining wave function overlap with the perturbation theory formulation can be made by use of the idea of damped dispersion forces (Tang and Toennies 1984; Meath and Koulis 1991). The partitioning of the interaction in eqn (2.10) has the advantage that the Hartree–Fock energy $V_{HF}(R)$ can be evaluated numerically with reasonable accuracy for many atomic combinations. For a pair of closed-shell atoms, $V_{HF}(R)$ is expected to be purely repulsive, as discussed above. Over a limited range of separations, the repulsive interaction can be represented as

$$V_{HF}(R) = V_0 \, e^{-\alpha R} \,, \tag{2.11}$$

where the exponential dependence essentially follows that of the wave function overlap. To this is added the change in the total correlation energy, assumed to have the form (Tang and Toennies 1984)

$$V_{corr}(R) = -f_N(\alpha R)\frac{C_6}{R^6} \tag{2.12}$$

with

$$f_N(x) = 1 - e^{-x} \sum_{n=0}^{N} \frac{x^n}{n!} \,. \tag{2.13}$$

The damping factor f_N vanishes as x^{N+1} as $x \to 0$ and tends to 1 for large x; it eliminates the small-R divergence of V_{vdW} if $N \geq 5$. When several terms are retained in the dispersion multipole series, the usual practice (Tang and Toennies 1986; Ahlrichs *et al.* 1988) is to have separate damping factors f_N for each inverse power law and set the corresponding value of N equal to that power, $N = 6$ for eqn (2.12).[*] The value of α is identified with the range parameter in the exponential repulsion,[†] although it may also be treated as an adjustable parameter (Aziz 1984; Meath and Koulis 1991). There is some plausibility to the Tang–Toennies prescription, because the departure of the correlation energy from its asymptotic form should reflect the degree of wave function overlap, which

---

[*]The choice $N = 5$ causes V_{corr} in eqn (2.12) to have a finite value as $R \to 0$ and amounts to a renormalization of the repulsive interaction in eqn (2.11). The convention for adatom–substrate potentials is to use f_2/z^3 (Nordlander and Harris 1984; Chizmeshya and Zaremba 1992).

[†]For an elaboration when the repulsion is more complex, see Tang and Toennies (1992).

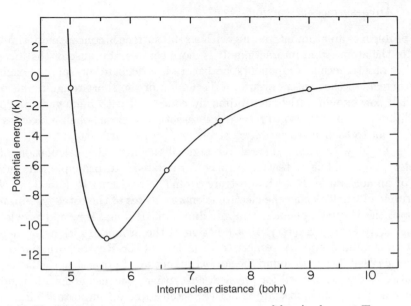

FIG. 2.2. He–He interatomic interaction, computed by Anderson, Traynor, and Boghosian (1993) using the 'exact' quantum Monte Carlo method. The result is consistent with the optimized semiempirical potential of Janzen and Aziz (1995).

also is dominant in determining the repulsion. It turns out that the prescription provides a very good representation of the interactions of inert gas pairs (Tang and Toennies 1986).

2.1.1.3 *He–He potential* The potential $V(R)$ of the helium dimer He_2 has been a demanding test case for the methods of this subsection. The potential well is about 11 K $\simeq$ 1 meV deep, while the total electronic energy at infinite separation is 157 eV, of which about 2 eV is correlation energy. To evaluate the depth with an accuracy of 10% by a total energy calculation requires a fractional accuracy of 1×10^{-6} in the total energy. Recently this requirement has been met and surpassed (Anderson *et al.* 1993) and the resulting potential energy curve is shown in Fig. 2.2. The potential is of thermodynamic accuracy (Aziz *et al.* 1995) and is sufficiently accurate to yield a He_2 binding energy consistent with the experimental value of $\sim$ 1 mK (Luo *et al.* 1993, 1996; Meyer *et al.* 1994; Schöllkopf and Toennies 1994, 1996). More generally, understanding of the physics needed to compute the well to 10% accuracy is developing, but has not yet been attained (Tang *et al.* 1995).

2.1.2 *Atom–surface interactions*

The problem of an atom interacting with a solid surface is conceptually similar to that of the atom–atom interaction. It is more complex because of the increased number of interacting components, and there is a delicate interplay of exchange and correlation that enters even in a discussion of the short-range repulsion, as we shall now explain. Although primarily concerned with inert gas adsorption, our discussion applies also to the weak interaction of closed-shell molecules such as H_2 prior to their dissociation on strongly reactive surfaces.[*]

We begin with some general remarks about how the calculation of the physisorption holding potential is to be formulated. In principle, the interaction of an adatom at $\mathbf{R}$ with substrate atoms at positions $\mathbf{R}_i$ ($i = 1, ..., N_a$) is determined by calculating the electronic energy states of the total system, in the spirit of the Born–Oppenheimer approximation. If done, it would provide the ground state energy $E(\{\mathbf{R}_i\})$ as a function of the atomic positions. When this function is minimized with respect to variations of the $\mathbf{R}_i$, the formulation contains the possibility of deriving the surface structure. In practice, the latter minimization is not performed for physical adsorption, because adsorption-induced reconstruction of the substrate is not expected to be an important effect when the atom–surface interaction is much weaker than the interactions within the substrate. Rather, one assumes that the substrate atoms reside at their equilibrium positions in the absence of the adsorbate. Nevertheless, for processes such as inelastic atom–surface scattering and surface diffusion, it is essential to allow for the motion of substrate atoms.

The total energy of the combined system is given by the analogue of eqn (2.2)

$$E(\mathbf{R}) = \langle \Psi | H(\mathbf{R}) | \Psi \rangle , \qquad (2.14)$$

where $H(\mathbf{R})$ is the Hamiltonian for position $\mathbf{R}$ of the adsorbate and $|\Psi\rangle$ is the total electronic wave function. As in eqn (2.3), the interaction energy is defined as

$$V(\mathbf{R}) = E(\mathbf{R}) - E_0^a - E_0^s , \qquad (2.15)$$

where E_0^a and E_0^s are the adsorbate and substrate ground state energies, respectively. At large separations where there is no overlap between the adsorbate and substrate wave functions, $V(\mathbf{R})$ is purely a correlation energy, the analogue of the van der Waals interaction for a pair of atoms, and is amenable to perturbation theory.

When the adatom and substrate wave functions overlap, the correlation energy is modified, and there are additional contributions to the energy which ultimately lead to a short-range repulsion. The net interaction can be formulated following the prescription of eqn (2.10). For example, this leads to the

---

[*]For a discussion of the relation between physisorption and chemisorption of H_2 on metals, see Nordlander *et al.* (1986).

potential wells for helium on noble metals shown in Fig. 2.3.* There is a good correspondence to the wells derived from scattering experiments.

However, an important difference from the atom–atom case is that correlation now may have a greater effect. For example, in the case of a metal, it leads to a modification of the work function and thus to a change in the spatial decay of the surface electron charge density. Thus a calculation of the overlap repulsion at the level of the HF approximation might be expected to be less reliable than in the atom–atom case. To avoid such an error, a more accurate description of the metal surface is needed. As we shall see, density functional theory is well-suited for this purpose.

In the next two subsections we give overviews of the perturbation theory of the attraction and the density functional theory of the short-range interaction. Then in Section 2.2 we describe the implementation of these two methods in more detail.

2.1.2.1 *Van der Waals interaction* The interaction of an atom with a substrate at large enough separations that there is negligible overlap can be treated with an extension of London's theory (1930a,b) of the van der Waals interaction between two closed-shell atoms. There is, in fact, a corresponding interaction between macroscopic objects (Hamaker 1937; Mysels and Scholten 1991; Israelachvili 1992). The interaction of one atom with a semi-infinite solid can even be derived as a special case of the interaction of two semi-infinite solids (Dzyaloshinskii *et al.* 1961). In Section 2.2.1 we present a systematic development of the adatom–substrate van der Waals interaction that includes the effects of the diffuseness of the surface and other refinements. We first give a more heuristic discussion of the interaction, which at large separations takes the form

$$V_{vdW}(Z) \simeq -C_3/Z^3 \ . \tag{2.16}$$

Several approaches[†] can be used to arrive at this result, with the van der Waals constant C_3 given by

$$C_3 = \frac{\hbar}{4\pi} \int_0^\infty d\omega \, \frac{[\epsilon(\imath\omega) - 1]}{[\epsilon(\imath\omega) + 1]} \alpha(\imath\omega) \ . \tag{2.17}$$

Here α is the electric dipole polarizability of the adatom, and ϵ is the long wavelength dielectric function of the substrate; both are analytically continued to pure imaginary frequencies. The function $\epsilon(\imath\omega)$ can be obtained from the

---

*In fact, one obtains the qualitative features of a physisorption holding potential for atomic hydrogen with a model in which the substrate has a sharp edge and the overlap repulsion is mimicked by the requirement that the adatom electron is excluded from the substrate (Bruch and Ruijgrok 1979; MacMillen and Landman 1984).

†These include a thorough treatment of the linear response of the solid (Zaremba and Kohn 1976), a simplified charged oscillator ('Drude') model of the linear response (Nijboer and Renne 1968, 1971), dispersion theory with an electrodynamic boundary condition for the substrate (Mavroyannis 1963; McLachlan 1964), and fluctuating electrodynamics (Lifshitz 1956).

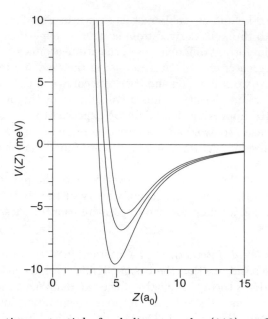

FIG. 2.3. Adsorption potentials for helium on the (110) surfaces of Cu, Ag, and Au (from top to bottom). This presentation by Chizmeshya (private communication) differs slightly from the results of Chizmeshya and Zaremba (1992) because the form of the dispersion has been taken from Tang and Toennies (1992). The distance Z is measured from the jellium edge, in units of Bohr radii a_0.

experimentally accessible absorptive part of the dielectric function $\epsilon_2(\omega)$ at real frequencies by the Kramers–Kronig relation

$$\epsilon(\imath\omega) = 1 + \frac{2}{\pi} \int_0^\infty d\omega' \, \frac{\omega' \epsilon_2(\omega')}{\omega'^2 + \omega^2} \ ; \tag{2.18}$$

expressions for the polarizability are given in Section 2.2.1.1.

Some insight into the origin of eqn (2.16) is provided by an elementary model first used by Lennard-Jones (1932b). Consider a neutral spherically symmetric atom at a distance Z from a metal surface. Because of the zero-point motion of the electrons in the atom, the atom has *instantaneous* multipole moments that generate dynamic long-range fields with which the metallic electrons interact. The metal is assumed to be a perfect conductor, so that its electrons respond instantaneously to the fields generated by the atomic charges and completely screen them. The leading term in the interaction at large Z is determined by the dipole fluctuations and can be evaluated by the method of *images*. An instantaneous dipole moment $\mathbf{p} = p_x \hat{x} + p_y \hat{y} + p_z \hat{z}$ has an image dipole $\mathbf{p}' =$

$-p_x\hat{x} - p_y\hat{y} + p_z\hat{z}$ situated at $-Z$ and the potential energy of interaction between the two is

$$V_p(Z) = \frac{1}{2} \left[\frac{\mathbf{p} \cdot \mathbf{p}' - 3p_z p_z'}{(2Z)^3} \right]. \tag{2.19}$$

The factor of one-half accounts for the fact that the image is induced by the external dipole, and the assumption of perfect screening eliminates any explicit reference to intrinsic properties of the metal. An application of first-order perturbation theory then leads to the result

$$V_p(Z) = -\frac{\langle p_z^2 \rangle}{4Z^3} = -\frac{\langle \mathbf{p}^2 \rangle}{12Z^3}. \tag{2.20}$$

The brackets denote an average in the atomic ground state. Equation (2.20) is consistent with eqn (2.17) since the perfect screening assumption corresponds to the limit $\epsilon \to \infty$ and[*]

$$\hbar \int_0^\infty d\omega\, \alpha(\imath\omega) = \pi \langle p_z^2 \rangle. \tag{2.21}$$

The analogous approximation for a dielectric which is able to respond instantaneously would have an additional factor of $[\epsilon - 1]/[\epsilon + 1]$ multiplying eqn (2.20).

We have thus shown that an attractive interaction arises from the correlation between the substrate charges and the zero-point fluctuations of the adatom charge density, even when the time-dependent dipole moment $\mathbf{p}(t)$ has zero average value. The root-mean-square dipole moment for a ground state inert gas atom is rather large: in debye units (1 debye = 10^{-18} esu cm), the values derived from the entries in Table A.2 range from 3.8 for He to 12.6 for Xe.

Retardation However, the idealization of perfect screening is too extreme, and a real metal substrate cannot completely screen the fluctuating dipole field of an inert gas adatom. The characteristic frequency of the external fluctuations amounts to an electronic excitation energy of 10–20 eV, which is higher than the plasma frequency of a typical metal.[†] Bardeen (1940) and Margenau and Pollard (1941) were the first to consider the retarded screening response of the substrate and found an interaction which is reduced in magnitude from the Lennard-Jones result. When all contributions to the screening response are included, the expression given in eqn (2.17) is obtained.[‡]

Relativistic retardation The domain of validity of eqn (2.16) is actually limited to a range where 'Z is large but not too large'. It must be large enough that wave functions of adatom and substrate electrons do not overlap. It must be small on

---

[*]In a single-frequency model of the atom, $\alpha(\imath\omega) = \alpha_0/[1 + (\omega/\omega_0)^2]$, eqns (2.20) and (2.21) yield $C_3 = \alpha_0\omega_0/8$, where ω_0 is the characteristic frequency of the atom. See also Appendix E for the result when the retarded response of the substrate is included.

[†]Thus, the physisorption wells of metals with the same density of s-electrons, matches which occur between some simple and noble metals, may be quite different (Nordlander *et al.* 1986).

[‡]The integral over imaginary frequencies here, and in eqns (2.8) and (2.61), results from using the Casimir–Polder identity to express convolutions of matrix elements of the interacting components as products of functions evaluated for the separate components.

the scale set by the wavelengths of the characteristic frequencies of the adatom. When Z becomes large on that scale, typically 100s of Å, effects of relativistic retardation become important. One way to express this is to note that eqn (2.19) and its generalization for retarded response of the substrate amount to using the 'near-zone' form of the field of an oscillating dipole. The extreme limit of relativistic retardation requires the use of fields in the 'radiation-zone'. Effects of relativistic retardation are most significant for the length scales involved with van der Waals forces between macroscopic objects and for thick films. They also lead to corrections in the analysis of certain geometries of atom–solid scattering.[§]

Relativistic retardation for the van der Waals energy was first treated by Casimir and Polder (1946, 1948), who showed that the $1/Z^3$ behavior of eqn (2.16) crosses over to a $1/Z^4$ dependence at very large Z. They took a rather different view of the interaction and formulated it in terms of the shift in the zero-point energy of the electromagnetic field, an approach that has been pursued by others of the Dutch school (van Kampen et al. 1968). Because the subject touches on such fundamental issues, the basic result has been re-derived several times (Lifshitz 1956; Dzyaloshinskii et al. 1961; McLachlan 1963; Renne 1971; Marvin and Toigo 1982). The following form encompasses the near-zone and radiation-zone limits

$$
\begin{aligned}
V_{rel}(Z) &= -\frac{1}{2\pi} \int_0^\infty du\, \alpha(\imath u)(\frac{u}{c})^3 \\
&\times \int_1^\infty d\eta\, [\Delta_s + (1 - 2\eta^2)\Delta_p] \exp[-2\eta(\frac{u}{c})Z]\,, \\
\sigma &= [\eta^2 + \epsilon(\imath u) - 1]^{1/2} \\
\Delta_s &= \frac{\sigma - \eta}{\sigma + \eta} \\
\Delta_p &= \frac{\sigma - \eta\epsilon(\imath u)}{\sigma + \eta\epsilon(\imath u)}\,,
\end{aligned}
\tag{2.22}
$$

where c is the speed of light.[*] In the limit $Z \to 0$, eqn (2.22) reduces to eqn (2.16), while for $Z \to \infty$ and $\epsilon(0) \to \infty$ the energy becomes

$$
V_{rel} = -\frac{3\hbar c\alpha(0)}{8\pi Z^4}\,,
\tag{2.23}
$$

which is the Casimir–Polder (1948) result for the interaction energy of an atom with a good conductor in the limit of extreme retardation.

2.1.2.2 *Density functional theory* We return now to the problem of determining the overlap repulsion. It is useful to base our discussion on density functional

[§] The retarded interaction must be included for a successful treatment of the ultra-low energy sticking of an H atom to liquid He (see Carraro and Cole 1995 and Section 4.1.2).

[*] Equation (2.22) is written in atomic units $\hbar = e = m_{el} = 1$, so that the numerical value of c must be adjusted accordingly to be $c \simeq 137$, the inverse of the fine structure constant.

theory (Hohenberg and Kohn 1964; Kohn and Sham 1965), which in principle is able to treat exchange and correlation on an equal footing. The method is widely used in molecular and solid state physics (Lundqvist and March 1983) and has proven to be versatile and accurate in many applications. More importantly, it provides a rigorous formulation for electronic structure calculations. We therefore begin with a brief review of density functional theory (Lang 1973, 1994; Parr and Yang 1989; Ellis 1995; Gross and Dreizler 1995).

There are two central theorems in the development (Hohenberg et al. 1990). First, it is proved (Hohenberg and Kohn 1964) that the ground state energy of the many-electron system is a functional of the single particle density $n(\mathbf{r})$. This itself is a startling result because it implies that knowledge of a function of 3 variables $n(\mathbf{r})$ suffices, in principle, to determine the ground state wave function of N electrons, a function of $3N$ variables. Second, there is a variational principle (Kohn and Sham 1965) with respect to a solution in terms of a set of single-particle Schrödinger equations. The latter leads to an approximate formulation similar to the Hartree equations (Kohn and Sham 1965; Lang 1973) in which the complexities are contained in an exchange-correlation energy functional $E_{xc}[n]$. Hohenberg et al. (1990) reviewed the conceptual developments and highlighted some of the subtle considerations in the formally rigorous reformulation of the quantum mechanical ground state problem in terms of the single particle density. Our purpose here is to outline the derivation of the Kohn–Sham equations and to state the leading approximations; we provide several references to sources where more complete accounts may be found.

Consider a Hamiltonian with coupling constant λ (Gunnarsson and Lundqvist 1976; Harris 1984)

$$H_\lambda = T + V_\lambda + \lambda U \,, \tag{2.24}$$

composed of the kinetic energy operator $T = \sum_i p_i^2/2m$, the electron–electron potential energy U

$$U = \frac{1}{2} \sum_{i \neq j} \frac{1}{|\mathbf{r}_i - \mathbf{r}_j|} \,, \tag{2.25}$$

and the interaction V_λ of the electrons with an external potential

$$V_\lambda = \int d\mathbf{r} \, v_\lambda(\mathbf{r}) \hat{n}(\mathbf{r}) \,. \tag{2.26}$$

v_λ is chosen in such a way that the ground state density at coupling constant λ, $\langle \Psi_\lambda | \hat{n}(\mathbf{r}) | \Psi_\lambda \rangle$, is equal to the true ground state density $n(\mathbf{r})$ at $\lambda = 1$ and that it then reduces to the physical potential $v(\mathbf{r}) \equiv v_{\lambda=1}(\mathbf{r})$. An application of the Hellmann–Feynman theorem then leads to an expression for the ground state energy in terms of an integration with respect to the coupling constant:

$$E = T_0[n] + \frac{1}{2} \int d\mathbf{r} \int d\mathbf{r}' \frac{n(\mathbf{r})n(\mathbf{r}')}{|\mathbf{r} - \mathbf{r}'|} + \int d\mathbf{r} \, n(\mathbf{r})v(\mathbf{r}) + E_{xc}[n] \,, \tag{2.27}$$

where $T_0[n]$ is the kinetic energy of a system of noninteracting electrons of density n and the exchange-correlation energy functional $E_{xc}[n]$ is defined as

$$E_{xc}[n] \equiv \int_0^1 d\lambda \langle \Psi_\lambda | U | \Psi_\lambda \rangle - \frac{1}{2} \int d\mathbf{r} \int d\mathbf{r}' \frac{n(\mathbf{r})n(\mathbf{r}')}{|\mathbf{r} - \mathbf{r}'|} . \tag{2.28}$$

It is important to recognize that eqns (2.27) and (2.28) together provide an exact expression for the ground state energy of the nonuniform interacting system. In fact, as proven by Hohenberg and Kohn (1964), $E = E[n]$ is a *functional* of the electronic density and is minimized by the true ground state density $n(\mathbf{r})$ at the true ground state energy, E_0.

Kohn and Sham (1965) showed, in an important development, that the interacting electron problem could be reduced to the self-consistent calculation of single-particle orbitals ψ_i in a potential derived from $E_{xc}[n]$. At $\lambda = 0$, the Hamiltonian eqn (2.24) corresponds to a collection of noninteracting fermions in the external potential $v_0(\mathbf{r})$. As such, $|\Psi_0\rangle$ is a single Slater determinant consisting of single-particle orbitals ψ_i defined by the Schrödinger equation

$$\left[-\frac{1}{2}\nabla^2 + v_0(\mathbf{r}) \right] \psi_i(\mathbf{r}) = \epsilon_i \psi_i(\mathbf{r}) , \tag{2.29}$$

and $v_0(\mathbf{r})$ is that potential which generates the exact ground state density according to*

$$n(\mathbf{r}) = \sum_{i=1}^N |\psi_i(\mathbf{r})|^2 . \tag{2.30}$$

Kohn and Sham showed that the energy functional is minimized when

$$v_0(\mathbf{r}) = v_{\lambda=1}(\mathbf{r}) + \phi(\mathbf{r}) + v_{xc}(\mathbf{r}) , \tag{2.31}$$

where $\phi(\mathbf{r}) = \int d\mathbf{r}' \, n(\mathbf{r}')/|\mathbf{r} - \mathbf{r}'|$ is the electrostatic potential of the electrons and $v_{xc}(\mathbf{r}) = \delta E_{xc}[n]/\delta n(\mathbf{r})$ is the exchange-correlation potential. That is, the orbitals ψ_i are obtained from the self-consistent solution of a closed set of equations, eqns (2.29)–(2.31). Since $E_{xc}[n]$ is not known exactly, approximations to this functional are required before explicit calculations can be performed. Fortunately, there are several approximations motivated by physical considerations which are relatively easy to use and lead to results of acceptable accuracy.

The total exchange-correlation energy $E_{xc}[n]$ can be divided into exchange and correlation contributions as follows (Harris and Jones 1974; Sahni *et al.* 1982; Sahni and Levy 1986):

$$E_{xc}[n] = E_x[n] + E_c[n] \tag{2.32}$$

---

*The index i includes a spin component; however we only treat cases for which the orbitals of the two spin orientations are degenerate.

with

$$E_x[n] = \langle \Psi_0 | U | \Psi_0 \rangle - \frac{1}{2} \int d\mathbf{r} \int d\mathbf{r}' \, \frac{n(\mathbf{r})n(\mathbf{r}')}{|\mathbf{r} - \mathbf{r}'|} \qquad (2.33)$$

and

$$E_c[n] = \int_0^1 d\lambda \langle \Psi_\lambda | U | \Psi_\lambda \rangle - \langle \Psi_0 | U | \Psi_0 \rangle . \qquad (2.34)$$

Note that we have *not* defined the correlation term E_c to be the difference between the true ground state energy and the HF approximation to it. To establish a formal relation with Hartree–Fock theory, it is useful to introduce the exchange-only (XO) energy functional

$$E_{XO}[n; v] = T_0[n] + \frac{1}{2} \int d\mathbf{r} \int d\mathbf{r}' \, \frac{n(\mathbf{r})n(\mathbf{r}')}{|\mathbf{r} - \mathbf{r}'|} + \int d\mathbf{r} \, n(\mathbf{r})v(\mathbf{r}) + E_x[n] , \quad (2.35)$$

so that the total energy functional is

$$E[n] - E_{XO}[n; v] + E_c[n] . \qquad (2.36)$$

As stated earlier, the minimization of $E[n]$ yields the true ground state density. On the other hand, minimization of the functional E_{XO} yields an approximate ground state density and energy. If a full variation of the single particle orbitals ψ_i used to define $|\Psi_0\rangle$ is performed, the usual set of HF equations is obtained. These orbitals generate the HF density $n_{HF}(\mathbf{r})$, which is different from $n(\mathbf{r})$; the corresponding energy $E_{XO}[n_{HF}; v]$ is the HF ground state energy. The energy E_{corr} defined by

$$E_{corr} = E[n] - E_{XO}[n_{HF}; v] \qquad (2.37)$$

is the more usual definition of correlation energy based on the HF reference point. Since $E[n]$ is stationary about the true ground state density n, replacing n by n_{HF} in eqn (2.36) leads to an error that is second order in the density difference $(n - n_{HF})$. To this level of accuracy, we can therefore write

$$E[n] \simeq E_{XO}[n_{HF}; v] + E_c[n_{HF}] . \qquad (2.38)$$

Thus, the conventional correlation energy E_{corr} is approximately equal to $E_c[n_{HF}]$ in the density functional setting. Equation (2.35) leads to a variant of HF theory with *local* potentials in the orbital equations, as in eqn (2.29).

We next describe the most commonly used approximation to $E_{xc}[n]$ in applications of density functional theory. In the *local density approximation* (LDA) the exchange-correlation energy functional is taken to be (Hohenberg and Kohn 1964)

$$E_{xc}^{LDA}[n] = \int d\mathbf{r} \, n(\mathbf{r}) \, \epsilon_{xc}[n(\mathbf{r})] , \qquad (2.39)$$

where $\epsilon_{xc}(n)$ is the exchange-correlation energy per particle of a uniform electron gas of density n. The essence of the approximation is to treat the inhomogeneous electronic system as locally homogeneous with regard to exchange and correlation. The function $\epsilon_{xc}(n)$ is subdivided into exchange and correlation parts,

$$\epsilon_{xc}(n) = \epsilon_x(n) + \epsilon_c(n)\,, \tag{2.40}$$

with the Hartree–Fock exchange energy given by (Mahan 1981)

$$\epsilon_x(n) = -\frac{3}{4}\left(\frac{3}{\pi}\right)^{1/3} n^{1/3}\,. \tag{2.41}$$

A similar closed-form expression for $\epsilon_c(n)$ is not available, although it is accurately known over a wide range of densities and has been fit to analytic expressions for computational purposes (Vosko *et al.* 1980). In most applications, $\epsilon_x(n)$ is the dominant contribution to the energy, but inclusion of $\epsilon_c(n)$ usually leads to improved agreement with experiment.

We emphasize that use of $\epsilon_x(n)$ by itself, i.e., dropping $\epsilon_c(n)$ from eqn (2.40), is not equivalent to a HF calculation even though it is nominally an 'exchange-only' approximation. The differences between LDA-exchange and HF-exchange are particularly evident in the calculation of interatomic potentials, see the review of Campbell *et al.* (1992).

With the LDA form of $E_{xc}[n]$, the XC potential is given by

$$v_{xc}^{LDA}(\mathbf{r}) = \frac{d}{dn}\Big(n\epsilon_{xc}(n)\Big)\Big|_{n=n(\mathbf{r})}\,. \tag{2.42}$$

This prescription, in principle, enables the Kohn–Sham orbital equations to be solved self-consistently to obtain an approximate ground state density and energy. Many calculations of this type have been performed for chemisorbed species. For physisorption, Lang and his coworkers (Lang 1981; Lang and Williams 1982; Lang and Nørskov 1983) have performed the most careful calculations of the interaction of inert gas atoms with metallic surfaces. Their results will be discussed more fully in Section 2.2.2.

The Hohenberg–Kohn theorem, that the ground–state of the nonuniform electron gas is a functional $E[n]$ of the density n, revived interest in statistical theories of electronic structure such as the Thomas–Fermi and Weizsäcker approximations. In that context, the Kohn–Sham orbital equations are a systematic way to include so-called *gradient corrections* to the kinetic energy term of statistical models. The density functional approach stimulated attempts to use the nonuniform electron density for estimates of interaction energies, particularly those likely to be dominated by overlap energies. We now describe two such approximations that have yielded useful estimates of the interaction energy without explicitly solving the Kohn–Sham orbital equations.

Gordon–Kim approximation   Gordon and Kim (1972) essentially adopted a Thomas–Fermi approximation in which the Kohn–Sham kinetic energy $T_0[n]$ is replaced by the kinetic energy density of the uniform electron gas of density n, $\epsilon_{kin}(n) = \frac{3}{10}(3\pi^2 n)^{2/3}$. In this way the energy functional becomes an explicit functional of the electron density, in contrast to the Kohn–Sham theory, which makes use of electron orbitals. Gordon and Kim expressed the interaction energies of inert gas pairs as the difference

$$E_{AB}(R) = V_{Coul} + \int d\mathbf{r}\, n(\mathbf{r})\{\epsilon_{stat}[n_A(\mathbf{r}) + n_B(\mathbf{R} - \mathbf{r})]$$
$$- \epsilon_{stat}(n_A) - \epsilon_{stat}(n_B)\}, \qquad\qquad (2.43)$$

where $\epsilon_{stat}(n) = \epsilon_{kin}(n) + \epsilon_{xc}(n)$ and V_{Coul} is the change in the Coulomb energy brought about by the overlap of the atomic charge densities. They further assumed that the total density n could be approximated by the atomic superposition $n_A + n_B$, where HF densities are used for the individual atoms and charge rearrangements of the interacting species have been omitted. Not only does their approximation reproduce the qualitative features of the overlap repulsion, but it also leads to reasonable magnitudes for the depth of the attractive potential well of inert gas pairs.

However, the asymptotic, large-R, dependence of the Gordon–Kim potential exhibits an exponential behavior rather than the power law that is characteristic of the van der Waals energy. In the Gordon–Kim calculation, the attractive interaction arises from the overlap of the atomic charge densities and its largest contribution is from the local *exchange* term in the energy functional. This serves to emphasize the point that a local approximation to the exchange is quite distinct from a calculation in the HF approximation, which would lead to purely repulsive interactions for inert gas pairs. Further, the nonlocal character of the van der Waals interaction has not been included. On the other hand, it appears that the local exchange approximation does, in a fashion not yet understood, include some of the attractive energy usually attributed to the van der Waals interaction.

A very similar approximation has been applied to the physisorption holding potential. Freeman (1975a) showed that it reproduces the main features of the holding potentials for inert gases on graphite, although the potential well depths he obtained for He, Ne, Ar, and Kr are only 25% of the observed values. Freeman (1975b) also estimated that the interatomic potential wells on graphite are 10–20% shallower than for the isolated pairs, due to adsorption-induced modifications of the adatom–adatom potential at short separations. Van Himbergen and Silbey (1977) treated the holding potential of Ar and of He on jellium using the substrate charge density from density functional theory (Lang and Kohn 1970) and Hartree–Fock charge densities for the atoms.* They obtained reasonable values for the physisorption well depths. However, the success of these

---

*There were prior estimates of the physisorption holding potential in which the short-range repulsion was obtained from the statistical model approximation for the kinetic energy (Kleiman and Landman 1973a,b, 1976). The work with the Gordon–Kim formulation obtained deeper potential wells. There were other simplifications made in the earlier work that had unexpectedly large effects in the numerical results.

rather primitive approximations in reproducing the energy scale[†] of the adsorption process again is linked to the contributions of local exchange to the potential well.

Embedded atom methods A rather different approach is that provided by the so-called *embedded atom methods*[*] first formulated by Nørskov and Lang (1980) and Stott and Zaremba (1980) and subsequently applied by Esbjerg and Nørskov (1980) to the scattering of helium atoms from metal surfaces. The general problem of interatomic interactions is viewed in terms of the process of embedding an atom into an electronic host, e.g., into the surface of a solid for adsorption. The success of the method depends on obtaining a reliable estimate of the embedding energy in terms of some set of properties characteristic of the atom and host.

The simplest embedding energy is that defined for a uniform extended medium of density $\bar{n}$, i.e., a jellium model: $\Delta E_{hom}(\bar{n})$. This embedding energy is a function of the atomic species and is determined by a self-consistent calculation of the screening of the nuclear charge in the electron gas. There are corrections to this result when the atom is embedded into an inhomogeneous host. One form of the correction replaces $\bar{n}$ by a sampled density and the embedding energy becomes

$$V(\mathbf{R}) \simeq \Delta E_{hom}(\bar{n}_0(\mathbf{R})), \qquad (2.44)$$

where $\bar{n}_0$ is an average of the host charge density weighted by the electrostatic potential of the atoms. In the context of adsorption, eqn (2.44) represents the potential of an atom interacting with a solid substrate.

For He and the other inert gases, $\Delta E_{hom}(\bar{n})$ is an approximately linear, monotonically increasing function of the density in the range of densities used for atom–surface scatering (Puska *et al.* 1981; Stott and Zaremba 1982), reflecting the expected repulsive nature of the interaction of these atoms with a homogeneous gas of electrons. For the specific case of He, we have $\Delta E_{hom}(\bar{n}) \simeq \alpha \bar{n}$ with $\alpha \simeq 300$ eV a_0^3, as determined by a density functional calculation in the LDA (Stott and Zaremba 1980; Puska *et al.* 1981). This value of α implicitly includes the interaction of the atom with the positive background of the electron gas. Outside a metal surface, this positive background does not exist and subtraction of the extraneous electrostatic interaction leads to an effective value of α given by (Lang and Nørskov 1983; Manninen *et al.* 1984)

$$\alpha_{eff} = \alpha - \int d\mathbf{r}\, \phi_a(\mathbf{r}), \qquad (2.45)$$

[†] Combinations of approximations can be found which do better. For instance, Ihm and Cole (1989) used the Gordon–Kim approximation for the kinetic energy and the Tang–Toennies damping for the asymptotic van der Waals energy.

[*] Various names are attached to this approximation, for example, *effective medium approximation* (Nørskov and Lang 1980) and *quasiatom theory* (Stott and Zaremba 1980). The term 'embedded atom method' is usually reserved for a variant introduced by Daw and Baskes (1984), but it will be used here since it conveys most succinctly the spirit of all these related approaches.

where $\phi_a(\mathbf{r})$ is the electrostatic potential of the screened atom. This correction reduces the effective value of α to about 160 eV a_0^3 in the LDA. We therefore arrive at the working equation

$$V_{rep}(\mathbf{R}) = \alpha_{eff}\, \bar{n}_0(\mathbf{R}),$$ (2.46)

which is of the form used by Esbjerg and Nørskov (1980) to analyze the scattering of He from metal surfaces. The addition of a 'hybridization' correction enabled Lang and Nørskov to construct a physisorption potential that was in qualitative agreement with the full LDA density functional calculation.

The 'best' value of α to be used in eqn (2.45) has had a somewhat confused history and has led to some controversy (Manninen et al. 1984; Celli 1992). As argued by Puska et al. (1981), the rigorous low-density limit of α_{eff} is $2\pi a$, in atomic units, where a is the s-wave scattering length. That this limit was not observed in their calculations is not a failure of the identification of α_{eff} with a but of the LDA, which fails to account for the effect of electron–atom polarization. In fact, the LDA α_{eff} values are in quite good agreement with the 'polarization-free' scattering lengths (Cole and Toigo 1985). By the same token, the LDA value differs from that determined within the HF approximation. The HF s-wave scattering length of e-He is 1.422 a_0 (Moiseiwitsch 1953), which gives $\alpha_{eff}^{HF} \simeq 250$ eV a_0^3. This value is also different from the HF-α_{eff} of about 500 eV a_0^3 as determined by Harris and Liebsch (1982) for a He atom at a metal surface. Thus even within the HF approximation, the value of α_{eff} depends on the physical situation and it does not have a 'universal' value.* While there is obvious appeal in having a direct relation between the interaction and the surface charge density, one must keep in mind that eqn (2.46) is but a simple approximation to the repulsion and has a limited validity. Values $\alpha_{eff}(He) \simeq 250 - 600$ eV a_0^3 are currently in use (Takada and Kohn 1988; Celli 1992).

Where eqn (2.46) is of particular interest is in the discussions of the corrugation of the atom–substrate potential which is probed directly in atomic diffraction experiments. There have been several attempts to account for the diffraction experiments on the basis of eqn (2.46) as a working hypothesis, together with the assumption that the surface charge density can be approximated as a superposition of atomic densities. Such work brings into sharper focus the following problem: the electron density at a metal surface is expected to differ from that obtained as a superposition of free atomic densities, but how much effect the difference has on the helium–metal potential is not decided. While some tests seem to support using the superposition of atomic densities (e.g., Batra et al. 1985 and Cortona et al. 1992a), the amplitude of the calculated corrugation is invariably too large (Celli 1992). We return to this question in Section 2.3.3.1, when we discuss the corrugation.

---

*Other issues include the fact that the scattered electrons are at negative energy, since they are bound in the surface, and that there is a dependence on the work function of the metal, because α_{eff} represents an average effect of the inhomogeneous substrate charge distribution (Takada and Kohn 1985, 1988).

2.1.2.3 *Qualitative considerations* Having considered the key physical processes that enter in the potential, it is worthwhile to look at empirical qualitative trends before beginning a detailed theoretical discussion. Figure 2.4 presents results for the van der Waals coefficient and welldepth of He and Ar adsorption on various surfaces and is an extension of the information already given in Table 1.1.

The most strongly adsorbing surface is graphite, and the most inert is cesium. Groups of other materials having similar properties lie in between. The strength of the graphite attraction arises from the fact that the π-bonded charge is localized relatively close to the plane of the carbon nuclei. This permits the atom to come close to this plane; the equilibrium distance is ~ 3 Å. In contrast, the electrons of the alkali metal spill far out from the surface, because the work function is so low, and the adatom is kept at a much larger equilibrium distance, ~ 7 Å. The ratio of the well depths on graphite and cesium is in the range 10–20 and can be understood straightforwardly as a consequence of the z^{-3} van der Waals attractions for these two cases. There is irony in the fact that the weakly bound electrons that are the origin of the chemical reactivity of the alkali surfaces are also the origin of the inertness of the same surfaces in physical adsorption. The same qualitative behavior is found in adsorption of the other inert gases.

2.2 Atom–surface potential theory

In the next two subsections, we obtain the Lifshitz coefficient C_3, the image plane location Z_0 for the van der Waals forces, and the short-range adatom–substrate repulsion. These pieces have been combined (Nordlander and Harris 1984; Chizmeshya and Zaremba 1992), following the ideas described in Section 2.1.1.2, eqns (2.10)–(2.13), to construct a net interaction of the form

$$V_{hold}(Z) = A_0\, e^{-\alpha Z} - \frac{C_3 f_2(\alpha[Z - Z_0])}{(Z - Z_0)^3}. \qquad (2.47)$$

This potential provides a semiquantitative model of the interactions of inert gases with noble metals. Results for helium were shown in Fig. 2.3 and are discussed in Section 2.4.3.

2.2.1 *The van der Waals interaction*

We shall derive the first corrections (Zaremba and Kohn 1976) to the Lifshitz energy formula, eqns (2.16) and (2.17). In so doing, we show where the effects of the diffuseness of the metal surface and of higher multipoles enter. The first steps are to identify the relevant second-order perturbation energy and to express it in terms of generalized susceptibilities; that part of the formulation is quite general. Then we adopt the jellium model of a metal and make use of the translational symmetry parallel to the surface to reduce these general forms to obtain more explicit results.

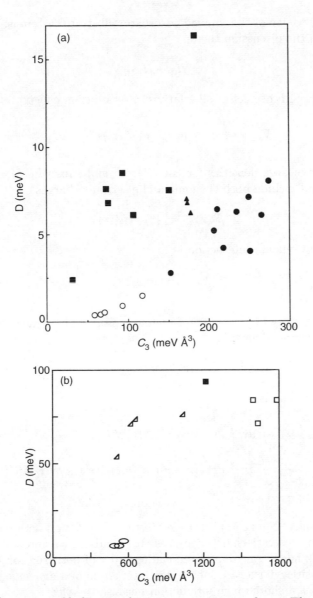

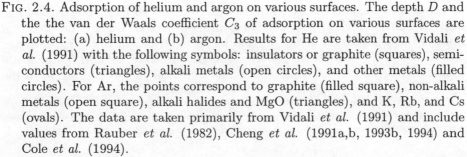

FIG. 2.4. Adsorption of helium and argon on various surfaces. The depth D and the the van der Waals coefficient C_3 of adsorption on various surfaces are plotted: (a) helium and (b) argon. Results for He are taken from Vidali *et al.* (1991) with the following symbols: insulators or graphite (squares), semiconductors (triangles), alkali metals (open circles), and other metals (filled circles). For Ar, the points correspond to graphite (filled square), non-alkali metals (open square), alkali halides and MgO (triangles), and K, Rb, and Cs (ovals). The data are taken primarily from Vidali *et al.* (1991) and include values from Rauber *et al.* (1982), Cheng *et al.* (1991a,b, 1993b, 1994) and Cole *et al.* (1994).

The Hamiltonian of the coupled adsorbate–substrate system is, in analogy to eqn (2.4) and the discussion there,

$$H = H_a + H_s + V_{sa} \, . \tag{2.48}$$

The coupling, with neglect of all relativistic retardation effects, is

$$V_{sa} = \int d\mathbf{r} \int d\mathbf{r}' \, \hat{\rho}^s(\mathbf{r}) \, \hat{\rho}^a(\mathbf{r}') \, v(\mathbf{r} - \mathbf{r}') \, , \tag{2.49}$$

where the net charge densities for the adatom and substrate ($j = a, s$) are in units of $|e|$ and include both the ionic and electronic charges*

$$\hat{\rho}^j(\mathbf{r}) = n^j_+(\mathbf{r}) - \hat{n}^j(\mathbf{r}) \, , \tag{2.50}$$

and the basic Coulomb interaction is

$$v(\mathbf{r} - \mathbf{r}') = \frac{e^2}{|\mathbf{r} - \mathbf{r}'|} \, . \tag{2.51}$$

The electron density fluctuation $\delta \hat{n}$ is defined to be the increment relative to the time-average charge density $\bar{\rho}$

$$\hat{\rho}^i(\mathbf{r}) = \bar{\rho}^i(\mathbf{r}) - \delta \hat{n}^i(\mathbf{r}), \quad i = a, s \, , \tag{2.52}$$

so that the coupling is

$$
\begin{aligned}
V_{as} &= \int d\mathbf{r} \int d\mathbf{r}' \, v(\mathbf{r} - \mathbf{r}') \bar{\rho}^s(\mathbf{r}) \, \bar{\rho}^a(\mathbf{r}') \\
&- \int d\mathbf{r} \int d\mathbf{r}' \, v(\mathbf{r} - \mathbf{r}') \, [\bar{\rho}^s(\mathbf{r}) \delta \hat{n}^a(\mathbf{r}') + \bar{\rho}^a(\mathbf{r}') \delta \hat{n}^s(\mathbf{r})] \\
&+ \ V_{disp} \, .
\end{aligned}
\tag{2.53}
$$

The first line of eqn (2.53) is the electrostatic interaction between the average adsorbate and substrate charge densities. If the adsorbate and substrate do not overlap, it still has a contribution in leading order of perturbation theory, using the 'bare' densities, for a case such as the interaction of a molecule with an ionic crystal. This is a short-range interaction because the electrostatic fields of the substrate decay exponentially with distance from the surface. The second line contains terms that give rise to induction energies at second order in perturbation theory: the polarization of the adatom by an electric field from the substrate and the screening charges (*image terms*) in the substrate induced by the static multipole moments of an adsorbate.[†] The third line is the interaction of the

---

*That is, n^j is an electron number density and is multiplied by $-|e|$ to be a charge density.

[†]For instance, the induction energy of a static quadrupole moment, as for a nitrogen molecule, contributes to the $1/Z^5$ term in the holding potential (Bruch 1983b).

charge fluctuations on the adatom and substrate; its explicit form is obtained by replacing $\hat{\rho}$ by $\delta\hat{n}$ in eqn (2.49). The third term has zero expectation value in the unperturbed ground state $|\Psi_0^a\rangle|\Psi_0^s\rangle$. We now state and then transform the second-order perturbation theory term for V_{disp}:

$$V_{disp}^{(2)} = -\sum_{mn} \frac{|\langle\Psi_0^s\Psi_0^a|V_{disp}|\Psi_m^s\Psi_n^a\rangle|^2}{E_m^s + E_n^a - E_0^s - E_0^a}. \tag{2.54}$$

Equation (2.54) can be expressed in terms of generalized susceptibilities of the adsorbate and substrate by using the spectral density function that determines the retarded response of a system to an external perturbation. For instance, if the adsorbate is subjected to the time-dependent electric potential ϕ_{ext}

$$\delta V_a^{ext}(t) = -\int d\mathbf{r}\,\hat{n}_a(\mathbf{r})\phi_a^{ext}(\mathbf{r},t), \tag{2.55}$$

the adsorbate polarization is given to first order in ϕ_{ext} by

$$\delta n_a(\mathbf{r},t) = \int dt' \int d\mathbf{r}'\,\chi_a(\mathbf{r},\mathbf{r}',t-t')\,\phi_a^{ext}(\mathbf{r}',t'). \tag{2.56}$$

The retarded response function in eqn (2.56) is given explicitly by (Martin 1968)

$$\chi_a(\mathbf{r},\mathbf{r}',t-t') = \imath\theta(t-t')\langle\Psi|[\delta\hat{n}_a(\mathbf{r},t),\delta\hat{n}_a(\mathbf{r}',t')]|\Psi\rangle, \tag{2.57}$$

where $[...,...]$ denotes a commutator, θ denotes the unit step function, and $|\Psi\rangle$ is the state before perturbation. The time-Fourier-transform is defined by

$$\chi_a(\mathbf{r},\mathbf{r}',\omega) = \int_0^\infty d\tau\,e^{\imath\omega\tau}\chi_a(\mathbf{r},\mathbf{r}',\tau), \tag{2.58}$$

and, when analytically continued to imaginary frequencies $\omega = \imath u$, the spectral representation takes the form

$$\chi_a(\mathbf{r},\mathbf{r}',\imath u) = \sum_n \frac{2(E_n^a - E_0^a)}{(E_n^a - E_0^a)^2 + u^2}\langle\Psi_0^a|\delta\hat{n}^a(\mathbf{r})|\Psi_n^a\rangle\langle\Psi_n^a|\delta\hat{n}^a(\mathbf{r}')|\Psi_0^a\rangle. \tag{2.59}$$

The corresponding expressions for the substrate s are obtained formally by the replacement $a \to s$. We now use these expressions and the Casimir–Polder identity

$$\frac{2}{\pi}\int_0^\infty du \frac{AB}{(A^2+u^2)(B^2+u^2)} = \frac{1}{A+B}, \tag{2.60}$$

to transform eqn (2.54).

The dispersion energy is then expressed in terms of a correlation of substrate and adsorbate charge fluctuations as follows:

$$
\begin{aligned}
V_{disp}^{(2)} &= -\int_0^\infty \frac{du}{2\pi} \int d\mathbf{r}_1 \int d\mathbf{r}_2 \int d\mathbf{r}_3 \int d\mathbf{r}_4 v(\mathbf{r}_1 - \mathbf{r}_2)\chi_s(\mathbf{r}_2,\mathbf{r}_3,iu) \\
&\quad \times v(\mathbf{r}_3 - \mathbf{r}_4)\chi_a(\mathbf{r}_4,\mathbf{r}_1,iu).
\end{aligned}
\tag{2.61}
$$

Equation (2.61) is the linear response, second-order-perturbation, energy[*] for the electrostatic Hamiltonian eqn (2.48). It does not rely on a multipole expansion for the adatom charge density or on a specific model for the substrate.

Equations (2.54) and (2.61) may also be regarded as the leading term in a more general strong-coupling theory (Casimir and Polder 1946, 1948; van Kampen *et al.* 1968; Renne 1971; Mahanty and Ninham 1976). That formulation replaces V_{sa} by λV_{sa} in eqn (2.48) and has a formal integration with respect to the coupling constant, as in eqn (2.28). Developing the theory in that fashion leads to an interpretation of the dispersion energy in terms of the shift in the total electromagnetic zero-point energy resulting from the coupling of the adsorbate and substrate. This was a major conceptual development that also enabled tests of the smallness of terms beyond the second order of perturbation theory (eqn (2.54)). It also helps maintain consistency in calculations which include both induction and dispersion energies.[†] However, all our ensuing analysis is based on eqn (2.61).

In the next subsections we develop the content of eqn (2.61) for an inert gas adatom and a jellium substrate. Two-dimensional Fourier decompositions are used to incorporate the symmetry of the planar semi-infinite substrate. The Coulomb potential is expressed as

$$
v(\mathbf{r}) = \int d\mathbf{q}\, \frac{2\pi}{q}\, \exp[\imath\mathbf{q}\cdot\boldsymbol{\rho} - q|z|],
\tag{2.62}
$$

where $\mathbf{q}$ is a 2D-wavevector and the 3D position vector is $\mathbf{r} = (\boldsymbol{\rho}, z)$. We also use a complex wavevector

$$
\boldsymbol{\kappa} = \mathbf{q} + \imath q\, \hat{z}.
\tag{2.63}
$$

We rewrite eqn (2.61) in terms of the 2D Fourier decomposition of the Coulomb potential, with the adsorbate coordinates $\mathbf{x}$ relative to the adatom center at $(\mathbf{R}, Z)$:

---

[*]Note that we may bias the answer and omit other leading processes such as charge transfer to or from the substrate by taking the perturbation expansion relative to a neutral adatom.

[†]For instance, in going to higher orders of perturbation theory, the question arises whether the average density $\bar{\rho}$ in eqn (2.53) is that for the unperturbed subsystem or the coupled system.

$$V_{disp}^{(2)} = -\int_0^\infty \frac{du}{2\pi} \int \frac{d^2q}{(2\pi)^2} \int \frac{d^2q'}{(2\pi)^2} \frac{2\pi}{q} \frac{2\pi}{q'} e^{-Z(q+q')} \exp[\imath\mathbf{R}\cdot(\mathbf{q}-\mathbf{q}')]$$

$$\times \int d\mathbf{r} \int d\mathbf{r}' \exp[-\imath\boldsymbol{\kappa}\cdot\mathbf{r} + \imath\boldsymbol{\kappa}'^*\cdot\mathbf{r}'] \chi_s(\mathbf{r},\mathbf{r}',iu)$$

$$\times \int d\mathbf{x} \int d\mathbf{x}' \exp[\imath\boldsymbol{\kappa}\cdot\mathbf{x} - \imath\boldsymbol{\kappa}'^*\cdot\mathbf{x}'] \chi_a(\mathbf{x},\mathbf{x}',iu). \qquad (2.64)$$

The definition of the complex wavevector $\boldsymbol{\kappa}$ means that the complex exponentials contain decaying, evanescent wave, components. In the further development based on eqn (2.64), we shall find the factorization of substrate and adsorbate properties represented by the last two lines to be very useful.

2.2.1.1 Adsorbate response function
Dipole approximation   We start by retaining only the first multipole moment for the adatom charge fluctuation in order to simplify the spatial integrals in eqn (2.61). The adatom is spatially localized on an atomic scale, and thus $\chi_a(\mathbf{r},\mathbf{r}')$ has appreciable strength only for values of $\mathbf{r}$ and $\mathbf{r}'$ close to the atomic center $\mathbf{R}$. Therefore, we change to variables $\mathbf{x} = \mathbf{r} - \mathbf{R}$ and $\mathbf{x}' = \mathbf{r}' - \mathbf{R}$ and expand the Coulomb interactions v in powers of $\mathbf{x}$ and $\mathbf{x}'$. The first nonvanishing term in this expansion of the spatial integrals in eqn (2.61) (denoted by angular brackets) is*

$$< v\chi_s v\chi_a > \simeq \sum_{\mu\nu} \alpha_{\mu\nu}(iu) \int d\mathbf{r} \int d\mathbf{r}' \nabla_\mu v(\mathbf{r}-\mathbf{R})\chi_s(\mathbf{r},\mathbf{r}',iu)\nabla'_\nu v(\mathbf{r}'-\mathbf{R}).$$

$$(2.65)$$

The dipole polarizability tensor for the adatom is defined by

$$\alpha_{\mu\nu}(iu) = \int d\mathbf{x} \int d\mathbf{x}' \, x_\mu \, x'_\nu \, \chi_a(\mathbf{x},\mathbf{x}',iu). \qquad (2.66)$$

In all that follows, we treat an S-state atom, in which case the tensor is diagonal, $\alpha_{\mu\nu}(iu) = \alpha(iu)\delta_{\mu\nu}$.

The double integration in eqn (2.65) can be rewritten as

$$D_{\mu\nu}^0(\mathbf{R},iu) \equiv \int d\mathbf{r} \int d\mathbf{r}' \, \nabla_\mu v(\mathbf{r}-\mathbf{R})\chi_s(\mathbf{r},\mathbf{r}',iu)\nabla'_\nu v(\mathbf{r}'-\mathbf{R}) \qquad (2.67)$$

$$= \nabla_{R_\mu}\nabla_{R'_\nu} \int d\mathbf{r} \int d\mathbf{r}' v(\mathbf{r}-\mathbf{R})\chi_s(\mathbf{r},\mathbf{r}',iu)v(\mathbf{r}'-\mathbf{R}')\Big|_{\mathbf{R}'=\mathbf{R}},$$

and has a simple physical interpretation. The integral in the last line is the electric potential at $\mathbf{R}$ from the substrate charge that is induced by an external

---

*The linear term vanishes because, by definition, the density fluctuation operator $\delta\hat{n}_a$ has a spatial integral equal to zero; then, for each Cartesian component μ and ν,

$$\int d\mathbf{x} \int d\mathbf{x}' \, x'_\mu\chi_a(\mathbf{x},\mathbf{x}',iu) = \int d\mathbf{x} \int d\mathbf{x}' \, x_\nu\chi_a(\mathbf{x},\mathbf{x}',iu) = 0$$

point charge located at $\mathbf{R}'$. The derivatives with respect to R_μ and R'_ν lead to a susceptibility tensor $D^0_{\mu\nu}(\mathbf{R}, iu)$ that represents the μ-component of the induced electric field at $\mathbf{R}$ due to a unit dipole moment at R pointing in the ν-direction.

Combining eqns (2.61), (2.65), and (2.67), we have

$$V^{(2)}_{disp} \simeq - \int_0^\infty \frac{du}{2\pi} \alpha(iu) \sum_\mu D^0_{\mu\mu}(\mathbf{R}, iu) \,. \tag{2.68}$$

Then the physical interpretation of the electric susceptibility, $\mathcal{D}^0(\mathbf{R}, iu)$, helps us see how eqn (2.20) arises. For an oscillating point dipole of unit amplitude situated at a distance Z from a perfectly conducting surface, the substrate charge that is induced can be represented by an image dipole of unit amplitude at distance Z below the surface. After integrating the frequency-dependent adatom polarizability, we recover eqn (2.20).

General multipole expansion In addition to the dipole fluctuations, there are fluctuations of the higher multipole moments that also contribute to the total dispersion energy. Although the dipolar component dominates the interaction at large adatom–substrate separations, the relative contributions of the higher multipoles become larger near the minimum of the holding potential. We now generalize the expansion eqn (2.65) by considering the atomic polarization function

$$h(q, iu) = \int d\mathbf{x} \int d\mathbf{x}' \exp[\imath \boldsymbol{\kappa} \cdot \mathbf{x} - \imath \boldsymbol{\kappa}^* \cdot \mathbf{x}'] \chi_a(\mathbf{x}, \mathbf{x}', iu) \,. \tag{2.69}$$

The function $h(q, iu)$ is purely an adsorbate property and has a relatively simple dependence on the wavevector $\mathbf{q}$.

The response function for a spherically symmetric atom has the following expansion in spherical harmonics Y_{lm}

$$\chi_a(\mathbf{x}, \mathbf{x}', iu) = \sum_{lm} \chi_l(x, x', iu) Y^*_{lm}(\hat{\mathbf{x}}) Y_{lm}(\hat{\mathbf{x}}') \,, \tag{2.70}$$

where the angular arguments are specified by the corresponding unit vectors $\hat{\mathbf{x}}$. The solid-angle integrations in eqn (2.69) can be done analytically with the result

$$h(q, iu) = \sum_{l=1}^\infty \frac{2^l \alpha_l(iu)}{l!(2l-1)!!} q^{2l} \,, \tag{2.71}$$

where the multipole polarizability $\alpha_l(iu)$ is defined by

$$\alpha_l(iu) = \frac{4\pi}{2l+1} \int_0^\infty dx\, x^{l+2} \int_0^\infty dx'\, x'^{l+2} \chi_l(x, x', iu) \,. \tag{2.72}$$

The case $l = 1$ corresponds to the dipole polarizability of eqn (2.66) and $l = 2$ to the quadrupole polarizability. Note that the expansion of the atomic polarization function contains only even powers of q. Empirically, the dynamic dipole polarizability for inert gas atoms is rather wellknown, but there are large uncertainties in modeling the dynamic quadrupole polarizability $\alpha_2(iu)$ (Standard and Certain 1985).

2.2.1.2 *Substrate response function* All the detailed information about the geometrical structure of the substrate and its response is buried within the substrate
response function χ_s. For example, the substrate may be insulating or metallic,
an ordered periodic solid or one with defects, each with its own characteristic
response. To begin, we consider a rather simple and idealized model, the *jellium*
model of a semi-infinite substrate with planar surface oriented perpendicular to
the z-axis of the coordinate system. No lattice structure is retained within this
model, and the substrate is translationally invariant along the surface. However,
the model does provide a realistic description of the electronic surface properties
in a direction perpendicular to the surface (Lang and Kohn 1970, 1971). The
density response function of the jellium has the property

$$\chi_s(\mathbf{r}, \mathbf{r}', iu) = \chi_{sJ}(z, z', \boldsymbol{\rho} - \boldsymbol{\rho}', iu), \tag{2.73}$$

and its 2D Fourier transform, defined by

$$\chi_s(z, z', q, iu) = \int d(\boldsymbol{\rho} - \boldsymbol{\rho}') \exp[-i\mathbf{q} \cdot (\boldsymbol{\rho} - \boldsymbol{\rho}')] \chi_{sJ}(z, z', \boldsymbol{\rho} - \boldsymbol{\rho}', iu), \tag{2.74}$$

is a function of only the magnitude of the 2D wavevector.

When the translational symmetry of the jellium response function eqn (2.73)
is used in eqn (2.64), the substrate integrals lead to a Dirac delta function
$\delta(\mathbf{q} - \mathbf{q}')$.* We thus arrive at the useful form

$$V_{disp}^{(2)} = -\int_0^\infty \frac{du}{2\pi} \int \frac{d^2q}{(2\pi)^2} \frac{2\pi}{q} e^{-2Zq} g(q, iu) h(q, iu). \tag{2.75}$$

The surface response function $g(q, iu)$ is the analogue of the atomic polarization function in eqn (2.69) and is given by

$$g(q, iu) = \frac{2\pi}{q} \int dz \int dz' e^{q(z+z')} \chi_s(z, z', q, iu). \tag{2.76}$$

It appears whenever the surface properties of the substrate are probed. For example, the inelastic scattering of electrons from surfaces provides a measure of
$g(q, \omega)$, the counterpart of $g(q, iu)$ at real frequencies (Ibach and Mills 1982).
That the 'same' function should appear in the context of the dispersion interaction is not too surprising since the substrate is responding to the fluctuating
charge distribution of the adsorbate. The g-function includes effects of the diffuseness of the surface charge density and of the nonlocal dielectric response of
both the surface and the bulk. A full determination of this function is a chal-
lenging task, but it has been evaluated in the case of jellium (Liebsch 1987) using
a time-dependent extension of density-functional theory.

---

*Lattice structure leads to Umklapp terms that give rise to terms in the dispersion energy
that depend exponentially on Z (Zaremba and Kohn 1976). Such terms must be included in
the theory of the potential corrugation, (Section 2.3.3).

We shall only require the coefficients in the small-q expansion of g,

$$g(q, iu) = g_0(iu) + g_1(iu)q + g_2(iu)q^2 + \cdots, \tag{2.77}$$

that enter in evaluating the large-Z behavior of the adatom–substrate potential. We demonstrate in Appendix C that g_0 is given in terms of the dielectric function ϵ by

$$g_0(iu) = \frac{\epsilon(iu) - 1}{\epsilon(iu) + 1}. \tag{2.78}$$

Because $g_0(iu)$ is a monotonic function of u, it may be approximated rather well with the form $g_0(iu) = g_0/[1 + (u/E_s)^2]$, see Appendix E. The function $g_1(iu)$ has been approximated using interpolation formulas that satisfy requirements on the limiting behavior at large and small argument (Zaremba and Kohn 1976; Persson and Zaremba 1984) and with time-dependent density functional theory (Liebsch 1986a,b). The modeling of $g_2(iu)$ is quite primitive (Hutson $et\ al.$ 1986).

2.2.1.3 $Asymptotic\ behavior\ of\ the\ interaction$ Using eqns (2.71) and (2.77) in eqn (2.75), we find a series for the adatom–substrate potential at large-Z

$$V_{disp}^{(2)} \simeq -\frac{C_3}{Z^3} - \frac{C_4}{Z^4} - \frac{C_5}{Z^5} + \cdots, \tag{2.79}$$

where

$$C_3 = \frac{1}{4\pi} \int_0^\infty du \alpha_1(iu) g_0(iu), \tag{2.80}$$

$$C_4 = \frac{3}{8\pi} \int_0^\infty du \alpha_1(iu) g_1(iu), \tag{2.81}$$

and

$$C_5 = \frac{3}{4\pi} \int_0^\infty du \left[\alpha_1(iu) g_2(iu) + \frac{1}{3} \alpha_2(iu) g_0(iu) \right]. \tag{2.82}$$

The leading term behaves as Z^{-3} and depends on the dipole polarizability of the adsorbate. Higher multipole polarizabilities of the adatom contribute terms with yet higher inverse powers of Z. For example, the quadrupole polarizability of the adsorbate enters in the definition of C_5. Note that the higher-order dispersion coefficients C_n are complex combinations of both adsorbate and substrate response properties. Thus, the C_5 coefficient depends on both the dipolar and quadrupolar polarizabilities, together with two different substrate response properties (X.-P. Jiang $et\ al.$ 1984; Hutson $et\ al.$ 1986). Since there are only even powers of q in the expansion of the atomic polarization function, the Z^{-4} term derives from the linear term $g_1(iu)$ in the expansion of $g(q, iu)$. We shall see that this term has a special physical significance.

As with the analogous interatomic dispersion series, eqn (2.79) is an asymptotic expansion and is not absolutely convergent. Further, even the truncated series must break down at small separations because of the unphysical divergence.

Thus, care must be exercised in applying eqn (2.79). Nevertheless, it displays the leading Z^{-3} dependence which is the hallmark of the atom–substrate van der Waals interaction. In Section 2.3.1, eqn (2.122), we show that the Z^{-3} dependence is obtained also by summing R^{-6} pair potentials over a half-space. The analysis here shows that the Z^{-3} dependence emerges even in cases which can not be formulated as a superposition of pair potentials.

The Lifshitz formula With the identification of g_0 in terms of the long wavelength dielectric function, eqn (2.78) and Appendix C, eqn (2.80) for the coefficient C_3 reduces to

$$C_3 = \frac{\hbar}{4\pi} \int_0^\infty du\, \alpha_1(iu) \frac{\epsilon(iu) - 1}{\epsilon(iu) + 1}. \tag{2.83}$$

Lifshitz (1956) originally derived this expression by using sophisticated field-theoretic techniques to treat the electromagnetic interaction between two continuous media. The present derivation demonstrates that the C_3 coefficient depends on the bulk, or macroscopic, dielectric function, i.e., the $q = 0$ limit of $\epsilon(q, \omega)$. Although we used a jellium model of the substrate, the arguments are quite general and carry over to crystalline materials as well (McLachlan 1964; Nijboer and Renne 1968, 1971; MacRury and Linder 1971, 1972; Zaremba and Kohn 1976). In the latter case, the macroscopic dielectric function must be determined from the appropriate limit of the microscopic dielectric function in order to properly account for *local field effects* in solids. Nevertheless, the Lifshitz formula with $\epsilon(\omega)$ equal to the macroscopic dielectric function is still correct. We emphasize that the Lifshitz formula *does not* depend on assuming that the continuous medium is described by a local dielectric function. Nonlocality of the bulk dielectric function, as determined by its explicit q-dependence, enters only in the higher-order dispersion coefficients.

The van der Waals reference plane An important consequence of the nonlocal dielectric response of the substrate is the definition of the van der Waals reference plane. The choice of the origin of coordinates to be used in the asymptotic expansion in eqn (2.79) is not obvious, but we would expect it to be close to the substrate surface, on physical grounds. If we write

$$V_{disp}^{(2)} \simeq -\frac{C_3}{(Z - Z_0)^3}, \tag{2.84}$$

the first two terms in the asymptotic expansion are reproduced by defining the reference plane position Z_0 to be

$$Z_0 \equiv \frac{C_4}{3C_3}. \tag{2.85}$$

Z_0 is specified by the response properties of the substrate and provides a precise definition for the origin of coordinates for the asymptotic van der Waals interaction.

To evaluate C_4, and hence Z_0, we require the coefficient $g_1(iu)$ in the small-q expansion of $g(q, iu)$. This coefficient can be shown to be given by (Persson and Zaremba 1984)*

$$g_1(iu) = 2g_0(iu)d_{IP}(iu), \tag{2.86}$$

where the dynamic image plane is

$$d_{IP}(iu) = \frac{d_{\parallel}(iu) + \epsilon(iu)d_{\perp}(iu)}{\epsilon(iu) + 1}, \tag{2.87}$$

with

$$d_{\perp}(iu) = \frac{\int dz\, z\delta n_s(z, 0, iu)}{\int dz\, \delta n_s(z, 0, iu)} \tag{2.88}$$

and

$$d_{\parallel}(iu) = \frac{\int dz\, z\, dj_{\parallel}(z, 0, iu)/dz}{\int dz\, dj_{\parallel}(z, 0, iu)/dz}. \tag{2.89}$$

In terms of $d_{IP}(iu)$ the van der Waals reference plane position is

$$Z_0 = \frac{1}{4\pi C_3} \int_0^\infty du\, \alpha_1(iu)\frac{\epsilon(iu) - 1}{\epsilon(iu) + 1}d_{IP}(iu). \tag{2.90}$$

The derivation of eqns (2.86)–(2.89) is quite involved and will not be reproduced here. However, all the quantities that appear have useful physical interpretations. $d_{\perp}(iu)$ is the centroid of the induced surface charge due to a uniform external electric field that is directed normal to the surface. For $u = 0$, the static limit, it defines the image-plane position for an external point charge (Lang and Kohn 1973). The other length parameter $d_{\parallel}(iu)$ reflects the spatial distribution of the currents induced parallel to the surface by a uniform tangential electric field.

In general, the dynamic image plane is determined by substrate responses to both parallel and perpendicular electric fields. However, $\epsilon(iu) \to \infty$ as $u \to 0$ for a metal, and then $d_{IP}(0) = d_{\perp}(0)$. Persson and Zaremba (1984) showed that the translational invariance along a jellium surface causes $j_{\parallel}(z, 0, iu)$ to be proportional to $n_0(z)$, the equilibrium density profile of the electrons. Then, eqn (2.89) becomes $d_{\parallel}(iu) = Z_B$, independent of frequency, where Z_B is the position of the edge of the uniform positive background of the jellium. $d_{IP}(iu)$ is essentially determined by $d_{\perp}(iu)$, and the centroid of the induced charge is the physical quantity of interest.

Several formal properties of the centroid as a function of frequency have been established (Zaremba and Kohn 1976; Feibelman 1982; Persson and Apell 1983; Persson and Zaremba 1984) which have guided a reasonably accurate evaluation of Z_0. Even in a more direct evaluation of the frequency-dependent centroids of

---

*Feibelman (1982) presents these results in a broader context and explains the role of the charge centroids $d_{\perp}$ and $d_{\parallel}$ in the microscopic theory of surface reflectivity.

the screening density (Liebsch 1986a,b) using time-dependent density functional theory, the sum rules provide tests of the internal consistency of the numerical results. Some representative values for the adsorption of He on metal surfaces are given in Table 2.2. Since Z_0 is a frequency average of $d_{IP}(iu)$, its value typically lies between the position $d_{IP}(0)$ of the static image plane (Lang and Kohn 1973) and the edge of the positive background, which is the high-frequency limit of $d_{IP}(iu)$. Although the displacement of the reference plane from the edge of the positive background is not large, it can be quantitatively significant in the calculation of adsorption potentials (Chizmeshya and Zaremba 1992). We show below, in eqn (2.123), that the analogous origin of coordinates in the case of a pairwise-summed potential lies one-half lattice constant outward from the top atomic plane.

2.2.2   Repulsive interaction

The short-range repulsive part of the adsorbate–surface interaction arises from wave function overlap, as discussed in Section 2.1.2. For adsorbed inert gases, it can be approximated closely by calculating the total HF energy. We shall describe the development of work initiated by Zaremba and Kohn (1977), who showed that the HF interaction energy of the adatom–metal complex can be expressed in terms of the shift in the single-particle density of states of the metal caused by electron scattering from the adatom, the *fragment formula* for weakly overlapping fragments. This line of work has led to He–metal potentials in good agreement with the holding potentials constructed from atomic scattering experiments.

As a preliminary, we discuss the relation of this approach to the solutions of the Kohn–Sham orbital equations within the *local density approximation*, LDA (Lang 1981; Lang and Williams 1982; Lang and Nørskov 1983). By solving the Kohn–Sham equations, the kinetic energy operator is treated correctly, in contrast to the local approximation in statistical (Thomas–Fermi) models such as the Gordon–Kim approximation. The solutions, which have been given for He, Ar, and Xe on jellium metals, show the rearrangement of charge which is driven by the presence of the adsorbed atom. There is no assumption that the overlap of adatom and metal electrons is small, although the resulting minimum of the holding potential does occur at rather large separations.

There is a conceptual difficulty with the use of the local approximation to the exchange and correlation energies, because the charge density is quite inhomogeneous in the spatial region of principal interest, and the van der Waals interaction is understood as a nonlocal correlation. The local approximation seems to mimic the effect of the van der Waals term near the potential minimum, but how that happens is unclear and the apparent agreement of calculations and experiments has been disputed (Nordlander and Harris 1984). However, the calculations of Lang and his coworkers are the only ones to show charge rearrangements for adsorbed Ar and Xe which correspond to the scale of observed adsorption-induced dipole moments and were the pioneering applications of density functional theory to the heavier inert gases.

The development of the Zaremba–Kohn analysis has mostly been for the helium–metal interaction, although there are recent applications to heavier inert gases. The combination in eqn (2.47) has been rationalized (Harris and Liebsch 1982; Nordlander and Harris 1984) as the beginning of a systematic expansion in the overlap of atomic orbitals and Bloch electron wave functions of the metal. In such terms the van der Waals energy is zeroth order in the overlap, while the Hartree–Fock repulsion has a term that is first order in the overlap. The leading omission is an overlap term in the atom–metal correlation energy (Harris and Liebsch 1982). The *fragment formula*, eqn (2.98), arises as a simplification of the Hartree–Fock interaction energy that retains terms that are first order in the overlap. Intuitively, such an analysis embodies the dominant features of the tightly bound atomic electrons of He, but it may be a less accurate picture for adsorbed Xe and other more extended adsorbates.

2.2.2.1 *The fragment formula* Zaremba and Kohn (1977) showed that the Hartree–Fock interaction energy can be approximated in terms of the shifts in the density of states of the Bloch electrons caused by scattering from the adsorbed atom. The requirement of self-consistency in the solution of the HF equations turns out to be of secondary importance, and the main perturbation is the presence of the atomic potential. We outline their derivation and then discuss the further steps that are needed to make estimates for the He–metal repulsion.

Hartree–Fock theory The Hartree–Fock approximation may be derived by using a Slater determinant of single-particle orbitals as the trial function in a variational calculation of the ground state energy of a many-electron system. A functional variation of the orbitals to minimize the trial energy leads to the following coupled set of equations

$$
\begin{aligned}
h\phi_i(\mathbf{r}) \quad &+ \quad \int d\mathbf{r}' \sum_j |\phi_j(\mathbf{r}')|^2 v(\mathbf{r} - \mathbf{r}')\phi_i(\mathbf{r}) \\
&- \quad \int d\mathbf{r}' \sum_j \phi_j^*(\mathbf{r}')\phi_i(\mathbf{r}')\delta_{\sigma_i \sigma_j} v(\mathbf{r} - \mathbf{r}')\phi_j(\mathbf{r}) \\
&= \quad \epsilon_i \phi_i(\mathbf{r}).
\end{aligned} \tag{2.91}
$$

The single-particle Hamiltonian $h = -\frac{1}{2}\nabla^2 + V_{ext}(\mathbf{r})$ is the sum of the kinetic energy operator and the Coulomb potential $V_{ext}(\mathbf{r})$ due to all external positive charges, and v is the Coulomb potential of two charges, eqn (2.51). The first integral in eqn (2.91) gives the direct electrostatic interaction of the electron in orbital ϕ_i with all electrons, while the second term is the exchange interaction, effective only between orbitals* with the same component of spin σ. The terms

---

*In general, the spatial orbitals $\phi_i(\mathbf{r})$ have a spin index, to allow for differences of 'up' and 'down' spin orbitals.

with $j = i$ in the two integrals cancel, so that there is no self-interaction of an electron with itself. It is convenient to write eqn (2.91) more formally as

$$(\hat{h} + \hat{V}_{HF})|i\rangle = \epsilon_i|i\rangle,\tag{2.92}$$

implicitly defining a nonlocal potential energy operator $\hat{V}_{HF}$. Then the HF energy is written as

$$E_{HF} = \sum_i \langle i|\hat{h}|i\rangle + \frac{1}{2}\sum_i \langle i|\hat{V}_{HF}|i\rangle + E_{NUC},\tag{2.93}$$

where the sum extends over all occupied orbitals i and includes spin. The term E_{NUC} denotes the Coulomb repulsion between all nuclear charges in the system. In the nonmagnetic case, when the spatial orbitals for up- and down-spin electrons are the same, the spin sum gives a factor of two and eqn (2.93) becomes

$$E_{HF} = 2\sum_i \langle i|\hat{h}|i\rangle + \sum_i \langle i|\hat{V}_{HF}|i\rangle + E_{NUC}.\tag{2.94}$$

Now, and in all that follows, the index i denotes distinct spatial orbitals; the spin sum has been removed. Finally, to obtain the repulsive interaction, we subtract from E_{HF} the HF energies of the adsorbate and substrate taken separately,

$$E_{HF}^a = 2\sum_a \langle a|\hat{h}_a|a\rangle + \sum_a \langle a|\hat{V}_{HF}^a|a\rangle + E_{NUC}^a\tag{? 95}$$

and

$$E_{HF}^s = 2\sum_s \langle s|\hat{h}_s|s\rangle + \sum_s \langle s|\hat{V}_{HF}^s|s\rangle + E_{NUC}^s.\tag{2.96}$$

The HF orbitals are denoted by $|a\rangle$ for the isolated adsorbate and $|s\rangle$ for the substrate, respectively.

The self-consistent solution of the HF equations is an extremely difficult problem, particularly in the low-symmetry case of an adsorbate interacting with a metallic surface. However, the task may be eased by taking into account the nature of the situation typical of physisorption. If the adsorbate levels do not overlap in energy with the valence band of the metal, they hybridize only weakly with the Bloch states and, to a first approximation, retain their localized atomic character.* Apart from the atomic-like adsorbate orbitals, the remaining occupied states are those of the metallic substrate. These are extended in character, a property that does not change with the addition of the adsorbate. However, substrate electron states are strongly perturbed in the vicinity of the adsorbate, an effect which cannot be treated by a straightforward application of perturbation theory.

---

*In typical cases of chemisorption, the atomic valence orbitals hybridize strongly with the substrate orbitals.

Fragments Thus, the picture we develop for physisorption is that the total self-consistent potential V_{HF} in eqn (2.92) is atomic-like in the vicinity of the adsorbate and similar to the unperturbed metallic potential further away. In other words, the superposition of the isolated adsorbate and substrate HF potentials

$$\hat{V}_{sup} \equiv \hat{V}_{HF}^a + \hat{V}_{HF}^s \tag{2.97}$$

should be a close approximation to the total HF potential $\hat{V}_{HF}$. Zaremba and Kohn (1977) used this separation and a series of approximations based on retaining the leading terms in the adsorbate–substrate overlap integrals to obtain an expression for the Hartree–Fock repulsion in terms of the shifts of the substrate energy levels:

$$V_{rep}(\mathbf{R}_A) = 2 \sum_{\mathbf{k}} [\epsilon_{\mathbf{k}} - \epsilon_{\mathbf{k}}^0], \tag{2.98}$$

where $\mathbf{R}_A$ denotes the position of nucleus of the adatom and $\epsilon_{\mathbf{k}}$ and $\epsilon_{\mathbf{k}}^0$ denote the HF-orbital energies of the Bloch electrons with and without the adatom present, respectively. Harris and Liebsch (1982) gave another derivation in the context of density functional theory. Their analysis led to a similar form, in which the energies ϵ are the Kohn–Sham orbital energies and the van der Waals energy is added. They identified the principal omission as terms in the adatom–metal correlation energy that depend on the overlap. We outline the derivation of eqn (2.98) in Appendix D.

Equation (2.98) is the central result to be used in the evaluation of the repulsive interaction between an inert gas atom and a metal surface. Although the atomic states do not appear explicitly, they enter implicitly via the HF potential governing the metallic electrons. By obtaining eqn (2.98), the HF problem has been simplified importantly, since it has been reduced to the evaluation of a single-particle property, namely the shift in the metallic eigenvalues induced by the adsorbate potential. Electron–electron interactions are included fully at the level of the HF approximation.

2.2.2.2 *Density of states* The formal sum-over-orbitals in eqn (2.98) actually runs over a continuous energy spectrum. The problem of evaluating the HF repulsion has been transformed to determining the shift in the substrate density of states caused by the presence of an external scattering center, the adatom. The symmetry is lower than for an impurity embedded in a bulk solid, but methods developed for the impurity problem have been used here as well. We now develop the description of the scattering problem.

Since a metallic state typically is distributed over the large volume Ω of the bulk substrate, the corresponding eigenvalue shift by a localized scattering center is of order Ω^{-1}. Each such shift is vanishingly small, but the smallness is compensated by summing over all electrons in the system. Hence, eqn (2.98) gives a finite value for the repulsion, as we make clearer by introducing the metallic density of states $\rho(\epsilon) = \sum_s \delta(\epsilon - \epsilon_s)$. The eigenvalue sum can be expressed as

$$\sum_s \epsilon_s = \int^{\epsilon_F} \epsilon\,\rho(\epsilon)\,d\epsilon, \qquad (2.99)$$

where the integration extends over all states up to the Fermi level ϵ_F. Equation (2.99) applies to the unperturbed metallic system in which the electrons have the energy ϵ_s. When the adsorbate potential is introduced, both the states and eigenvalues change:

$$\sum_s \tilde{\epsilon}_s = \int^{\tilde{\epsilon}_F} \epsilon\,\tilde{\rho}(\epsilon)\,d\epsilon. \qquad (2.100)$$

The new density of states $\tilde{\rho}(\epsilon)$ differs from the original by the amount

$$\Delta\rho(\epsilon) = \tilde{\rho}(\epsilon) - \rho(\epsilon). \qquad (2.101)$$

The shift in the Fermi energy is determined by the conservation of the total number of electrons,

$$\int^{\tilde{\epsilon}_F} \tilde{\rho}(\epsilon)\,d\epsilon = \int^{\epsilon_F} \rho(\epsilon)\,d\epsilon. \qquad (2.102)$$

Then the sum of eigenvalue shifts is given by

$$\sum_s \delta\tilde{\epsilon}_s = \int^{\epsilon_F} (\epsilon - \epsilon_F)\,\Delta\rho(\epsilon)\,d\epsilon. \qquad (2.103)$$

If the change in density of states at some energy ϵ below the Fermi level is positive, the states in this energy range make an *attractive* contribution to the atom–surface interaction. Conversely, a decrease in the density of states leads to a *repulsion*. The latter is expected to dominate for the physisorbed systems.

Zaremba and Kohn (1977) showed how to relate the change in density of states to the solution of a scattering problem. The 'incident' wave is a state of the bare substrate, and the scattering potential is the electron–adsorbate potential. The density of states formally depends on the solution to a multiple scattering problem, but the fact that the adsorbate is at a distance where the overlap with substrate orbitals is small leads to a simplification again. This final step reduces the calculation to a single-scattering problem. Technically, summing the eigenvalue shifts reduces to evaluating the scattering T-matrix. Computation then involves either the solution of the scattering problem (Zaremba and Kohn 1977; Takada and Kohn 1988; Chizmeshya and Zaremba 1992) or an approximation of the strong scattering potential by a pseudopotential (Harris and Liebsch 1982; Nordlander and Harris 1984; Chizmeshya and Zaremba 1989).

2.2.2.3 *Scattering theory* By repeatedly exploiting the smallness of the overlap of the substrate orbitals with the adatom orbitals, the evaluation of the HF repulsion has been reduced to the solution of a single-scattering problem. The incident wave is an unperturbed metallic orbital that experiences the full influence of the substrate potential, including the effects of the surface. The scattering center is the adatom charge distribution. The potential it creates is not weak; an approximation such as the first Born approximation is inadequate. The whole development has been based on the smallness of the overlap, so it is crucial that realistic asymptotic forms for the wave functions be used in the computations.

The adatom The adatom charge distribution is expressed in terms of the HF orbitals. While this is unambiguous in principle, several varieties of 'HF orbitals' are available in practice, because of differences in how thorough the solution for self-consistency is. For instance, there are orbitals, based on limited variational searches, that give good approximations to the HF energy but have significant errors in the asymptotic tails. The repulsion V_{rep} depends sensitively on the tails. This causes serious enough uncertainties in applications to inert gases heavier than helium that numerical HF orbitals are used (Chizmeshya and Zaremba 1989).

The adatom presents a strong scattering potential for an incident electron mainly because of the overlap of that electron with adatom electrons. There is a systematic procedure, the *pseudopotential method*, for parameterizing the effect of a strong potential or an excluded volume by a weak scattering potential that is treated by perturbation theory. The idea arises from the fact that many dense Fermi systems show independent-particle behavior in conditions where the interactions are strong. The resolution to this apparent paradox is that the interactions are *effectively* weak in the sense that they do not scatter the fermions very strongly. In the context of physisorption, pseudopotentials were introduced by Zaremba and Kohn (1977) and developed extensively by Harris and Liebsch (1982).

The pseudopotential method is extremely accurate in the case of He (Nordlander and Harris 1984; Chizmeshya and Zaremba 1989). The atomic $1s$ electrons are essentially frozen. Harris and Liebsch (1982) showed that the Bloch electrons experience a negative electrostatic potential and a strong orthogonalization energy, with the net effect being a strong repulsive scattering potential. This identification is based on a partial wave analysis of the nonlocal pseudopotential constructed by the technique that we now describe. The analysis has been extended to heavier inert gases by Chizmeshya and Zaremba (1989).

The pseudopotential approach provides an approximate solution of the equation for a substrate electron orbital, eqn (D.1), which we write here as

$$(\hat{T} + \hat{V}_s + \hat{V}_a)|\tilde{s}\rangle = \tilde{\epsilon}|\tilde{s}\rangle \,. \tag{2.104}$$

The unperturbed metallic states satisfy the analogous equation

$$(\hat{T} + \hat{V}_s)|s\rangle = \epsilon_s|s\rangle \,. \tag{2.105}$$

We define the pseudowavefunction $|s_{ps}\rangle$ by

$$|\tilde{s}\rangle = N_s \left[|s_{ps}\rangle - \sum_{nLM} |\psi_{nLM}\rangle - \langle\psi_{nLM}|s_{ps}\rangle \right] \,. \tag{2.106}$$

The orbitals $\tilde{s}$ are now automatically orthogonal to all the adsorbate core states, by construction, i.e., $\langle\psi_{nLM}|\tilde{s}\rangle = 0$. The normalization constant N_s is given by

$$N_s = \left[\langle s_{ps}|s_{ps}\rangle - \sum_{nLM}|\langle\psi_{nLM}|s_{ps}\rangle|^2\right]^{-1/2}. \qquad (2.107)$$

The pseudostates s_{ps} extend throughout the metal, so that N_s^2 deviates from unity only by terms of order Ω^{-1} that therefore can be neglected. Substituting eqn (2.106) into eqn (2.104) leads to

$$(\hat{T} + \hat{V}_s + \hat{V}_{ps})|s_{ps}\rangle = \tilde{\epsilon}_s|s_{ps}\rangle, \qquad (2.108)$$

where the pseudopotential is defined as

$$\hat{V}_{ps} = \hat{V}_a + \sum_{nLM}(\epsilon_s - \epsilon_{nL} - \hat{V}_s)|\psi_{nLM}\rangle\langle\psi_{nLM}|. \qquad (2.109)$$

The second term in eqn (2.109) is a nonlocal operator that offsets the strong attractive atomic potential $\hat{V}_a$. If the cancellation is sufficiently complete, leading orders of a perturbation series in $\hat{V}_{ps}$ may give an accurate solution for the orbital energy $\tilde{\epsilon}_s$ in eqn (2.108).

Since the lowest order pseudowavefunction is the unperturbed state $|s\rangle$, the first-order shift in the Bloch orbital energy is

$$\begin{aligned}\Delta\epsilon_s^{(1)} &= \langle s|\hat{V}_{ps}|s\rangle = \langle s|\hat{V}_a|s\rangle \\ &+ \sum_{nLM}[\epsilon_s - \epsilon_{nL} - V_s(\mathbf{R})]|\langle\psi_{nLM}|s\rangle|^2.\end{aligned} \qquad (2.110)$$

The last term* is the pseudopotential estimate of the 'correction' to the first-order perturbation theory for the bare potential $\hat{V}_a$. Nordlander and Harris (1984) carried the series in the pseudopotential to second order and found a 20% correction to eqn (2.110). That the series is reliable has been tested (Harris and Liebsch 1982; Chizmeshya and Zaremba 1992) by comparison with the full scattering theory solution.

The principal advantage of the pseudopotential method is that the interaction is simply determined by evaluating a matrix element of a nonlocal operator. Because the dependence on the substrate wave functions is quite transparent, this method has been used to analyze the foundations of the *effective medium approximation* (Harris and Liebsch 1982).

---

*We have assumed that the matrix element $\langle s|\hat{V}_s|\psi_{nLM}\rangle$ can be approximated by $V_s(\mathbf{R})\langle s|\psi_{nLM}\rangle$.

The substrate   We require the 'tail' of the substrate charge distribution, i.e., the evanescent states, that extends to the vicinity of the adatom. The dominant contribution to this tail arises from the valence electrons of the metal, so that electrons with energy close to the Fermi level dominate. Thus, a jellium model for the substrate orbitals should embody the dominant physical characteristics of the evanescent wave functions. We now describe the adjustments that are made to the available models of the surface charge density in order to have a good estimate of the overlap with the adatom.

The first consideration is that a strict *Hartree–Fock* solution to the substrate surface charge density is known to be inadequate, e.g., in reproducing work functions and surface potential barriers (Juretschke 1953; Lang and Kohn 1971). Therefore, Zaremba and Kohn (1977) were led to base their calculations on the solutions of Lang and Kohn (1970) for the metallic charge density 'far' from the surface. Since the solutions of Lang and Kohn are based on the local density approximation to exchange and correlation, there arose a question about how much correlation effect is contained in the Zaremba–Kohn 'Hartree–Fock repulsion'. Harris and Liebsch (1982) concluded that correlation terms in the overlap are omitted.

In the jellium model, the substrate potential $V_s(z)$ is only a function of the perpendicular coordinate z. As a result, the metallic states have the form

$$\psi_s(\mathbf{r}) = \sqrt{\frac{2}{\Omega}}\,\exp[\imath\mathbf{k}_\| \cdot \boldsymbol{\rho}]\,u_{k_z}(z)\,, \qquad (2.111)$$

where $u_{k_z}(z)$ is the solution of

$$\left[-\frac{1}{2}\frac{d^2}{dz^2} + V_s(z)\right] u_{k_z}(z) = \frac{1}{2}k_z^2 u_{k_z}(z)\,. \qquad (2.112)$$

The potential V_s is a local potential calculated self-consistently with density functional theory rather than the HF potential.*

To treat a specific metal substrate, three further steps are taken. (1) The metal is characterized by an average electron density, and the Lang–Kohn solutions are interpolated to give solutions for the corresponding density. (2) The potential barrier in the Lang–Kohn solution is rescaled (Zaremba and Kohn 1977) to fit the experimental value for the work function, a procedure essential for constructing physically realistic evanescent tails. (3) The edge of the jellium background is shifted (Nordlander and Harris 1984) to reestablish the charge neutrality of the substrate. The outcome of this procedure is believed to be a reasonably accurate substrate density of states in the region of the adatom.

---

*Although, the HF potential is nonlocal and is cumbersome in some applications, that is not the deciding issue because quite satisfactory *local* approximations to the HF potential are available.

The wavevector $(\mathbf{k}_\parallel, k_z)$ of a state in the metal corresponds to an evanescent wave with decay constant $k_\perp$ related to the orbital energy ϵ_s and the asymptotes $V_s(-\infty)$ and $V_s(\infty)$ of the substrate potential by

$$
\begin{aligned}
\epsilon_s &= \frac{1}{2}[k_\parallel^2 + k_z^2] + V_s(-\infty) \\
&= \frac{1}{2}[k_\parallel^2 - k_\perp^2] + V_s(\infty).
\end{aligned}
\tag{2.113}
$$

The work function of the metal is given in terms of the asymptotic value of the substrate potential and the Fermi energy by $\Phi = V_s(\infty) - \epsilon_F$. This determines the smallest magnitude of $k_\perp$ and hence the most persistent of the evanescent states

$$
\psi_s(\mathbf{r}) = A \exp[\imath(\mathbf{k}_\parallel \cdot \boldsymbol{\rho}) - k_\perp z].
\tag{2.114}
$$

Nordlander and Harris (1984) found that the substrate potential reaches its asymptotic value sufficiently slowly that the Bloch orbitals have substantial deviations from this asymptotic form in the region of the physisorption potential minimum. A semiclassical approximation for the wave functions suffices to correct for this effect.

2.2.2.4 *Scattering solutions* The low symmetry of the adatom–metal potential causes the solution of the scattering problem to become quite detailed. The scattering problem is already unconventional because of the role of the evanescent substrate electron states as the 'incident' wave. We must defer to the literature for the full description of the solutions that are obtained when using a formulation in terms of the cylindrical symmetry of the problem (Zaremba and Kohn 1977) or in terms of a (coupled) spherical partial wave decomposition which exploits some formal similarities to other scattering problems (Takada and Kohn 1988).

2.3   Semiempirical methods

Most of the adsorbates treated in physical adsorption are complex enough that first principles calculations of the type described in the Section 2.2 are not yet capable of providing realistic adsorbate–substrate interactions. Thus, we must usually rely on semiempirical constructions, especially when trying to understand processes which involve a balance between adsorbate–adsorbate and adsorbate–substrate forces. Two productive lines of approach have been to systematize empirical data into families that have similar shapes for the potential function (*universal functions*) and to construct the potential functions from models for the interactions of the constituents. The first of these is mostly the work of Vidali *et al.* (1983a,b; 1991) and is described in Section 2.4.1. The second was developed by Steele (1973, 1974) and has been particularly useful for dielectric substrates. For monolayers, the *holding potentials* that result from such constructions are supplemented by adsorbate–adsorbate interactions and by adsorption-induced interactions, as discussed in the following subsections. For molecular adsorbates,

Table 2.1 *Parameters of the Lennard-Jones (12,6) potentials between adatoms and carbon sites of a graphite substrate.*[†]

Adatom[‡]	ϵ	σ	source
He	16.2	2.74	Carlos and Cole (1980a)
Ne	26.5	3.00	Bruch *et al.* (1984)
Ar	57.9	3.40	Cheng and Steele (1989)
Kr	75	3.42	Steele (1978)
Xe	107	3.40	Klein *et al.* (1984)
H_2	42.8	2.97	Wang *et al.* (1980)
$N[N_2]$	31.9	3.36	Steele (1977)
$O[O_2]$	38.2	3.20	Bhethanabotla and Steele (1987)
$O[CO_2]$	45.8	3.16	Bottani *et al.* (1994b)
$C[CH_4]$	47.7	3.30	Severin and Tildesley (1980)
$C[CO_2]$	27.1	3.11	Bottani *et al.* (1994b)
$H[CH_4]$	17.0	2.98	Severin and Tildesley (1980)
$H[NH_3]$	24.3	2.82	Cheng and Steele (1990)

effects of the electrostatic multipole moments are included as supplementary terms. We begin with an account of constructions following Steele's procedures and then discuss some of the modifications that have been made.

2.3.1 *Atom–atom sums*

The starting point is to assume that the interaction of an adsorbate with atomic components α and a substrate with atomic components β at sites j_β can be written as the sum of pairwise interactions

$$V(\{\mathbf{R}_\alpha\}) = \sum_{\alpha, j_\beta} \phi_{\alpha,\beta}(\mathbf{R}_\alpha - \mathbf{R}_{j_\beta}) \,. \tag{2.115}$$

We shall discuss this assumption in the next paragraph. First, we note that in the applications one proceeds by assuming a functional form for $\phi_{\alpha,\beta}(\mathbf{r})$ and by adopting some rules for determining the parameters that enter in the functional form. Most important, the use of eqn (2.115) requires a knowledge of the substrate structure, especially of the relaxations of the lattice near the surface, but

---

[†]The energy scale ϵ is given in Kelvin and the length scale σ in angstrom. The entries are values used in recent applications. The overall trends in the Table reflect the original constructions of Steele (1973) where values $\epsilon_{C-C} = 28$ K and $\sigma_{C-C} = 3.40$ Å were used for the nominal interactions in the graphite and the values for mixed pairs were derived from combining rules. The interplanar spacing in the graphite is taken to be $d = 3.37$ Å.

[‡]The entries for H_2 are for the spherical molecule approximation. The notation X[Y] denotes the interaction of adatom X when bound in a molecule Y. The C and H entries for CH_4 are also used in modeling C_2H_4 (Cheng and Klein 1992) and C_2H_6 (Moller and Klein 1989).

the most common procedure is simply to assume a sharp truncation of the bulk structure.

The pairwise sum must only be suggestive of the interaction of the adsorbate with the extended composite object which is the substrate, because even for the relatively simple dense phases of inert gases the potential energy contains many-body terms (Barker 1986, 1989, 1993). There are substantial corrections when the dense phase is composed of a molecular species (Elrud and Saykally 1994). Even if we were able to determine the interaction of an α–β pair, the function entering in eqn (2.115) would be different because of modifications of the species β in the dense substrate environment and of the species α when bound in various molecular environments. However, for physical adsorption, the substrate is only weakly perturbed, so that effective interaction parameters may be assigned for the substrate species. Further, it is found that there are series of admolecules where the constants for a given species α have small variations across the series. Thus, with judicious use, eqn (2.115) may be applied to systematize a great range of data and, effectively, to extrapolate and interpolate to a variety of molecular adsorbates.

The most extensive use of eqn (2.115) is with a Lennard-Jones (12,6) potential for ϕ:

$$\phi(r) = 4\epsilon[(\sigma/r)^{12} - (\sigma/r)^6] . \tag{2.116}$$

The $1/r^6$ term represents the London–van der Waals dispersion energy, while the $1/r^{12}$ term is intended to model the strong short-range overlap repulsions. An exponential dependence for the repulsion is more realistic and has been used in some cases (Bclak et al. 1985; Kuchta and Etters 1987a; Kumar and Etters 1991). However, eqn (2.116) is a two-parameter form that has been applied systematically to many cases. We analyze the holding potential that results from using eqn (2.116) in eqn (2.115). The implicit assumption of central forces is relaxed in eqn (2.134).

The simplest series is the inert gases interacting with a graphite substrate. Historically, there were typically two pieces of information to determine the parameters ϵ and σ for each case, an adsorption energy and an estimate of the perpendicular vibration frequency (Constabaris et al. 1959, 1961; Sams et al. 1960). While just sufficient, there was in fact some ambiguity in the fits. A systematic procedure following trends across the inert gas series and based on combining rules proved to be more satisfactory (Steele 1973, 1974). The combining rules (Hirschfelder et al. 1964) for the parameters of a mixed pair XY in terms of like pairs XX and YY are:[*]

$$\epsilon_{XY} = \sqrt{\epsilon_{XX}\epsilon_{YY}}$$
$$\sigma_{XY} = \frac{1}{2}(\sigma_{XX} + \sigma_{YY}) . \tag{2.117}$$

---

[*]As noted by Ihm et al. (1990) and Scoles (1990), the combining rules tend to be accurate when the ionization energies of the atoms are similar, but are not accurate otherwise.

Parameters for several adsorbate species interacting with graphite are listed in Table 2.1. Those for the inert gases are refinements of Steele's estimates while those for molecular components still are quite directly based on eqn (2.117). For more complex substrates, an even larger body of experimental data is needed to establish systematic trends. Girardet and his coworkers have constructed sets for ionic crystals (Girard and Girardet 1987; Lakhlifi and Girardet 1991). The situation is much more favorable for calculating the interaction of an inert gas atom with a graphite substrate that has been preplated with a monolayer or several layers of another inert gas. The interactions among inert gas atoms are pair-additive as a good leading approximation, and 3D pair potentials have been summed to give rather good accounts of the interactions for argon and xenon platings (Chung *et al.* 1986a; Jónsson *et al.* 1987; Aziz *et al.* 1989).

The periodicity of the substrate lattice, which enters in eqn (2.115) through the positions $\mathbf{R}_{j\beta}$, has the consequence that the potential sum for an adatom α above a planar semi-infinite substrate is periodic for displacements parallel to the planar surface. As explained in Section 3.2 the periodicity can be expressed using the reciprocal lattice vectors $\mathbf{g}$ of the substrate surface:

$$V(\mathbf{r}, z) = V_0(z) + \sum_g V_g(z) \exp(\imath \mathbf{g} \cdot \mathbf{r}) \, , \qquad (2.118)$$

where $\mathbf{r}$ and z are the components of the adatom position $\mathbf{R}$ parallel and perpendicular to the surface, respectively. The coordinate z is measured relative to the topmost substrate plane and $\mathbf{r}$ is taken relative to a point in the Bravais lattice of the surface.

The functions V_0 and $V_{\mathbf{g}}$ for adatom α are given in terms of the basis vectors $\mathbf{b}_{j\beta}$ of substrate atoms in a unit cell of area A_c in the j-th layer (interplanar spacing d) by*

$$
\begin{aligned}
V_g(z)|_\alpha &= \frac{1}{A_c} \sum_\beta \sum_{j=0}^\infty \exp(-\imath \mathbf{g} \cdot \mathbf{b}_{j\beta}) \int d^2r \, \exp(-\imath \mathbf{g} \cdot \mathbf{r}) \\
&\times \quad \phi_{\alpha,\beta}(\sqrt{r^2 + (z + jd)^2}) \, .
\end{aligned}
\qquad (2.119)
$$

The laterally averaged term is, for the Lennard-Jones (12,6) potential

$$V_0(z)|_\alpha = \frac{1}{A_c} \sum_{j=0}^\infty \sum_\beta \int d^2r \, \phi_{\alpha,\beta}(\sqrt{r^2 + (z + jd)^2})$$

---

*Implicit in this notation is the assumption that successive planes contain the same species $\{\beta\}$. The basis vectors $\mathbf{b}_\beta$, Section 3.1, may depend on the layer index j because of lateral displacements of the unit cell from one layer to the next. Such displacements have been tabulated by Bruch and Venables (1984).

$$= \frac{2\pi}{A_c} \sum_{j=0}^{\infty} \sum_{\beta} \epsilon_{\alpha,\beta}\, \sigma_{\alpha,\beta}^2 \left[\frac{2}{5} \Big(\frac{\sigma_{\alpha,\beta}}{z+jd} \Big)^{10} - \Big(\frac{\sigma_{\alpha,\beta}}{z+jd} \Big)^4 \right]. \qquad (2.120)$$

The corresponding amplitudes of the spatially periodic terms are primarily determined by the surface layer and are given by[†]

$$V_{\mathbf{g}}(z)|_\alpha = \frac{4\pi}{A_c} \sum_{\beta} \exp(-\imath \mathbf{g} \cdot \mathbf{b}_\beta)\, \epsilon_{\alpha,\beta}\, \sigma_{\alpha,\beta}^2$$

$$\times \left[\frac{1}{60} \Big(\frac{g\sigma_{\alpha,\beta}^2}{2z} \Big)^5 K_5(gz) - \Big(\frac{g\sigma_{\alpha,\beta}^2}{2z} \Big)^2 K_2(gz) \right], \qquad (2.121)$$

where the K_n are modified Bessel functions which vary as $\exp(-gz)$ for large argument. Except for quite small values of z, the series in eqn (2.118) converges rapidly. Generalizations of eqns (2.120) and (2.121) to other inverse power laws, an exponential repulsive potential, and the Tang–Toennies damping function are presented in Appendix F.

At large z, the j sum in eqn (2.120) may be replaced by an integration, and we have

$$V_0(z)|_\alpha \simeq - \left[\frac{2\pi}{3A_c d} \sum_{\beta} \epsilon_{\alpha,\beta}\, \sigma_{\alpha,\beta}^6 \right] /z^3, \quad z \to \infty. \qquad (2.122)$$

Thus, as remarked in Section 1.2, the $1/z^3$ dependence is reproduced by a model with pairwise interactions, although this power law really arises from much more general considerations, Section 2.2.1. Equation (2.122) provides an expression for C_3 in terms of the parameters listed in Table 2.1. When those parameters are used, the values for C_3 are of the same order of magnitude as those calculated with the empirical dielectric response function (Appendix E). A more detailed agreement between the two approaches is not to be expected, and is not obtained, because the parameters in Table 2.1 were set to reproduce the interaction near the minimum of V_0 and not in the asymptotic $1/r^6$ region.

2.3.1.1 *Continuum edge* The atom–atom model for the holding potential may be used to derive the $d/2$ rule for the position of the edge of the continuum substrate model: the edge is one-half interplanar spacing above the top–most plane of atoms. This is the location adopted for jellium models of metals, subject to modest corrections for the diffuseness of the surface which were discussed in Section 2.2.1.3. In the case of the explicit sum over layers j in eqn (2.120), the $d/2$ rule arises from the first correction in the Euler–Maclaurin summation formula when a sum is approximated by an integral:

$$\sum_{k=1}^{\infty} f_k \;=\; \int_0^{\infty} f(k)\,dk - \frac{1}{2}f(0) - \frac{1}{12}f'(0) + \dots,$$

[†]The approximation of retaining only the topmost layer omits terms which are responsible for the small energy difference between the fcc and hcp stacking sites of an fcc(111) surface.

$$\sum_{k=1}^{\infty} f_k \; \simeq \; \int_{\frac{1}{2}}^{\infty} f(k)\, dk + \frac{1}{24} f'(0) + \dots \qquad (2.123)$$

That is, use of the $d/2$ rule includes the first correction to the integral replacement. It arises as a piece of applied mathematics and is only weakly dependent on the form of the summand through the magnitude of further correction terms.

2.3.1.2 *Molecular moments: distributed point charges* The $1/r^6$ attraction in eqn (2.116) is the dominant long-range interaction for adsorbates without permanent multipole moments. However, molecules have nonspherical charge distributions and net electrostatic moments that lead to long-range molecule–molecule interactions. While the interaction can be evaluated explicitly in terms of the Coulomb potential between two specified charge densities, there are occasions when other representations are useful. In computer simulations, there is a premium for rapid-to-evaluate expressions. For a given molecule, the detailed charge distribution may be imperfectly known, but a few electrostatic moments may have been measured. The multipole expansion of the Coulomb potential for well-separated molecules has a leading component determined by the first non-vanishing electrostatic moment, but the relevant separations in a dense phase may not be large enough for the point moment approximation to be accurate.

This leads to the use of approximate forms similar to eqn (2.115) with sums taken over a small number of discrete charges that have been assigned to intramolecular positions in such a way as to mimic the first few electrostatic moments of the isolated molecule. Procedures for assigning distributed charges and, more generally, distributed dipoles and quadrupoles have been reviewed by Dykstra (1993). In physical adsorption most applications have used distributed point charges. Some unusual cases, such as C_2H_6 (Hansen *et al.* 1985), have been encountered where the distributed monopoles correspond to a situation that violates chemical intuition and the distribution is more plausibly given by a small number of distributed dipoles. For molecular adsorption on ionic crystals, the substrate electric field is so nonuniform that it is essential to include the distributed character of the molecular charge when modeling the holding potential (Picaud and Girardet 1993). In fact, as the assigned point moment distributions become more elaborate, the actual reporting of the parameters of the interaction model becomes more implicit, so that it is not a simple matter to reproduce the starting point of a calculation.

2.3.1.3 *Electrostatic energies* There are important special cases of eqns (2.115) and (2.118) for which explicit results are available: the electrostatic potentials which arise from monopole, dipole, and quadrupole distributions at substrate lattice sites. Then the potential external to the substrate satisfies Laplace's equation

$$\nabla^2 V_{el}(R) = 0 \, , \quad z > 0 \, , \qquad (2.124)$$

and the coefficients V_g vary as e^{-gz}.

The potentials to be used in eqn (2.115) are, for a test charge q,[*]

$$\phi(\mathbf{R}) = q\boldsymbol{\mu} \cdot \mathbf{R}/R^3 \quad \text{dipole} \tag{2.125}$$

$$= q\mathbf{R} \cdot \boldsymbol{\Theta} \cdot \mathbf{R}/R^5 \quad \text{quadrupole} . \tag{2.126}$$

The functions V_0 and V_g that arise from a sum over one plane of dipole moments μ parallel to $\hat{z}$, perpendicular to the plane, are[†]

$$V_0(z) = \frac{2\pi q}{A_c} [\sum_\beta \mu_\beta] \frac{z}{|z|}$$

$$V_g(z) = \frac{2\pi q}{A_c} \exp(-gz) \sum_\beta \mu_\beta \exp(-\imath \mathbf{g} \cdot \mathbf{b}_\beta) . \tag{2.127}$$

The corresponding planar sum for a lattice of axially symmetric quadrupoles aligned along the z-axis, $\theta = \Theta_{zz}$, has $V_0 = 0$ and

$$V_g(z) = \frac{\pi q}{A_c} \exp(-gz) \sum_\beta \theta_\beta \exp(-\imath \mathbf{g} \cdot \mathbf{b}_\beta) . \tag{2.128}$$

The coefficient for an array of charges Q_β is (Lennard-Jones and Dent 1928; Gready *et al.* 1978)

$$V_g(z) = \frac{4\pi q}{gA_c} \exp(-gz) \sum_\beta Q_\beta \exp(-\imath \mathbf{g} \cdot \mathbf{b}_\beta) . \tag{2.129}$$

The results for V_g all have the anticipated e^{-gz} dependence, and the net electric fields are appreciable only very close to the layer. Close to the surface of an ionic crystal, corresponding to the use of eqn (2.129), the electric fields and the electric field gradients are large. The large gradients have led to surprises in the analysis of a monolayer of water adsorbed on NaCl(100) (Bruch *et al.* 1995).

The cancellation of attractive and repulsive Coulomb fields that is implicit in eqn (2.129) has a serious consequence for the modeling of inert gas adsorption on ionic crystals using eqn (2.115). There are data, principally derived from the mobility of ions in dilute inert gases, for the pair potentials between ions and inert gas atoms. Such data and calculations have led to models for the interactions of inert gas atoms with alkali and halide ions (Ahlrichs *et al.* 1988). However, the attraction is dominated by the induction energy $-\alpha Q^2/2R^4$ (Hirschfelder *et al.* 1964) that arises from the polarization of the inert gas atom by the Coulomb field of the ion-charge Q. The turning point in an ion–inert gas atom collision

---

[*]The zz component of the quadrupole tensor for a charge distribution $\rho(\mathbf{r})$ is $\Theta_{zz} = \int d^3r\, \rho(\mathbf{r})[\frac{3}{2}z^2 - \frac{1}{2}r^2]$.

[†]The first part of eqn (2.127) leads to eqn (6.1) for the work function change due to a dipole layer.

is smaller than it would be in the absence of the induction energy, so that the mobility data probe rather different parts of the interaction than are needed to construct the net interaction of inert gas atoms with the corresponding ionic crystals. A further complication is that the crystal environment also changes the ionic response that enters in the evaluation of the C_6 coefficients (Fowler and Hutson 1986; Hutson and Fowler 1986; Fowler and Tole 1988). However, a realistic potential for He with LiF(100), for which a wide range of data is available, has been constructed by Celli *et al.* (1985).

We return to the electrostatic contributions to V_g when we discuss models for the corrugation terms in the holding potential for molecules in Section 2.3.3.

2.3.2 Adsorption-induced interactions

2.3.2.1 *Dielectric screening* We repeat some of the considerations that were discussed in connection with the theory of the dispersion forces in Section 2.2.1. The role of the semi-infinite planar substrate can be approximated by an electro-magnetic boundary condition (McLachlan 1964) in which the substrate response to an external charge Q is an image charge of strength $-Q[\epsilon - 1]/[\epsilon + 1]$ at the mirror position. For a static charge, ϵ is the static dielectric constant while for charge fluctuations it is a frequency-dependent quantity as given in Section 2.2.1 and Appendix E. The mirror position is calculated with respect to somewhat different positions in the two cases, although it is basically the continuum edge given by the $d/2$ rule, Section 2.3.1.1.

The relevant dielectric constant for dielectrics with planar anisotropy is a geometric average of the components parallel and perpendicular to the surface,

$$\epsilon = \sqrt{\epsilon_\parallel \epsilon_\perp}. \tag{2.130}$$

This result is readily obtained for the screening response to a static point charge and has been shown to give the appropriate average for van der Waals forces (Kihara and Honda 1965; Okano 1965). It was also obtained from the Drude model of adatom and substrate responses (Bruch and Watanabe 1977).

The most direct evidence for the static screening is the observation of energy shifts in photoelectron emission from adsorbed inert gases. The shifts reflect final state interactions of the remaining ion with the screening charge of the substrate (Kaindl *et al.* 1980; Lang *et al.* 1982; Kaindl and Mandel 1987) and are consistent with mirror plane locations which include the shift relative to the jellium edge calculated by Lang and Kohn (1973) for a static point charge.

The screening of molecular moments is to be included in modeling the adsorbate–substrate potential as a term supplementing the van der Waals interaction derived from eqn (2.115). Lang *et al.* (1985) included a self-consistent screening of the adsorbate charges in a calculation of the lateral interactions of chemisorbed species.

2.3.2.2 *McLachlan interaction* The screening response also modifies the $1/r^6$ attraction between inert gas atoms. The substrate plays the role of an extended

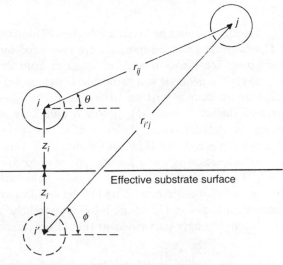

FIG. 2.5. Geometry of the McLachlan interaction. Geometry and coordinates related to the McLachlan substrate-mediated interaction between adatoms, as discussed in the text.

third body, and many-body interactions among the two adatoms and the substrate atoms contribute to the net interaction between the adatoms (Sınanoğlu and Pitzer 1960; MacRury and Linder 1971, 1972). McLachlan (1964) formulated the interaction using the electromagnetic boundary condition in such a way that empirical data may be used to evaluate the strength coefficients. For two atoms at lateral separation r and heights z_1 and z_2 above the continuum edge or *mirror plane* as sketched in Fig. 2.5, the McLachlan interaction is:

$$
\begin{aligned}
\phi_{McL} = \quad & C_{s1} f(r, z_1, z_2) / ([r^2 + (z_1 + z_2)^2][r^2 + (z_1 - z_2)^2])^{3/2} \\
& - C_{s2} / [r^2 + (z_1 + z_2)^2]^3 \\
f(r, z_1, z_2) = \quad & \frac{4}{3} - \frac{(z_1 + z_2)^2}{r^2 + (z_1 + z_2)^2} - \frac{(z_1 - z_2)^2}{r^2 + (z_1 - z_2)^2} ,
\end{aligned}
\tag{2.131}
$$

where the coefficients are

$$
\begin{aligned}
C_{s1} &= \frac{3\hbar}{\pi} \int_0^\infty d\omega \, \frac{[\epsilon(\imath\omega) - 1]}{[\epsilon(\imath\omega) + 1]} \, [\alpha(\imath\omega)]^2 \\
C_{s2} &= \frac{3\hbar}{\pi} \int_0^\infty d\omega \left(\frac{[\epsilon(\imath\omega) - 1]}{[\epsilon(\imath\omega) + 1]} \right)^2 [\alpha(\imath\omega)]^2 .
\end{aligned}
\tag{2.132}
$$

There is a small distinction (W.-K. Liu 1985) between the positions of the mirror plane for the McLachlan interaction and for the single molecule dispersion energy $-C_3/z^3$, but that has not been incorporated in the applications of the McLachlan interaction to inert gas monolayers. H.-Y. Kim *et al.* (1988)

generalized eqn (2.131) to the case of linear molecules. Tabulations of the coefficients C_{s1} and C_{s2} for inert gases on metals were presented by Bruch (1983a). Estimates for many more combinations can be obtained from the approximation of Rauber *et al.* (1982); the necessary information is assembled in Appendix E.

The McLachlan interaction is written in terms of the $q \to 0$ limit of the wavevector-dependent dielectric constant (Persson and Zaremba 1984). This omits the screening oscillations, i.e., Friedel oscillations, associated with the Fermi surface of a metallic substrate (Lau and Kohn 1978; Einstein 1997). The oscillatory terms in the adatom–adatom interaction have longer range than the $1/r^6$ terms in the McLachlan interaction, but thus far there are no data which isolate their effect for physisorbed atoms. The Friedel oscillation terms are more prominent for chemisorbed species (Einstein 1997), apparently because the substrate electrons are more strongly perturbed in the adsorption.

2.3.2.3 *Dipole* arrays

We evaluated the potential arising from a lattice of specified dipole moments in eqn (2.127). Now we examine two related questions: adsorption-induced dipole moments and depolarization effects. In the general scheme of intermolecular force theory, these are 'properties' to be derived from accurate ground state wave functions of the many-body system. An actual evaluation is more difficult than the calculation of the energy because, roughly, wave function errors have first-order effects in the 'properties' and second-order effects in the energy.

Adsorption-induced moments That there are adsorption-induced dipole moments is a consequence of the charge rearrangements that occur as the adatom–substrate complex adjusts its configuration to reduce the energy. The major question is how large the moments are. In cases such as alkali adsorption, the rearrangement is so large that it is characterized qualitatively as charge transfer between the alkali atom and the substrate (Aruga and Murata 1989). For adsorbed inert gases, the most likely candidate for a charge transfer instability is xenon, but detailed studies (Lang *et al.* 1982) appear to exclude it. In the next paragraph, we describe a mechanism for the formation of an adsorption-induced dipole moment that is analogous to the mechanism for the van der Waals energy. The mechanism should be present for adsorption both on graphite and on metals. However, while work function changes and dipole moments are readily detected for inert gas adsorption on metals, the corresponding effects are much smaller for adsorption on graphite.* Thus, there are questions about the quantitative importance of this process, even though a physical mechanism is easily recognized.

In Section 2.1.2, we evaluated the dispersion energy for a fluctuating dipole moment on the adatom if the substrate has an ideal complete electrodynamic

---

*Mandel *et al.* (1985) detected a work function change of 0.1 eV for Xe/graphite, corresponding to a dipole moment 0.04 D per adatom; this is 5–10 times smaller than the dipole moments inferred for Xe/metals. They were unable to detect a work function change for Ar or Kr adsorbed on graphite.

screening of the dipole field. The energy, eqn (2.20), is proportional to the mean-square dipole moment of the adatom. However, the interaction of the fluctuating dipole moment with the screening charge can drive a net displacement of the adatom charge distribution; a similar process for a mixed pair of inert gas atoms leads to a net dipole moment that depends on the interatomic spacing as $1/R^7$ (Feynman 1937). Antoniewicz (1974) developed the perfect-screening theory of the adsorption dipole, that varies as $1/Z^4$, and Zaremba (1976) and Linder and Kromhout (1976) showed how to include the retarded electrodynamic response of the substrate. Later, a pair-sum model of the collision-induced dipoles and quadrupoles on pairs and triples of atoms, analogous to eqn (2.115) was used to display terms corresponding to the polarization of the substrate as well as that of the adatom (Bruch and Osawa 1980). However, Kromhout and Linder (1979) concluded that the magnitudes of the *dispersion dipoles* are too small to account for the work function changes of inert gases adsorbed on metals. There does appear to be reasonable agreement between photoemission measurements and self-consistent calculations of the dipole moment of an inert gas atom interacting with jellium (Lang 1981; Lang and Williams 1982).

Depolarizing fields The net dipole moment for a monolayer of atoms is not simply equal to the sum of the adsorption-induced dipoles of well-separated adatoms. This arises from the finite polarizability of the adatoms. The electric field from the other adsorption dipoles changes the net moment: in the common case where the moments are oriented perpendicular to the plane of the mono-layer, the contribution is negative, i.e., depolarization. The problem just stated is a relatively straightforward self-consistent field calculation if the polarizable moments form an isolated 2D lattice. If the 2D system is a fluid, there is a complex self-consistency problem related to that encountered in modeling polar fluids, such as water. For a 2D lattice in the presence of a substrate that has electrostatic screening charges, the modeling must include the contribution of *image dipoles* in the substrate. The screening charges increase the electrostatic forces between the dipoles (Kohn and Lau 1976) and change the effective polar-izability in the self-consistent solution (Meixner and Antoniewicz 1976). Data for the work function change of an adsorbed gas reflect some of the trends expected from this discussion, but there is not a detailed confirmation – see, e.g., the discussion of Moog and Webb (1984). For recent modeling of electrodynamic screening terms and dipole arrays, see Maschhoff and Cowin (1994).

As the dipole–dipole energy varies as $1/r^3$, it has a longer range than the McLachlan interaction. The latter, though, has a larger effect in calculations of the monolayer equation of state. This happens for adsorbed inert gases, because the root-mean-square dipole moment is much larger than the mean dipole moment. A rather large net dipole moment for adsorbed Xe is in the range 0.5–0.7 debye, while the root-mean-square dipole moment for Xe derived from Table A.2 is 12 debye.

2.3.2.4 *Elastic distortion of the substrate* Thus far we have described adsorption-induced interactions that depend on the electronic properties of the substrate and have implicitly regarded the substrate as being mechanically rigid. There are small elastic distortions of the substrate driven by the stress applied by the adlayer. The interaction of the elastic distortions from various adatoms leads to an indirect interaction of the adatoms.

The indirect interaction of two adatoms, acting as force dipoles at a planar surface, is a long-range potential which varies as $1/r^3$ at large lateral distances. This has been evaluated using a continuum elasticity theory for the response of an isotropic substrate by Lau and Kohn (1977), Stoneham (1977), and Maradudin and Wallis (1980). When the elastic anisotropy of materials used as substrates is included (Lau 1978; Kappus 1978), the interaction too is anisotropic and is attractive along some azimuths and repulsive along others. Later work was reviewed by Kern and Krohn (1989). To evaluate the interaction at separations r on the order of a few substrate lattice constants requires a lattice dynamical treatment of the substrate response rather than a continuum approximation; Tiersten *et al.* (1989) performed such an analysis for chemisorbed species.

Although the $1/r^3$ power law of the indirect elastic interaction means that it is a much longer range term than the McLachlan interaction, its strength is rather weak, and it makes only small contributions in equation of state calculations for physisorbed monolayers (Bruch 1983a; Gottlieb and Bruch 1993). Even so, there are situations where the elastic response of the substrate makes a dominant contribution. It is responsible for the damping of certain adlayer normal modes at small wavevectors (Hall *et al.* 1985, 1989) where anharmonic damping within the adlayer is very small. Its long range makes it an important term for processes involving large distance scales such as the interaction of steps on surfaces (Alerhand *et al.* 1988; Narasimhan and Vanderbilt 1992). The anisotropy of the interaction may also be evident in some field ion microscope studies (Watanabe and Ehrlich 1992) for chemisorbed species.

2.3.3 *Corrugation models*

It is remarkable how far modeling of the adatom–substrate potential can go without taking explicit account of the discrete atomic structure of the substrate. To a considerable extent, a typical substrate in physical adsorption can be treated as a continuum that is a source of a featureless potential surface, $V_0(z)$ of eqn (2.118), that determines the plane of a monolayer. However, the corrugation terms $V_g(z)$ determine the diffracted intensities for an atomic beam at 'positive energies', and they are responsible for the existence of commensurate and modulated monolayer structures at 'negative energies' near the minimum of V_0. To the extent that the holding potential is derived from jellium models, there is little information about the corrugation terms from *a priori* calculations. For most physical adsorption systems, the energy corrugation has variations on the scale of a few tens of meV. These rather small energy variations are accessible through quantitative analyses of intensities in atomic diffraction and from data for the structure and dynamics of nearly commensurate monolayers. Corrugation effects are larger for molecular adsorbates on ionic crystals and may be responsible for making hysteresis effects more troublesome in such cases than in most physisorbed systems.

The corrugation can be represented in nominally equivalent ways, according to the phenomenon under consideration. A presentation as contour plots of $V(\mathbf{R})$ in the $x - y$ plane is a direct display of the surface topography and shows the barriers to lateral motion of an adsorbed particle. Alternatively, specifying the height z_E at which $V(\mathbf{R}) = E$ defines a function $z_E(x, y)$ which corresponds to the classical turning point of an incoming positive energy atom. The corrugation is a combination of contributions from the repulsion and the attraction. In Hartree–Fock approximation, the corrugation of the repulsion is determined by the overlap of the adsorbate and substrate wave functions. Thus, it is directly associated with the underlying atomic structure. The corrugation of the attraction depends on the overall screening response of the substrate.

We now describe the insights into the scale and sources of the corrugation that have been developed from the study of several examples. The holding potential for He/graphite may be the most extensively analyzed case. We show the potential energy as a function of the height z above three sites of the graphite in Fig. 2.6, from Carlos and Cole (1980a), to illustrate two ways of characterizing the corrugation. The difference in the depth of the potential minima shown there gives an energy barrier of 3 meV for lateral motion across the surface.* A second view of the diagram is in terms of the difference in the turning points z_0 at a specified collision energy: for Fig. 2.6 the difference in the zero energy turning points is 0.23 Å.

2.3.3.1 *Positive energies* The most primitive atom–substrate potential model in a scattering experiment is a hard wall.[†] Then the surface topography is specified by the lateral variation of the height $z(x, y)$. For a helium beam of 63 meV the total range of z is about 0.2 Å on graphite and about 0.02 Å on Ag(111) (Boato *et al.* 1978).[‡] Since the fitted corrugation generally depends on the energy of the incident beam, diffraction data are sensitive to energy dependence of the adatom–substrate repulsion. We may use the effective medium approximation given in eqn (2.46) for the repulsive interaction of a helium atom with an electron gas of density $\rho(\mathbf{R})$ to estimate that the turning point of the helium atom is in a region of charge density $2 \times 10^{-4}\, a_0^{-3}$ at $3 - 4$ Å from the topmost layer of

---

*Notice that the difference between the minimum energies at the saddle-point SP and the atop site A is much smaller than the differences between those sites and the hexagon center S. This is characteristic of the topography of a physisorption potential surface with a hexagonal array of adsorption sites.

[†]An explicit model for the hard wall is obtained using billiard balls for the incident (a) and substrate (s) atoms. The van der Waals radii for the two species, r_a and r_s, are taken to be $\frac{1}{2}$ of the nearest-neighbor spacings in the corresponding 3D solids. For an fcc(111) surface, the height at an atop site is then $r_a + r_s$ and at a three-fold-hollow is $\sqrt{r_a^2 + 2r_s r_a - (r_s^2/3)}$.

[‡]The 0.02 Å scale also is the range that would fit the diffracted intensity for 18 meV He/Pt(111) found by Kern *et al.* (1986a). While estimates of the corrugation may use observed peak intensities, there are several corrections needed for quantitative analysis of diffracted intensities (Horne *et al.* 1980).

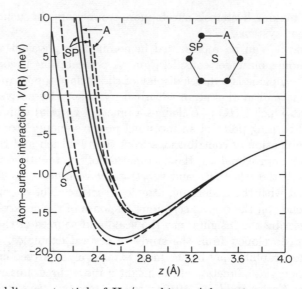

FIG. 2.6. Holding potential of He/graphite. Adsorption potentials calculated by Carlos and Cole (1980a) for He interacting with graphite. The solid and dashed curves represent results for alternative models. The labels A, S, and SP refer to sites above the hexagonal graphite surface indicated at the upper right.

atoms. The helium atom does not penetrate much into the substrate charge density and the repulsion may be viewed as mostly arising from a small overlap of the adatom charge density with the tail of the surface charge density.

While eqn (2.46) is a useful approximation for estimating V_0 (Cortona *et al.* 1992a), it overestimates the effective corrugation and the diffraction intensities for helium scattering from metal surfaces (Celli 1992). The most extensively studied case is He/Cu(110), where there is a surprising result that the charge density in the region of interest is nearly a simple superposition of the atomic charge densities (Batra 1984; Batra *et al.* 1985). Frigo *et al.* (1986) found for He/LiF(001), too, that the use of the effective medium approximation led to an overestimate of the corrugation in the repulsion relative to that inferred from diffraction intensities. Cortona *et al.* (1992b) have shown that the diffracted intensities for He/Ag(110) and He/Cu(110) can be fit with a model that allows for a rearrangement of the atomic electron clouds at the surface.

There have been several proposals (Takada and Kohn 1985; Annett and Haydock 1984a,b, 1986) of mechanisms that might reduce the calculated excess corrugation, but there is not agreement that they will suffice (Harris and Zaremba 1985). However, for Xe/Pt(111) there is a construction (Barker and Rettner 1992) using noncentral potentials that gives a potential with large corrugation at eV energies but rather small corrugation at low scattering energies.

Thus, the corrugation which determines the diffracted intensities might have been expected to be dominated by the repulsive term, but the use of pair-sum approximations for the charge density leads to an overestimate of the corrugation in the low energy scattering. That is, the contours of constant small charge density appear to have too much corrugation. This occurs both for metals and for ionic crystals.

2.3.3.2 *Negative energies*

The corrugation at negative energies, e.g., the region of the potential well in Fig. 2.6, shifts the bound states of the potential well (Cole and Toigo 1981) and enters in the balance between adsorbate–adsorbate and adsorbate–substrate energies that determines the structures of modulated and commensurate monolayers. The corrugation in the region of the potential minimum has significant contributions from both the attractive and repulsive components, and it is even more difficult to derive from *a priori* calculations than the corrugation for a scattering experiment. The greatest progress for a physisorption potential well has been for the graphite substrate, where the adequacy of several proposed sources of corrugation has been tested by comparison to experimental data. Most of our presentation treats this case, but we conclude with an example for a metal substrate.

Inert gases on graphite   The starting point is to calculate the amplitudes $V_g(z)$ of eqn (2.118) from the pairwise sum of eqn (2.115). If the Lennard Jones (12,6) potential parameters of Table 2.1 are used with eqn (2.116), it is found that the calculated corrugation is smaller in magnitude than is required to fit crucial experimental observations. This is true for the observed band structure effects in low coverage helium (Carlos and Cole 1980a,b), for the known stability of the commensurate monolayer lattice of krypton (Gooding *et al.* 1983; Shrimpton and Steele 1991), and for the frequency gap in the lattice dynamics of a commensurate monolayer solid (Bruch 1991a).*

A first step to cope with this discrepancy is to make a purely empirical adjustment of the calculated amplitudes

$$V_g(z) \rightarrow s\, V_g(z)\,; \qquad (2.133)$$

the choice $s \simeq 1.5$ for the scaling factor appears to suffice (Shrimpton and Steele 1991). The next step is to identify a mechanism which leads to a modification of this magnitude.

Carlos and Cole (1980a) proposed to incorporate effects of the dielectric anisotropy of the graphite substrate by use of a noncentral pair potential:

---

*The frequency gap at zero wavevector for lateral motions of the commensurate adlayer may be observed by inelastic scattering and in the low-temperature specific heat. It is mostly determined by the corrugation, see the next paragraphs.

$$\phi(\mathbf{R}) = \quad 4\epsilon\{(\frac{\sigma}{R})^{12}[1 + \gamma_R(1 - \frac{6}{5}\cos^2\theta)]$$

$$- (\frac{\sigma}{R})^6[1 + \gamma_A(1 - \frac{3}{2}\cos^2\theta)]\}, \qquad (2.134)$$

where θ is the angle between the surface normal and the vector $\mathbf{R}$ from the adatom to a substrate carbon atom. Equation (2.134) generalizes an earlier model of Bonino et $al.$ (1975). The value $\gamma_A = 0.4$ is set by graphite dielectric data and is nearly constant across the inert gas series. In the repulsion, $\gamma_R = -0.54$ is fit to $atomic$ band structure effects in the selective adsorption resonances of He/graphite (Boato et $al.$ 1979; Derry et $al.$ 1980). As formulated in eqn (2.134), the γ_R and γ_A terms do not contribute to the lateral average potential $V_0(z)$.

This modification is able to account for many observed corrugation effects. Band structure effects in the low coverage specific heat data of He/graphite (Silva-Moreira et $al.$ 1980) are reproduced with these parameters (Cole et $al.$ 1981). The same anisotropy parameters also bring the calculated corrugations for Ne/graphite and Kr/graphite into the range needed to reproduce observations (Bruch 1991a).

$Molecules$ on $graphite$ The most-studied case of linear-molecule adsorption is N_2/graphite (Section 6.1.7.3). There is a commensurate monolayer solid and apparently only one fluid phase (Migone et $al.$ 1983a), a situation attributed to the corrugation of the N_2–graphite potential well. In an early attempt to reproduce the phase diagram, Joshi and Tildesley (1985) found it necessary to increase the calculated corrugation beyond what they obtained using the parameters of Table 2.1. They adopted the anisotropic substrate mechanism of Carlos and Cole, but adjusted the parameter of the repulsion to $\gamma_R = -1.04$.

Inelastic neutron scattering measurements of the excitation spectrum of the commensurate N_2/graphite monolayer solid demonstrated that there is an energy gap of 19 K (0.4 THz) at the 2D Brillouin zone center at low temperatures. The corresponding normal mode has an in-phase vibration of all the molecular centers of mass, so that the force constant that determines the frequency is primarily due to the corrugation. Thus, the measured frequency provides a very direct measure of corrugation models. For the parameters of Table 2.1 the calculated frequency gap is 8.0 K, with the anisotropy parameters of Carlos and Cole it is 9.5 K, and with the parameters of Joshi and Tildesley it is 9.7 K (Bruch 1991a). The factor of 2 discrepancy between the calculation and the experiment is actually a factor of 4 in the underlying corrugation. There is a serious omission in the corrugation model for the molecule.

Vernov and Steele (1992) proposed that there is an additional source of corrugation for an adsorbed molecule with a permanent electrostatic moment. The carbon atoms have nonspherical charge distributions as a result of bonding in the graphite, so that there is a corrugated electrostatic potential external to the substrate, as in eqn (2.128). Hansen et $al.$ (1992) adopted a value $\theta = 1.0 \times 10^{-26}$ esu cm^2 for the effective quadrupole moment of each carbon atom. With this value the calculated frequency was 50% larger. Hansen et $al.$ suggested that the adjustment of γ_R by Joshi and Tildesley might have been driven partially by the omission of the substrate electrostatic field and retained the value assigned by Carlos and Cole (1980a). This opened the possibility that the corrugation for physisorbed molecules could be based on a combination of van der Waals terms identified for the inert gases and electrostatic interactions between substrate fields and molecular moments.

The present situation is inconclusive. With the sign of θ adopted by Hansen *et al.*, the electrostatic corrugation term and the van der Waals term add together for N_2/graphite; that sign was based on X-ray structure factors of Chen *et al.* (1977). However, Vernov and Steele assigned the opposite sign, using an extrapolation of quadrupole moments of small molecules, and Whitehouse and Buckingham (1993), in a measurement of the total quadrupole moment of a graphite sample, also arrived at a sign opposite to the X-ray data. With a value $\theta \simeq -1.0 \times 10^{-26}$ esu cm^2, the two contributions to the corrugation offset each other and the frequency gap is 6 K, i.e., 0.125 THz (Hansen and Bruch 1995). However, it is established that a rather small nonspherical component to the charge density at lattice sites in graphite will generate a significant electrostatic contribution to the corrugation of adsorbed molecules.

Xenon on platinum (111)　　The commensurate and nearly commensurate monolayer lattices of Xe/Pt(111) have been the subject of intense investigation, partially because of what may be unusually large or different mechanisms for the corrugation. Analogy to the minimum-energy site at the hexagon center S in Fig. 2.6 for inert gases on graphite led to an expectation that the corresponding site on an fcc(111) surface would be at a three-fold-hollow site. It was difficult to construct interaction models based on this idea that reproduced the observations (Black and Janzen 1989a). Then it was realized that the observation (Kern 1987a) of a particular modulation satellite for incommensurate Xe/Pt(111) could be used to argue* (Gottlieb 1990a) that the minimum energy site is the atop site A instead.

The atop site as the minimum energy site can be given a basis in local density functional theory in terms of hybridization of orbitals (Müller 1990, 1993). This amounts to invoking the possibility of xenon 'chemistry' (Ishi and Viswanathan 1991). The energy corrugation from the atop site to the hollow site is $\simeq 25$ meV in a model that fits a wide range of Xe/Pt(111) data (Barker and Rettner 1992). Such a variation is three times as large as the S-to-A variation of recent models of Kr/graphite (Shrimpton and Steele 1991), and its magnitude is an indication that novel mechanisms may be involved. Barker and Rettner were led to a rather complex interaction to reconcile the large corrugation in the attractive well with evidence for a much smaller corrugation at intermediate scattering energies. Their potential model has an interaction with delocalized conduction electrons and pair potentials to the platinum ion cores and also to force centers displaced from the cores.

---

*'An argument persuades a skeptical person, a proof a stubborn one.' There are other possibilities; see Zeppenfeld *et al.* (1992b).

Thus, the Xe/Pt(111) system has stimulated several lines of development for analysis of the corrugation. It indicates some of the possibilities for more systematic future studies of adsorption on transition metals.

2.4   Examples

We present several examples to illustrate the concepts and methods of the previous sections. Phenomena are selected to display trends for the holding potential and for adsorption-induced interactions. The monolayer phase diagrams for several of these systems are reviewed in Chapter 6.

2.4.1   *Universal functions*

There is an extensive body of information on V_0 for physical adsorption. A recent review (Vidali *et al.* 1991) treated over 250 adsorbate–substrate combinations. One response has been to look for unifying features in the data similar to the *Law of Corresponding States* for 3D pair potentials of inert gases (Hirschfelder *et al.* 1964). The experience with the pair potentials encourages such an effort, since the potential minima for the like-pairs have a very similar shape and differ principally in the scales for energy and length (Scoles 1980; Aziz 1984). The domain of physical adsorption is so broad that reduction to one shape function for $V_0(z)$ will probably not be achieved. Rather, there will be systematic departures reflecting trends in the underlying mechanisms and 'outliers' reflecting ambiguities in the interpretation of experimental signals (Vidali *et al.* 1983b).

To begin, consider an empirical model for V_0 obtained by superposing the overlap repulsion* and the dispersion energy (Vidali *et al.* 1983a), without refinements such as Tang–Toennies damping or higher multipole terms:

$$V_{emp} = Ae^{-\alpha z} - (C_3/z^3). \eqno(2.135)$$

There are three parameters A, α and C_3. Equation (2.135) may be recast as a universal functional shape with the reduced curvature κ, depth D and separation z_m at the potential minimum as the three parameters. Analogy to the 3D cases suggests that there are families with different curvatures arising, e.g., from different surface diffuseness.

Vidali *et al.* (1983b) generalized the considerations slightly by examining the correlation of data that is achieved by assuming the following three-parameter form

$$V_0(z) = Dg(z^*)$$

---

*Equation (2.135) has an exponential dependence for the repulsion, rather than the inverse power law in eqn (2.120). At the empirical level, this so-called *Born–Mayer* form is a familiar option to mimic the rapid spatial variation of the repulsion term (Hirschfelder *et al.* 1964). For the adsorption problem, the exponential is the more realistic form, as shown in Section 2.2.2.

$$z^* = (z - z_m)/\ell, \tag{2.136}$$

with a characteristic length ℓ in place of the reduced curvature. If eqn (2.136) is further required to reproduce the dispersion term at large z^*, the length ℓ is specified

$$g(z^*) \simeq -1/z^{*3} \; z \to \infty$$
$$\ell = (C_3/D)^{1/3}, \tag{2.137}$$

and the data should be reduced with only two scaling parameters.

A first application is fitting the energy levels observed for atomic hydrogen, helium and molecular hydrogen in selective adsorption resonances (Hoinkes 1980). Because the bound states are well approximated using the Sommerfeld quantization rule of semiclassical quantum mechanics, the levels are insensitive to the value of z_m. Thus, if eqn (2.137) is valid, the levels should follow a one-parameter scaling with the depth D. Vidali et al. found the scaling holds for the overall trend of the data, but could recognize some systematic deviations for the noble metals. That is, there is evidence for different functional shapes or, at least, different curvatures. They also found that the constraint imposed on the reduced curvature by eqn (2.137) is too strong for the potentials of the heavier inert gases, for which the curvature is available from the frequencies measured by inelastic atom scattering and from the Debye–Waller factor in diffraction experiments.

The modeling of inert gas adsorption on graphite, Section 2.3.1, leads one to expect that there may be a common shape for V_0 in that series. Vidali et al. (1983b) showed that this expectation is fulfilled, by demonstrating that the data for Ne, Ar, Kr, and Xe all fall on one curve for a scaled gas–surface virial coefficient B_{AS}, eqn (4.35), as a function of reduced temperature.

Another hypothesis that arises from considering eqns (2.136) and (2.137) is that the length scales ℓ and z_m may be proportional (Ihm et al. 1987) when z_m is measured relative to the continuum edge. As a test, Ihm et al. (1987) plotted ℓ as a function of the equilibrium distance z_m. The z_m data have considerable uncertainties, but the plot is roughly a straight line with slope 1.2. Such an analysis does not impose a requirement that the same variation occur in different adsorption families, but the assembled data show departures are only on the scale of 0.2–0.3 Å. Thus, there is an apparent contradiction between a value $z_m \simeq 3.5$ Å for Xe/Ag(111) (Stoner et al. 1978) and $z_m \simeq 4.2$ Å for Xe/Pt(111) (Potthoff et al. 1995), since the correlation found by Ihm et al. leads to the expectation that the value for Pt(111) would be smaller.

Using the *Law of Corresponding States* to systematize the large database helps to identify general trends and to identify data which deviate sharply from the trend. Such data may, indeed, signal the presence of quite different interaction mechanisms. Systematic departures from the trend may correspond to subfamilies already recognized by other considerations, such as the grouping into adsorption on insulators, noble metals and alkali metals discussed in Section

2.4.3 (Vidali *et al.* 1991). In the next subsections, we examine a few examples in more detail.

2.4.2 *Potentials from selective adsorption resonances*

The phenomenon of *selective adsorption* was observed already in early diffraction experiments that demonstrated the wave character of atoms (Estermann and Stern 1930; Estermann *et al.* 1932; Frisch and Stern 1933). There are minima in the reflected intensities as a function of angle of incidence, e.g., for He/LiF(001), that correspond to transitions of incident particles into bound surface states (Hoinkes 1980). The bound states are basically eigenstates of motion of the adatom of mass m in the laterally averaged potential V_0, although band structure effects of the adatom motion in the spatially periodic potential arising from the substrate corrugation are sometimes observed (Boato *et al.* 1979; Derry *et al.* 1980).

For the cases of He, Ne, H, or H_2 the technique of scattering has proved to be an invaluable spectroscopic tool thanks to the phenomenon of bound state resonances. As illustrated in Fig. 2.7, the incident particle is diffracted into a bound state whose energy is actually positive because of significant translational kinetic energy across the surface. The resonance occurs when

$$\begin{aligned} E_{inc} &= \hbar^2(K^2 + k_z^2)/2m \\ &= E_{bound} \sim E_j + \hbar^2(\mathbf{K} + \mathbf{G})^2/2m, \end{aligned} \qquad (2.138)$$

where $\mathbf{G}$ is the diffraction vector associated with the periodic lattice, $\mathbf{K}$ is the projection of the incident wavevector onto the surface, and E_j is the energy of the j-th bound state. Implicit in this equation is an assumption that the adsorbed particle propagates freely across the surface. A clear exception to the assumption is the behavior near a Brillouin zone boundary depicted in the figure, where the corrugation has large effects as in the analogous circumstances of electrons in solids. By monitoring the resonance position as a function of the incident conditions, one can derive quantitative information about both the energy levels of the laterally averaged potential and the band structure effects. The technique is limited to the quantum scattering case in which the wavelength is of the order of the lattice constant and there are relatively few relevant bound states, otherwise the scattering becomes too complicated to decipher unambiguously. In the case of light atomic and molecular scatterers, beams can be produced with sufficiently narrow energy distributions that energy levels can be determined with uncertainties of only 0.1 meV (Hoinkes 1980; Vidali *et al.* 1991). We shall describe the use of observed bound state energies E_j to determine V_0.

First, as in the case of diatomic molecules, there is a test to confirm that the potential has the anticipated asymptotic form

$$V_0 \simeq -C_3/z^3, \quad z \to \text{large}. \qquad (2.139)$$

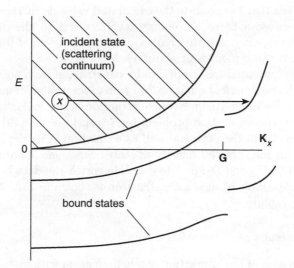

FIG. 2.7. Bound state resonances. Schematic depiction of bound state resonance transition of an atom between its incident state (circled x) and adsorbed state. $\mathbf{G}$ denotes the reciprocal lattice vector associated with the diffraction into the bound state. The energy–momentum relation for the bound state is shown in the extended zone scheme and is not continuous because of band structure effects. A kinematic description of the transition is presented in the text.

What is needed are the bound state energies for weakly bound ('shallow') states with turning points in the large-z range where eqn (2.139) is accurate. Le Roy (1976) adapted his analysis of the spectra of diatomic molecules to show that near the dissociation limit, a plot of $|E_j|^{\frac{1}{6}}$ varies linearly with the index j and the slope is proportional to $C_3^{-\frac{1}{3}}$.

$$|E_j|^{1/6} \simeq 0.9922(j_D - j)\hbar/\sqrt{m}C_3^{1/3}, \qquad (2.140)$$

where $\hbar$ is the reduced Planck constant and m is the mass of the adatom. The accuracy of the derived values of C_3 depends sensitively on the data for the most weakly bound energy levels. There is good agreement between the derived and calculated values of C_3 for H/LiF(001) and He/LiF(001) (Bruch and Watanabe 1977). This is valuable confirmation of the idea that the long-range attraction is based on correlated charge fluctuations of the adsorbate and substrate.

For later data for He/graphite (Chung *et al.* 1986b; Ruiz *et al.* 1986), the Le Roy plot is linear, but the slope corresponds to $C_3 = 245 - 255$ meV Å^3, which is distinctly larger than the value 180 calculated from the dynamic dielectric constant. However, the discrepancy is not considered to be critical, because model potentials that incorporate the calculated value do fit the observed energy levels, and because a fit to the energy levels closest to the dissociation limit reduces the discrepancy. For the shallowest observed level, at 0.13 meV (Ruiz *et al.* 1986), the turning point is at about 11 Å.

Second, the potential well itself can be constructed from the observed eigenstates using the Sommerfeld quantization rule of semiclassical quantum mechanics. This is the Rydberg–Klein–Rees construction of molecular spectroscopy and was adapted to the adsorption problem by Schwartz *et al.* (1978). What is obtained is the width of the potential well as a function of its depth in terms of an integral over an interpolated density-of-states function. Isotopic substitutions such as H and D, H_2 and D_2, or ^{3}He and ^{4}He enrich the data base for the interpolations. An example of such a construction is shown in Fig. 2.8 for molecular hydrogen on graphite.

2.4.3   *Helium–substrate potentials*

Several parameters of the interaction of a helium atom with metals and insulators are listed in Table 2.2. The coefficient C_3 gives a measure of the strength of the

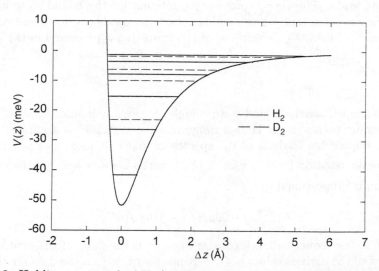

FIG. 2.8. Holding potential of H_2/graphite. The laterally averaged adsorption potential of a (rotationally averaged) hydrogen molecule interacting with graphite. The energy levels shown are those measured by Mattera *et al.* (1980) for the two isotopes and the potential is constructed semiempirically from these. Note that the two deepest levels of D_2 were not detected. This presentation is by G. Vidali (private communication).

Table 2.2 *Helium–substrate potentials.*
The van der Waals coefficient C_3 in meV Å^3
and depth D of the potential well in meV
are listed for several substrates.[†]

Substrate	C_3	D	d_c[‡]	Z_{vdW}
Cu(110)	235	6.3	1.28	0.17
Ag(110)	249	6.0	1.44	0.15
Au(110)	274	9.7	1.44	0.12
Ni(110)	218	4.2	1.24	–
Pd(110)	211	8.0	1.38	–
Pt(110)	251	8.5	1.38	–
Al(110)	202	≃5	1.43	0.38
Na	92	0.6	1.49	0.31
K	70	0.4	1.77	0.30
Rb	65	0.4	1.98	–
Cs	58	0.4	2.14	–
Graphite	180	16.6	1.68	
LiF(001)	93	8.5	1.00	
NaCl(001)	106	6.1	1.40	
MgO(001)	151	7.5	1.05	–

van der Waals attraction, while D is the depth of the potential minimum. There appear to be three families in the table: the noble and transition metals, the alkalis, and the dielectrics.

The values of C_3 for the transition metals tend to be the largest in the table, a consequence both of the polarizability of the metal and of the interband transitions that contribute to the dynamic screening in the 10–20 eV range. A first surprise is that the depths D for the insulators are in the same range as for these metals and that the deepest well is for graphite. This has been understood (Liebsch *et al.* 1984) in terms of the greater diffuseness of the electron distribution at the surface of the metal. The tighter binding of electrons in the insulator reduces the 'spill-out' of the electron density at the surface, and the helium atom is able to approach the substrate more closely.

Such an argument was pursued in the opposite direction and led to the prediction that the large diffuseness of the electron density at the surface of the heavier alkali metals would result in values of D so small that the helium would

[†]Values are taken from the review of Vidali *et al.* (1991). Also listed (in Å) are the height of the equivalent continuum edge d_c and, for the metals, the shift in the effective continuum edge Z_{vdW} calculated from the centroid of the electron density. The values of D and Z_{vdW} for the alkali metals are taken from calculations with jellium models; the corresponding entries for d_c are $\frac{1}{2}$ of the interplanar spacing for a bcc(110) surface.

[‡]$d_c = d/2$, where d is the interplanar spacing. Values for the centroid Z_{vdw} are taken from Nordlander and Harris (1984) and Chizmeshya and Zaremba (1989).

not wet the metal (Cheng *et al.* 1991a,b). This has been observed for He and also for H_2 (Cole 1995; Hallock 1995).

Also included in Table 2.2 are entries for the position of the jellium edge d_c and the displacement Z_{vdW} from that edge to the origin of coordinates of the dispersion energy. The values of Z_{vdW} are only half the size of the corresponding estimates for the displacement of the mirror plane of a static point charge (Lang and Kohn 1973). This reflects the ineffectiveness of the low-density portion of the surface charge density in screening the high-frequency charge fluctuations that determine the van der Waals interaction.

2.4.4   *Monolayer equation of state*

The principal tests of models for adsorbate–adsorbate interactions are based on information for the monolayer equation of state. As for the 3D inert gases, this carries with it questions of nonuniqueness and sensitivity. Our goal here is to examine evidence for the contributions of adsorption-induced interactions. For the examples we discuss, the basis of the interaction model is the interaction of an isolated pair, assumed to be transferable from 3D determinations (Scoles 1980; Aziz 1984). The McLachlan interaction is found to be the largest of the additional terms for the equation of state and raises the net lateral energy by about 15% (Unguris *et al.* 1981).

To evaluate the McLachlan energies, we need the height z above the mirror plane, Fig. 2.5. It is derived from information on the height z' of the monolayer above the topmost layer of substrate atoms.* For ^{4}He/graphite, the height is $z' = 2.85 \pm 0.05$ Å (Carneiro *et al.* 1981), a value that includes an average over considerable zero-point motion.[†] For Xe/Ag(111), the value is 3.45 ± 0.1 Å (Stoner *et al.* 1978). The Xe/Ag(111) value is close to the value which is estimated using the packing of spheres with radii set by the nearest-neighbor spacing in the 3D solids (Cohen *et al.* 1976) and the slightly smaller value of z' for Kr/Ag(111) (Unguris *et al.* 1981) matches to these packing ideas. For models of Xe/Pt(111), the values are in the range $z' = 3.1 - 3.35$ Å (Barker and Rettner 1992; Rejto and Andersen 1993), which also are in accord with the packing of hard spheres. However, photoelectron polarization experiments have been interpreted (Potthoff *et al.* 1995) to give a value $z' = 4.2$ Å for Xe/Pt(111). Thus, there is uncertainty in the structural information on which the calculations are based, so that discussion of the substrate-mediated energies becomes a matter of weighing evidence.

For He/graphite, a comparison of a calculated monolayer equation of state with data for the chemical potential as a function of coverage is shown in Fig. B.1. Fair agreement with the data is obtained for a zero-temperature calculation

---

*The height z' is derived from interferences in the multiple scattering of waves by the adlayer and the substrate. The intensity of a diffraction peak of the adlayer or the substrate, including the specular beam, depends on the perpendicular momentum transfer, and analysis of this dependence leads to a value for z'.

[†]The rms value is ~ 0.25 Å (Cole *et al.* 1981).

on a smooth surface ($V_g = 0$) which does not include any adsorption-induced interactions (Ni and Bruch 1986). A sensitive test of the interaction model is the question of whether the ground state of ^{4}He/graphite is a commensurate lattice or a liquid. The energies of the two states are sufficiently close that a 15% change in the He–He interaction changes the conclusion. With the McLachlan term included, the ground state is a liquid; otherwise it is a commensurate solid (Gottlieb and Bruch 1993). The recent experiment of Greywall (1993) indicates the ground state is liquid, a finding that provides support for the role of the McLachlan interaction.

For Xe/Ag(111), the monolayer solid appears to 'float' on the substrate and, at monolayer condensation, the lattice constant is 2% larger than in the 3D Xe solid at the same temperature (Cohen *et al.* 1976; Unguris *et al.* 1979). Geometric effects lead to an expansion of 1% between 3D and 2D, and the remaining 1% can be derived from the McLachlan term (Bruch *et al.* 1976a).

The Xe/Pt(111) monolayer has large effects from the substrate corrugation, and the modeling begins by fitting the corrugation. It appears (Rejto and Andersen 1993) that an energy term on the scale of the McLachlan interaction must be included to reproduce the observed phase diagram.

On the other hand, the situation is less compelling for adsorbed molecules. Bhethanabotla and Steele (1987, 1988a) were able to reproduce the monolayer melting of O_2/graphite without including any substrate-mediated interactions. Also, the submonolayer melting of N_2/graphite has been fit better with a model that omitted the McLachlan interaction (Joshi and Tildesley 1985) than with a model that included it (Hansen and Bruch 1995). There is evidence in low-density monolayers that the potential well between adsorbed molecules is 20% shallower than in the 3D phases. Bojan and Steele (1987a,b) interpreted their data for adsorption virials for N_2, CO, and O_2 on graphite as showing such a reduction. This is on the scale of the McLachlan term, but it has not yet been reconciled with the modeling of the monolayer solid just described. There also is a reduction of 20% or more in the molecule–molecule potential of CH_4/Ag(110) (Elliott *et al.* 1993).

3

MONOLAYER STRUCTURES

Physically adsorbed layers are readily compressed by increasing the pressure of the coexisting 3D gas.* The lattice constants of the monolayer solids are on the scale of a few angstroms, the compressions are on the scale of tenths of angstroms, and the lengths derived in high resolution diffraction experiments typically have precisions 0.002–0.02 Å.

A small number of lattice classes suffices to describe the average lattices of most monolayer solids. As for 3D solids, the primitive vectors, basis vectors, and reciprocal lattice vectors provide a convenient set of coordinates with which to express the periodicity of the lattice. Further, the spatial periodicity of the potential arising from the substrate is conveniently formulated with such coordinates. Finally, the periodicities of the modulated structures that arise from the competition between the intrinsic ordering of the monolayer solid and the periodicity set by the substrate are easily expressed in terms of differences of reciprocal lattice vectors of the adlayer and the substrate.

Sections 3.1 and 3.2 contain the formulation of the geometry of the average monolayer lattices. Sections 3.3 and 3.4 contain the definition of commensurate structures and the specification of modulated lattices derived from them. Section 3.5 contains a discussion of diffraction from adsorbed layers. Finally, Section 3.6 contains a description of orientational alignment, the *Novaco–McTague rotation*, which is a consequence of the modulation energies (Novaco and McTague 1977a,b).

3.1 Monolayer packing geometry

We introduce the 2D Bravais lattices that are basic to the discussion of the geometries of the adlayer and the substrate surface and then describe the specification of the components in a lattice with a basis.[†] There is an additional terminology (Wood 1964) for commensurate layers with a basis (e.g., c(2 × 2) and p(2 × 2)) which we do not develop here.

For disks in two dimensions there is one close-packed lattice, the triangular lattice shown in Fig. 3.1 with primitive vectors a_1 and a_2 of equal length a and

---

*The number of adsorbed particles on a given surface then increases enough to keep the chemical potential of the 3D gas equal to that of the adsorbed layer, the relation in the Gibbs adsorption isotherm eqn (B.7).

[†]These concepts are developed in standard texts, e.g., N. W. Ashcroft and N. D. Mermin, *Solid State Physics*, (Saunders, Philadelphia, 1976), Chaps. 4 and 5 and C. Kittel, *Introduction to Solid State Physics*, 6th edition (Wiley, New York, 1986), Chaps. 1 and 2.

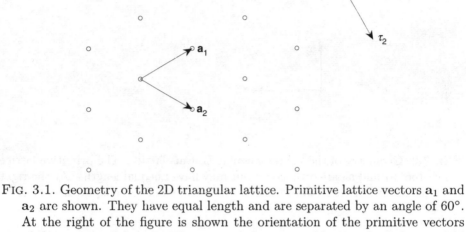

FIG. 3.1. Geometry of the 2D triangular lattice. Primitive lattice vectors a_1 and a_2 are shown. They have equal length and are separated by an angle of 60°. At the right of the figure is shown the orientation of the primitive vectors of the reciprocal lattice. The space and reciprocal lattices both have 6-fold rotation symmetry.

at an angle of 60° or 120°. It may be intuitively obvious that inert gas atoms on a smooth surface solidify in the triangular lattice, to maximize the packing. However, there is no rigorous proof setting out the conditions on the interatomic potential for this to be the structure of minimum potential energy. The triangular lattice is firmly established as the two-dimensional ground state for only a few highly idealized models (Radin 1986); more extensive results are available for 1D chains (Ventevogel 1978; Radin 1984; Nijboer and Ruijgrok 1985). The triangular lattice (also frequently termed the hexagonal lattice because of the configuration of the nearest neighbors) is in fact the most commonly encountered lattice in monolayer solids.

The remaining 2D Bravais lattices are the rectangular lattice, the centered rectangular lattice, and the oblique lattice. The rectangular lattice, Fig. 3.2, has primitive vectors a_1 and a_2 of unequal length and at an angle of 90°. The centered rectangular lattice, Fig.(3.3), has equal length primitive vectors a_1 and a_2 at an angle not equal to 60° or 90°. It reduces to the triangular lattice when the angle is 60° and to the square lattice when the angle is 90°. The most general

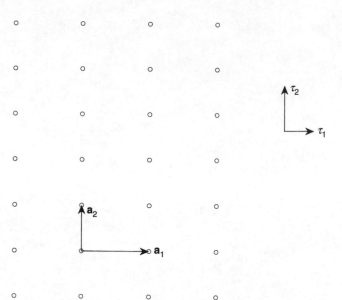

FIG. 3.2. Geometry of the 2D rectangular Bravais lattice. The primitive lattice
vectors a_1 and a_2 are orthogonal but may have unequal lengths. At the right
is shown the orientation of the primitive vectors of the reciprocal lattice.

2D Bravais lattice, which we do not treat in any detail, is the oblique lattice with
primitive vectors of unequal length and at an angle other than 90°. Points $R_{m,n}$
of the Bravais lattice are integer superpositions of the primitive lattice vectors:

$$R_{m,n} = ma_1 + na_2. \tag{3.1}$$

A more general lattice consists of a Bravais lattice with a basis. For example,
the basal plane graphite surface, Fig. 3.4, has a triangular lattice; its unit cell
has a basis of two carbon atoms at positions b_1 and b_2 given by

$$b_1 = (a_1 + a_2)/3 \; ; \quad b_2 = 2b_1 \quad \text{(graphite)}. \tag{3.2}$$

Another example is the herringbone lattice (Eckert *et al.* 1979; Diehl and Fain
1983) of low-temperature nitrogen adsorbed on basal plane graphite, Fig. 3.5.
It is a rectangular Bravais lattice with a basis of two molecules at positions

$$b_1 = 0 \; ; \quad b_2 = (a_1 + a_2)/2 \quad (N_2/\text{graphite}). \tag{3.3}$$

The number of sites j in a basis may be very large. An example that occurs for
monolayer solids is the high-order coincidence lattice with the substrate discussed

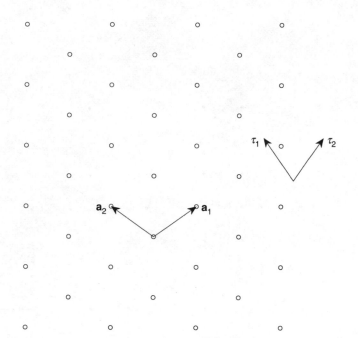

FIG. 3.3. Geometry of the centered rectangular lattice. The equal length primitive vectors a_1 and a_2 are at an angle other than $60°$, $90°$, or $120°$. In the example shown, which arises for an fcc(110) surface, the angle is $109.4°$. At the right is shown the orientation of the primitive vectors of the reciprocal lattice.

in Section 5.2 (Fuselier *et al.* 1980). In general, the sites of a Bravais lattice with a basis are generated by

$$\mathbf{R}_{m,n,j} = \mathbf{R}_{m,n} + \mathbf{b}_j. \tag{3.4}$$

Addition of a constant vector on the right-hand side of eqn (3.4) is a way to allow for a shift in the origin of the adlayer relative to the substrate lattice.

3.2   Reciprocal lattice

The periodicity of functions which arise as sums over sites of a Bravais lattice is readily expressed using reciprocal lattice vectors. This is the analogue of Fourier transformation of a periodic function. The Fourier decomposition of the adatom–substrate potential in terms of reciprocal lattice vectors was given in eqn (2.118) and is derived here at eqn (3.9) (Steele 1973, 1974). We adopt the convention that reciprocal lattice vectors of the substrate surface are denoted $\mathbf{g}$ and those of the adlayer are denoted τ.

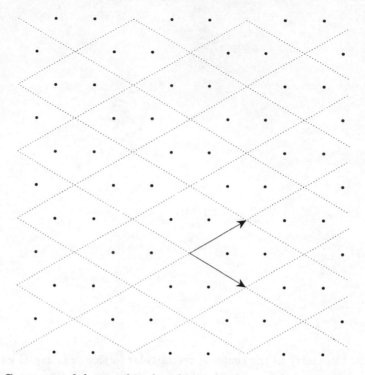

FIG. 3.4. Geometry of the graphite basal plane. The triangular Bravais lattice is
outlined by dotted lines and has the same scale and orientation as in Fig.(3.1).
The carbon atom positions are denoted by solid dots. Note that the vertices
of the Bravais lattice are situated at centers of hexagons of carbon atoms.

The 2D reciprocal lattice is generated by two primitive vectors $\boldsymbol{\tau}_1$ and $\boldsymbol{\tau}_2$
that are related to the primitive vectors $\mathbf{a}_1$ and $\mathbf{a}_2$ by

$$\boldsymbol{\tau}_i \cdot \mathbf{a}_j = \delta_{i,j} \, , \tag{3.5}$$

using the Kronecker delta notation (1 for equal susbcripts, zero otherwise). The
orientations of the primitive reciprocal lattice vectors relative to the space lattice
were shown for the triangular, rectangular, and centered rectangular Bravais
lattices in Figs. 3.1–3.3, respectively.

An explicit Cartesian representation of the vectors for the centered rectangular lattice is

$$\begin{aligned}
\mathbf{a}_1 &= a[+\sin(\psi/2)\hat{x} + \cos(\psi/2)\hat{y}] \\
\mathbf{a}_2 &= a[-\sin(\psi/2)\hat{x} + \cos(\psi/2)\hat{y}] \\
\boldsymbol{\tau}_1 &= \tau_o[+\cos(\psi/2)\hat{x} + \sin(\psi/2)\hat{y}]
\end{aligned}$$

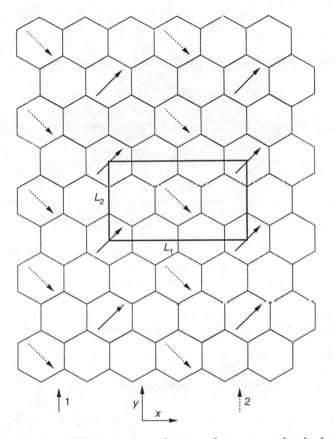

FIG. 3.5. Commensurate herringbone lattice of nitrogen adsorbed on the basal plane surface of graphite. Monolayer herringbone lattice showing rectangular unit cell and two sublattices denoted by solid and dotted arrows,respectively, from Roosevelt and Bruch (1990). The case shown is commensurate and has $L_1 = \sqrt{3}L_2 = 3 \times 2.46$ Å. There are also uniaxially incommensurate herringbone lattices for this system (Diehl and Fain 1983).

$$\tau_2 = \tau_o[-\cos(\psi/2)\hat{x} + \sin(\psi/2)\hat{y}] \tag{3.6}$$

with $\tau_o = 2\pi/a\sin\psi$. In Fig. 3.3, the angle is $\psi = 109.4°$. Equation (3.6) is used in Section 3.3.

For any vector $\mathbf{T}$ generated by integer sums of the τ_i,

$$\mathbf{T} = \mu\tau_1 + \nu\tau_2 , \tag{3.7}$$

and an arbitrary vector $\mathbf{R}_{m,n}$ of the Bravais lattice, eqn (3.1), we have

$$\exp[\imath\mathbf{T}\cdot(\mathbf{r} + \mathbf{R}_{m,n})] = \exp[\imath\mathbf{T}\cdot\mathbf{r}] . \tag{3.8}$$

Thus a Fourier representation of the spatial periodicity of the adatom–substrate potential is

$$V(\mathbf{r}, z) = V_0(z) + \sum_{\mathbf{g}} V_g(z)\exp[\imath\mathbf{g}\cdot\mathbf{r}] , \tag{3.9}$$

in terms of the reciprocal lattice vectors $\mathbf{g}$ of the substrate surface. For the graphite basal plane shown in Fig. 3.4, the reciprocal lattice vectors are oriented as shown in Fig. 3.1.

The Brillouin zone is a further construction based on the reciprocal lattice. The first Brillouin zone is the area around the origin of the reciprocal lattice that is enclosed by the nearest perpendicular bisectors to the reciprocal lattice vectors. The first Brillouin zones for the triangular, rectangular, and centered rectangular lattices are shown in Fig. 3.6. From the definition, the Fourier integral representation of a property defined at sites of the Bravais lattice, such as atomic displacement,

$$\mathbf{D}_{m,n} = \int d\mathbf{k}\,\mathbf{D}(\mathbf{k})\exp[\imath\mathbf{k}\cdot\mathbf{R}_{m,n}] , \tag{3.10}$$

can be restricted to wavevectors contained in the first Brillouin zone. When boundary conditions are imposed on a large but finite lattice, the Fourier integral becomes a discrete sum and the number of $\mathbf{k}$s is equal to the number of sites N_s of the Bravais lattice.

3.3 Commensurate lattices

Commensurate and registry lattices are defined in terms of the degree of overlap of the reciprocal lattices of the adlayer and the substrate. A commensurate monolayer has the same Bravais lattice as the substrate surface, and its orientation and lattice constant a are such that all points of the reciprocal lattice of the substrate surface are points in the reciprocal lattice of the adlayer. For example the 1×1 commensurate monolayer has its Bravais lattice aligned with that of the substrate and its lattice constant is equal to the surface lattice constant L;

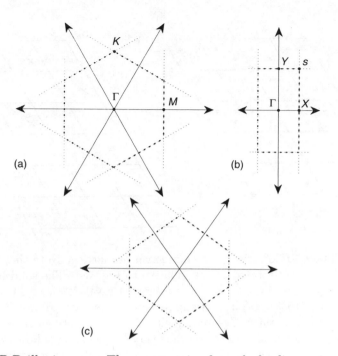

FIG. 3.6. 2D Brillouin zones. The construction from the leading reciprocal lattice vectors is shown. The perpendicular bisectors are drawn as dotted lines and the zone boundary is marked by bold-face dotted lines. (a) Triangular lattice. The zone is a regular hexagon with 6-fold rotation symmetry. The Γ, M and K points are noted. (b) Rectangular lattice. The zone is a rectangle and the Γ, X, Y, and s points are noted. (c) Centered rectangular lattice. The zone is a hexagon with four sides of equal length.

the 2×2 case has a similar orientation and has a lattice constant equal to twice L.

More complex ratios arise for the triangular lattice. The $\sqrt{3}\,R\,30°$ commensurate structure shown in Fig. 3.7 has lattice constant $a = \sqrt{3}L$ and its axes are rotated (R) by $30°$ relative to the underlying triangular lattice. To emphasize the fact that there are three equivalent such structures, we show in Fig. 3.8 the three superlattices that may be placed on the triangular Bravais lattice of Fig. 3.1. The $\sqrt{3}\,R\,30°$ lattice is observed in monolayer adsorption of He, H_2, Kr, and Xe on graphite, but other commensurate lattices also occur. The $\sqrt{7}\,R\,19.1°$ lattice of Ne/graphite has a Bravais lattice with a four-atom basis. It is aligned at $19.1°$ to the graphite Bravais lattice and has lattice constant $a = \sqrt{7}L$ (Calisti et al. 1982; Tiby et al. 1982).

The Bravais lattice of a registry lattice is in a distinct class from that of the

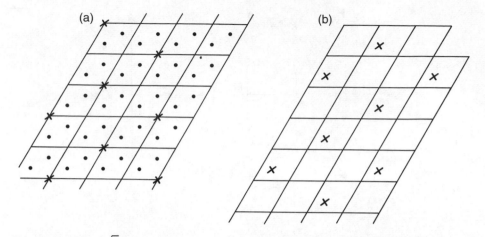

FIG. 3.7. The $\sqrt{3}R30°$ triangular commensurate lattice. Note that the axes of the substrate are rotated by 30° relative to Fig. 3.4. The solid lines outline several unit cells of the triangular substrate Bravais lattice for two examples, from Bruch (1988). (a) The triangular commensurate adlayer lattice (x) on the basal plane surface of graphite, with the dots denoting carbon atoms. (b) Adatoms in threefold sites (x) on the (111) face of an fcc solid, with substrate atoms at the unit cell vertices. The nearest neighbor spacing in the adlayer lattice is $\sqrt{3}$ times the surface lattice constant and the axes of the adlayer are rotated by 30° relative to the surface axes for both examples.

substrate surface. However, its orientation and lattice constant are such that some points of the substrate reciprocal lattice are points in the adlayer reciprocal lattice (Bruch and Venables 1984). An example is the uniaxially incommensurate (UIC) lattice treated later in this section.

For the commensurate lattice, all Fourier components of the adatom–substrate potential, eqn (3.9), add coherently to the energy of the monolayer solid, while for a registry lattice only some of the Fourier components add coherently. For each substrate reciprocal lattice vector that is in the adlayer reciprocal lattice, the sum over sites of the adlayer Bravais lattice of N_s sites

$$\sum_{m,n} V_{\mathbf{g}} \exp[i\,\mathbf{g}\cdot\mathbf{R}_{m,n}] \tag{3.11}$$

leads to a contribution of order N_s in the adlayer energy; this is called the registry energy. The sum is zero for those vectors $\mathbf{g}'$ not in the adlayer reciprocal lattice.

Misfit wavevectors $\mathbf{q}_i$ are defined as the differences between the substrate and adlayer reciprocal lattice vectors, translated if necessary into the first Brillouin zone of the adlayer. For those $\mathbf{g}$ for which $\mathbf{q} = 0$, the sum in eqn (3.11) is of order N_s.

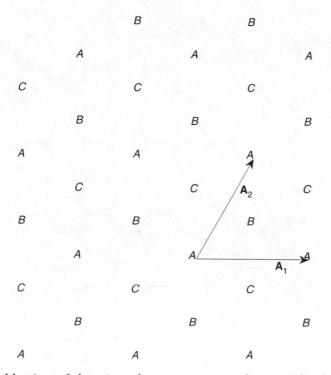

FIG. 3.8. Sublattices of the triangular commensurate lattice. The three equivalent sublattices of the $\sqrt{3}R30°$ commensurate lattice are shown, denoted A, B, and C. The scale and orientation of the underlying triangular Bravais lattice are the same as in Figs. 3.1 and 3.4. Primitive vectors $\mathbf{A}_1$ and $\mathbf{A}_2$ of the superlattice are shown.

An important special case (Cui *et al.* 1988) that illustrates these concepts and which is central to the theory of continuous commensurate–incommensurate transitions is the registry lattice, denoted UIC, formed by uniaxial compression of the $\sqrt{3}R30°$ commensurate monolayer. The primitive vectors of the space and reciprocal lattices are those for the centered rectangular lattice, eqn (3.6). Primitive vectors of the reciprocal lattice for the triangular surface for the lattice orientation shown in Figs. 3.8 and 3.9 are

$$\mathbf{g}_1 = g_0\hat{x}, \quad \mathbf{g}_2 = g_0[\frac{1}{2}\hat{x} + \frac{\sqrt{3}}{2}\hat{y}] \, . \tag{3.12}$$

with $g_o = 4\pi/L\sqrt{3}$; the vector $\mathbf{g}_3 = \mathbf{g}_2 - \mathbf{g}_1$ also has length g_0. The primitive vectors $\mathbf{A}_1$ and $\mathbf{A}_2$ of the reference $\sqrt{3}$ commensurate lattice have length $a_0 = \sqrt{3}L$ and are oriented as shown in Fig. 3.8, i.e., they make an angle of 60° and $\mathbf{A}_1$ is parallel to the x-axis. All vectors of the reciprocal lattice of the substrate surface lie in the $\sqrt{3}$ reciprocal lattice because

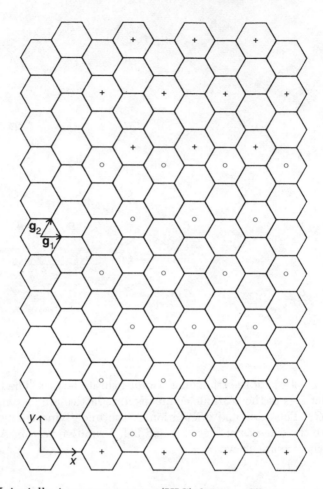

FIG. 3.9. Uniaxially incommensurate (UIC) lattice. Illustration of a six-row
UIC cell for a compressed monolayer with superheavy domain walls on the
basal plane of graphite. The $\circ$ and $+$ denote adsorption sites on the various
domains, from Gottlieb and Bruch (1989). The reciprocal lattice vectors
$\mathbf{g}_i$ of the graphite surface lattice are shown. A similar structure arises for
compressed H_2/graphite. The illustration corresponds to a limit in which the
domain walls are sharp, the adatoms are all at centers of carbon hexagons,
and the monolayer is a periodic succession of commensurate domains.

$$\mathbf{g}_1 = \boldsymbol{\tau}_1 - \boldsymbol{\tau}_2; \quad \mathbf{g}_2 = 2\boldsymbol{\tau}_1 + \boldsymbol{\tau}_2 \quad (\sqrt{3}R30°) . \tag{3.13}$$

The UIC lattice has $a\sin(\psi/2) = a_0/2$ in eqn (3.6). Define the fractional
misfit $\overline{m}$ of the space lattice by

$$a\cos(\psi/2) = a_0 \frac{\sqrt{3}}{2} \times (1 - \overline{m}) = \frac{3}{2} \times L \times (1 - \overline{m}) \,. \tag{3.14}$$

Then the area per adatom relative to that in the commensurate layer is

$$A/A_{\sqrt{3}} = 1 - \overline{m} \,. \tag{3.15}$$

Misfit wavevectors formed from $\mathbf{g}_1$, $\mathbf{g}_2$, and $\mathbf{g}_3$ are defined by

$$
\begin{aligned}
\mathbf{q}_1 &= \mathbf{g}_1 - (\boldsymbol{\tau}_1 - \boldsymbol{\tau}_2) = 0 \\
\mathbf{q}_2 &= \mathbf{g}_2 - (2\boldsymbol{\tau}_1 + \boldsymbol{\tau}_2) = -\left(\frac{2\pi}{L}\right) \times \frac{\overline{m}}{(1 - \overline{m})} \hat{y} \\
\mathbf{q}_3 &= \mathbf{g}_3 - (\boldsymbol{\tau}_1 + 2\boldsymbol{\tau}_2) = \mathbf{q}_2 \,.
\end{aligned}
\tag{3.16}
$$

Thus, two ($\pm\mathbf{g}_1$) of the six shortest reciprocal lattice vectors of the substrate have zero misfit wavevector and remain members of the adlayer reciprocal lattice; the corresponding two amplitudes V_{go} contribute linearly to the registry energy. The uniaxial compression along $\hat{y}$, perpendicular to the rows as in Fig. 3.9, preserves this registry energy; a compression parallel to the rows, along $\hat{x}$, would not do so.

There are similar constructions for the misfit wavevectors arising under uniaxial compression from other commensurate geometries. The identification of the compression axis which leaves some reciprocal lattice vectors of the substrate coincident with those of the adlayer has to be done in a case-by-case manner. For a 2×2 commensurate lattice on a triangular substrate that has been discussed for CF_4/graphite, the axis which does this is parallel rather than perpendicular to the rows (Kjaer et al. 1982).

For a Bravais lattice with a basis, the construction of the misfit wavevectors is similar, but the identification of the registry energy must be generalized when the sum in eqn (3.11) includes a sum over members of the basis. Positions in the basis are determined by balancing energies from the substrate corrugation (V_g) and from the adsorbate–adsorbate interactions. The registry energy now includes contributions from the relaxations of the positions in the basis. In perturbation theory terms, the latter are second-order and higher in V_g while the registry energy for the commensurate lattice with a one-atom basis, eqn (3.11), is first order in V_g. When the first substrate reciprocal lattice vector that lies in the adlayer reciprocal lattice is not a primitive vector, the first-order energy arising from its Fourier amplitude may be smaller than the higher-order perturbation energy arising from the leading Fourier amplitude (Theodorou and Rice 1978; Bruch 1988). A specific example of the relaxations in a Bravais cell with a large basis is treated in the next section.

3.4 Domains

Consider the $\sqrt{3}$ commensurate lattice indicated in Fig. 3.8. Only one-third of the sites of the substrate Bravais lattice are occupied and there are three equivalent ways of situating the adlayer on the substrate. This count of degenerate configurations is an essential part of determining the relevant universality class for critical phenomena in the adlayer (Section 5.1.2). An extended region of the adlayer that is situated in one of these configurations is called a domain, and the interface between domains is called a domain wall. The domains can be arranged so that the average adlayer density is either increased or decreased from the commensurate density. In the analogous terminology for magnetic phenomena, domains consist of large regions in which magnetic moments are aligned (e.g., 'up' or 'down'), and the domain walls are the relatively narrow regions in which the average direction of the moments changes orientation. In this section, the UIC lattice is used to illustrate the application of such concepts for the monolayer solid.

The UIC lattice shown in Fig. 3.9 has an average spacing $d = (3L/2) \times (1-\overline{m})$ between rows and the commensurate spacing $\sqrt{3}L$ within rows. However, at small misfits $\overline{m}$ or large corrugation V_g (there is an interplay in these variables), the character of the UIC lattice is drastically different than that indicated by the average lattice. Then, instead of achieving the compression or expansion by equal increments on the interrow spacing, the misfit is distributed very nonuniformly. For the geometry shown in Fig. 3.9, one domain can be translated into another by a 'slip' of $-L\hat{y}$. The language of domains and domain walls is most appropriate when the regions in which the slips occur are well-separated and only a few rows wide. Such a structure of the UIC lattice is termed a *striped lattice* because the domain walls are parallel and well spaced. At large misfits, the domain walls are so close to each other that a description in terms of the properties of individual walls loses accuracy.

To make the discussion explicit, we treat the case of the superheavy wall (SHW) (Kardar and Berker 1982; Houlrik and Landau 1991) that is observed for H_2/graphite and for Xe/Pt(111).[*] This is a compression in which the displacements can be described by a scalar function, as can the superlight wall with displacement $L\hat{y}$. While a slip by $L(\sqrt{3}/2\hat{x} + (1/2)\hat{y})$ also translates one domain into another, there has been no observation of the corresponding heavy and light walls, HW and LW. We refer to the literature, (Gooding *et al.* 1983; Gordon 1987; Houlrik and Landau 1991; Bruch 1993), for treatments of those cases.

The algebraic representation of the domain wall phenomenon is based on expressing the adlayer positions in terms of displacements $\boldsymbol{\eta}_j$ from the commensurate lattice positions $\mathbf{R}_{m,n}(C)$

$$\mathbf{r}_j = \mathbf{R}_{m,n}(C) + \boldsymbol{\eta}_j . \tag{3.17}$$

For the compressed UIC lattice with aligned parallel domain walls, which is the observed situation, the displacement $\boldsymbol{\eta}$ depends only on the row index j of the commensurate lattice and has the periodicity

$$\boldsymbol{\eta}_{j+\mathcal{L}} = \boldsymbol{\eta}_j - L\hat{y}, \tag{3.18}$$

---

[*]Masuda *et al.* (1992) observe such a structure for K/Co.

where $\mathcal{L}$ is the number of rows in the domain. Because of the symmetry of the adatom–substrate potential for an adlayer positioned as in Fig. 3.9, the vector displacement function is oriented along the $\hat{y}$-axis, $\boldsymbol{\eta} = \eta\,\hat{y}$. For the uniform UIC lattice, the function η is linear in the row-index j, while the limit of sharp domain walls is reached when η changes by L over ranges Δj that are small compared to $\mathcal{L}$. For intermediate cases, η is expressed as a sum of the uniform variation, linear in j, and sinusoidal terms with wavevectors that are multiples (harmonics) of the misfit wavevector. The adlayer is then said to be modulated, and the leading modulation amplitude is usually that of the basic misfit wavevector.

The presentation in eqns (3.17) and (3.18) is for the aligned superheavy wall (SHW); expressions for the superlight wall (SLW) are obtained by changing the sign of L. Pokrovsky and Talapov (1980) analyzed a generalization with parallel domain walls oriented at an angle to the slip axis ($\hat{y}$); such an orientation has yet not been observed. For the situation where the average adlayer lattice is triangular and incommensurate, the domain walls form a honeycomb net with important energy terms from the wall–wall intersections. Apart from qualitative arguments, reviewed in Section 5.2.2, on the significance of the sign of the wall intersection energy (Bak et al. 1979) and on the entropy of *breathing modes* of the honeycomb net (Villain 1980b,c), information on domain walls in the triangular incommensurate lattice comes from detailed solutions for specific interaction models.

3.5 Diffraction

The diffraction conditions for scattering by a 2D lattice are

$$[\mathbf{k}_f - \mathbf{k}_i]|_{\parallel} = \boldsymbol{\tau} \quad ; \quad |\mathbf{k}_f| = |\mathbf{k}_i| \, . \tag{3.19}$$

That is, the scattering is elastic, and the difference between the parallel components of the initial and final wavevectors is equal to a reciprocal lattice vector of the 2D lattice. The Ewald construction (Webb and Lagally 1973) for identifying solutions of this pair of equations is shown in Fig. 3.10. The scattered wavevector must terminate on a Bragg rod oriented perpendicular to the plane of the lattice; this is an expression of the lack of periodicity perpendicular to the plane. In fact, diffraction experiments on multilayer lattices include a scan parallel to the Bragg rod because the intensity variation then gives information on the stacking sequence of the layers (Ignatiev and Rhodin 1973; Larese et al. 1988c, 1989; Nuttall et al. 1993; Dai et al. 1994).

Another variant on the diffraction condition is to examine the interference between waves scattered from the substrate surface and from the adlayer. The intensity then depends on the perpendicular momentum transfer. Effects of such multiple scattering in the diffracted beams for the substrate or adlayer have been used to determine the interplanar spacing between the substrate surface and the adlayer, with LEED (Stoner et al. 1978; Shaw et al. 1980), neutron diffraction (Carneiro et al. 1981; Smalley et al. 1981), and X-ray diffraction

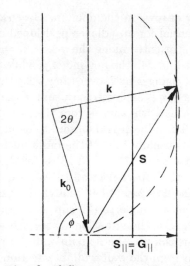

FIG. 3.10. Ewald construction for diffraction by a 2D array, such as a crystalline surface. The vertical lines represent reciprocal lattice rods, lines drawn perpendicular to the surface through the 2D array of reciprocal lattice points. The incident wave vector is denoted $\mathbf{k}_0$ and the scattered wave vector $\mathbf{k}$. The elastic scattering condition is represented by the dashed spherical surface of radius k_0. Coherent scattering occurs in directions for which this Ewald sphere intersects a reciprocal lattice rod. The condition is that the parallel component of $\mathbf{k} - \mathbf{k}_0$ be equal to a reciprocal lattice vector of the surface, here denoted by $\mathbf{G}_{\parallel}$ and in the text by τ. From Barnes (1967).

(Hong *et al.* 1986). A related phenomenon is *double diffraction* at wave vectors involving differences between the reciprocal lattice vectors of the adlayer and the substrate. At times it has been difficult to distinguish double diffraction from a signal arising from the modulation of the adlayer; helium atomic scattering, with its selective sensitivity to the adlayer, helps to settle such questions.

The diffraction from the lattice is analyzed in terms of the thermal average of the structure factor

$$S(\mathbf{k}) = \langle |\sum_{j,\alpha} \exp[i\,\mathbf{k} \cdot \mathbf{R}_{j,\alpha}]|^2 \rangle \,, \tag{3.20}$$

where the indices j, α on the sum denote the cells of the Bravais lattice and the members of the basis, eqn (3.4), respectively. The structure factor of a static lattice evaluated at a reciprocal lattice vector τ is proportional to N_s^2, reflecting the constructive interference. For a fluid, $S(\mathbf{k})$ is proportional to N_s, the number of particles in the scattering volume. As discussed in Section 4.2.3, the structure

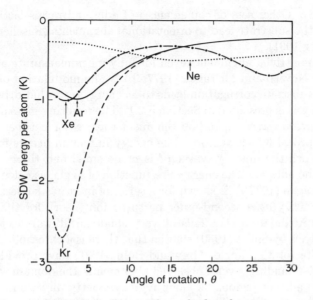

FIG. 3.11. Novaco–McTague effect. The calculated energy per atom that the adlayer gains by the formation of static distortion waves (SDW) is shown as a function of rotation angle for various gases on graphite, from Novaco and McTague (1977b). The absolute minima correspond to the equilibrium angles for orientational epitaxy. The undistorted lattice constant has been used in the calculation.

factor for the finite temperature 2D harmonic solid is not as sharply peaked as for the 3D solid, and peak heights at reciprocal lattice vectors are not as large as the N_s^2-dependence obtained for the static lattice (Imry and Gunther 1971; Nelson and Halperin 1979).

The structure factor, eqn (3.20), is proportional to the intensity of the Bragg diffraction peaks in the quasi-elastic approximation. Thus it applies to the analysis of X-ray diffraction and LEED. Additional analysis is required for neutron scattering (Webb and Lagally 1973; Imry 1978).

3.6 Orientational alignment

The commensurate and registry adlayer lattices have specific simple alignments relative to the substrate axes. Additional named cases that arise for triangular lattices on centered-rectangular surface lattices (e.g., the bcc(110) surface) are the Nishiyama–Wasserman and Kurdjumov–Sachs epitaxial orientations (Bauer 1982). They also can be understood in terms of coincidences of substrate reciprocal lattice vectors with vectors of the adlayer reciprocal lattice (Bruch and

Venables 1984). Other sets of coincidences of adlayer Bravais lattices with superlattices of the substrate lead to orientational alignments (Fuselier *et al.* 1980; Grey and Bohr 1991).

There are nontrivial alignments even for fully incommensurate adlayers, first recognized by Novaco and McTague (1977a,b).* The modulation of the adlayer lattice by the substrate corrugation leads to an energy term that drives the alignment; the analysis is presented in Section 5.2.1. The resulting *mass density waves* or *static distortion waves* depend on the misfit wavevector between adlayer and substrate reciprocal lattice vectors. The energy minimum occurs for a rotation angle where both the misfit wavevector is made small and there is a shearing distortion of the adlayer. The energy as a function of angle, as calculated by Novaco and McTague (1977b), is shown for a series of inert gas lattices on graphite in Fig. 3.11. They used second-order perturbation theory for the modulation energy, an approximation that fails at very small misfit wavevector. A more thorough analysis by Shiba (1980) showed that there is a threshold misfit for the onset of a finite rotation angle. Grey and Bohr (1991) have provided a 'rule of thumb' to understanding several observed alignments: the domain walls between commensurate adlayer regions lie along high symmetry directions of either the substrate or the adsorbate. The rule of thumb includes the alignment observed (Cui and Fain 1989) for the γ phase of D_2/graphite, which is an example where more sophisticated atomic-scale theory is required to reproduce the experimental data.†

---

*We discuss orientational epitaxy in the context of physically adsorbed layers. The phenomenon is also prominent in the observations of alkali atom monolayers (Doering and Semancik 1986; Diehl 1991) and of reconstructed metal surfaces (Abernathy *et al.* 1992).

†The rule is most applicable for systems with quite narrow domain walls; van der Merwe (1982 a,b,c) developed a 'rigid model' of ideal epitaxy for this limit. For more general cases, the relaxations of atomic positions and details of the interactions become significant in determining the alignment.

4

MONOLAYER FILMS: THEORETICAL CONTEXT

The development of scientific concepts usually proceeds from the simple to the complex. We now follow such a route in presenting the theory of monolayer adsorption. This chapter provides an introduction to the theory; Chapters 5 and 6 provide a more detailed description of the theory and a discussion of illustrative examples of film behavior, respectively. Three principal lines of conceptual development have strongly influenced the goals of theoretical efforts for monolayers and also have set a context in which most of the experiments have been analyzed.

1. Lattice gas models are a schematic description of monolayer formation and have provided a language and theoretical framework in which monolayer phase transitions are analyzed. We treat the most elementary of such models, the Langmuir and BET adsorption isotherms, in Section 4.1 and develop the analysis for models with phase transitions in Section 5.1.

2. Knowledge of interactions among inert gases and the available computational resources have led to quantitative theories of the 3D bulk phases. The corresponding analysis for the monolayer is developed in Section 5.2. Here, in Section 4.2, the monolayers are treated at the level of the two-particle clusters for the gas and the harmonic approximation to the solid. The analysis of 2D gas data in terms of the second virial coefficient was a historically important step in achieving quantitative understanding of a physically adsorbed layer. The effects of reduced dimensionality in the subtle character of ordering of the 2D solid are evident already in the harmonic approximation.

3. The theory of topological defects provides a scenario for the loss of order in reduced dimension with increasing temperature. We outline the application of these concepts to monolayer melting and to 2D superfluidity in Section 4.3. The most striking prediction of the topological defect theory is that melting in 2D may be a continuous transition. In 3D, melting is always discontinuous; the difference in 2D is the possibility of thermally exciting edge dislocations that destroy the resistance of the solid to static shear stresses.

We begin by treating several statistical mechanical models that yield estimates of thermodynamic conditions appropriate to a monolayer film. Although the alternative descriptions discussed in Sections 4.1 and 4.2.1 are extreme opposites in their assumptions about how the adsorption potential varies across the surface, i.e., localized *versus* delocalized adsorption, some common conclusions are derived that remain valid in a more realistic analysis. Section 4.2.2 presents a general discussion, based on the virial expansion, from which the behavior in all regimes of low coverage, high temperature, and/or weak interaction can be derived. Section 4.2.3 addresses the dynamics of a monolayer film by developing the simple harmonic (phonon) model. Section 4.3 continues the discussion of the monolayer solid and discusses the thermal excitation of topological defects. Finally, in Section 4.4, we treat briefly the question of persistent correlations in fluids, i.e., *long-time tails* in correlation functions, a phenomenon present in computer simulations of 2D fluids but whose relevance to the monolayer is not yet established.

The simplest models make drastic assumptions about the geometry and interactions. They are useful insofar as nature is qualitatively represented or the experimental design is appropriately chosen to probe a specific regime of behavior. The virial expansion is an example of the latter instance. As for 3D systems, a low-density expansion can provide both an indication of when interactions play a role and a reasonable estimate of the conditions for phase transitions at high density.[*] In the case of adsorption, an even more important role is played by the virial approach because the leading one-body term in the series is determined by the adsorbate–substrate holding potential $V(\mathbf{r})$. In fact, low-density adsorption isotherm studies are often the most convenient way to learn about the well depth of this potential.

In the following, we evaluate the properties of systems that conform to a set of idealized assumptions. The results serve to outline the qualitative features of monolayer phenomena, but they can be applied quantitatively only with caution. These limitations may be appreciated when comparison is made with data in this chapter and the next.

We focus on the thermodynamic properties of the film when it is in equilibrium with the vapor.[†] Then the chemical potential of the film, μ, is equal to that of the 3D vapor, which we assume to be ideal ($n = P\beta$):

$$\mu = k_B T \ln(n\lambda^3/g) , \tag{4.1}$$

where k_B is Boltzmann's constant, n is the number density of the vapor, g is the spin degeneracy of the atoms (often assumed implicitly to be one), and the DeBroglie thermal wavelength is defined as

$$\lambda = \sqrt{2\pi\beta\hbar^2/m} , \tag{4.2}$$

where m is the atomic mass and the inverse temperature is

$$\beta = 1/(k_B T) . \tag{4.3}$$

---

[*]We have in mind the van der Waals theory of condensation that is developed in Section 4.2.2 by using terms identified in the second virial coefficient to evaluate parameters in the van der Waals equation of state.

[†]This has been exploited by Taborek and Goodstein (1979), who proposed a manometer based on a detailed knowledge of the chemical potential of the ^{4}He/graphite monolayer as a function of temperature and coverage.

4.1 Noninteracting lattice gas models

4.1.1 Langmuir isotherm

Some of the simplest and earliest studies of adsorption are based on the idea that the surface consists of an array of N_s possible adsorption sites, usually assumed to be identical. The surface binding energy is denoted ϵ_o, a negative energy when referenced to the potential energy of the adatom at infinite separation. Further simplification follows if the adatoms are assumed to be mutually noninteracting, apart from the restriction that no more than one atom may reside at a given site. We solve here this simpler model, which reduces to a set of N_s identical independent site problems, and treat more complicated models in Chapter 5. In the grand canonical ensemble, each site's grand partition function becomes

$$\Xi_s = 1 + z \exp[\beta(\mu - \epsilon_o)] , \tag{4.4}$$

where z is the partition function associated with possible *internal* degrees of freedom at each site; z may be taken as unity in some cases. The two terms correspond to the absence and presence of a particle at the site, respectively. The grand canonical free energy is

$$\Omega = -(1/\beta) \ln \Xi = -(N_s/\beta) \ln \Xi_s . \tag{4.5}$$

Since the mean number of particles in this ensemble satisfies

$$N = -(\partial \Omega / \partial \mu)_\beta , \tag{4.6}$$

the fractional occupation is given by

$$\theta = N/N_s = P/(P + P_L) , \tag{4.7}$$

where the characteristic scale of pressure is

$$P_L = [g/(z\beta\lambda^3)] \exp(\beta\epsilon_o) . \tag{4.8}$$

We have used the ideal gas law, $P = n/\beta$. Equation (4.7) is the venerable Langmuir (1918) isotherm. It indicates that the coverage rises linearly at low P according to Henry's law and saturates eventually at values of P that are much greater than P_L. This characteristic pressure is particularly low if the substrate binding energy is strong relative to the temperature.

   An equally convenient model that extends the lattice gas model to the case of multilayer films is that of Brunauer, Emmett, and Teller (1938). The BET model allows particles to occupy a fully three-dimensional array of sites above the surface. The interactions between sites are neglected, but the site closest to the substrate is assumed to experience an additional attraction V_1. The relative probability that there are exactly N sites occupied above a given surface site is proportional to the corresponding term in the grand partition function for that site,

$$\Xi_s = 1 + c \sum_{N=1}^{\infty} z^N \exp(N\beta\mu) , \qquad (4.9)$$

where $c = \exp(-\beta V_1)$ and z is the internal partition function per site of the bulk adsorbate. After evaluating eqns (4.5) and (4.6), we obtain the isotherm, which may be written

$$xN_s/[(1-x)N] = [1 + x(c-1)]/c \ \text{(BET)} , \qquad (4.10)$$

where x is the ratio of the pressure to the saturated pressure, the value at which N diverges. Although eqn (4.10) is only qualitatively reliable, it provides an estimate of the monolayer completion phenomenon manifested by the vertical riser in the isotherm shown in Fig. 4.1. As we discuss in Chapter 5, this approach has been largely superseded by the model of De Oliveira and Griffiths (1978). Their model also suffers from some necessary approximations implicit in the lattice gas approach, but it provides a more accurate treatment of both the substrate potential and intralayer interactions.

A widely used approach to the determination of surface area for heterogeneous materials involves fitting the low coverage data to the BET relation. In such a case, N is measured as a function of x so as to deduce the unknowns, N_s and c.

4.1.2 Kinetic Langmuir model

Langmuir's original derivation (1918) of the isotherm, eqn (4.7), took a form quite distinct from that in Section 4.1.1. Instead of evaluating the thermodynamic properties of the adatoms, he considered the film's kinetic properties. The argument traces the consequences of detailed balance between the film and its vapor.

The rate at which gas atoms impinge on the surface is given by classical kinetic theory as $P[\beta/(2\pi m)]^{1/2}$. Denoting by s the sticking coefficient on a bare surface and assuming that atoms do not stick at sites that are already occupied, the sticking probability is assumed to be $s(1 - \theta)$. The product of the incident flux and the sticking probability is then set equal to the desorption rate:

$$P[\beta/(2\pi m)]^{1/2} s(1 - \theta) = \theta N_m/(A\tau) , \qquad (4.11)$$

where N_m/A is the monolayer density and τ is the adsorption lifetime. Solving for P, we find a relation identical to eqn (4.7), with the quantity

$$P_L = sN_m/(A\tau[\beta/(2\pi m)]^{-1/2}) . \qquad (4.12)$$

Combining this derivation and the one in Section 4.1.1 leads to an expression for the lifetime by equating the two formulas for P_L. Not surprisingly, the resulting desorption rate has an Arrhenius form.

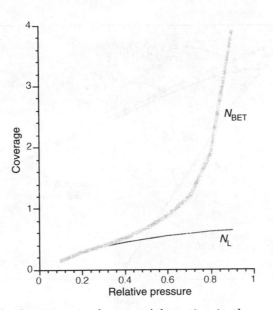

FIG. 4.1. Model adsorption isotherms. Adsorption isotherms computed from the Langmuir model, eqn (4.7), and the BET model, eqn (4.10), with c = 2. N_L and N_{BET} are the coverages, in layers, as a function of the pressure relative to the saturation value. The curves have been defined to agree at low P.

Since this kinetic derivation makes no explicit assumption about the states of the adatoms, the result might appear to be quite general. Such is not the case, as is made obvious by the fact that the Langmuir isotherm does not coincide with the isotherm calculated for a smooth surface, alternative versions of which are derived in the next section. The flaw in this argument is the assumption that sticking decreases in a simple linear way with increasing coverage, all the way to full monolayer coverage. In general, the variation is not linear over the complete range of coverage.

Having raised the subject of desorption, we turn briefly to the more general subject of adsorption kinetics, although it is not a principal concern of this book. The problem is difficult because it involves the coupling between the adsorbate and the dynamical properties of the substrate, including possible electronic excitation. Among the numerous topics of interest are atomic sticking to surfaces, diffusing along surfaces, and desorbing from surfaces. However, surface imperfection often plays an important role in determining these properties; usually it is neither reproducible nor comprehensible in a straightforward way. Even well-characterized and regular surfaces exhibit behavior that is difficult to predict from first principles. Nevertheless there has been significant progress. We refer the reader to a set of relatively recent reviews that treat physisorption kinet-

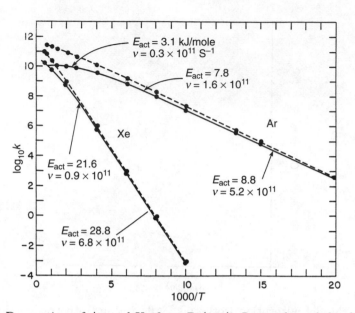

FIG. 4.2. Desorption of Ar and Xe from Pt(111). Logarithm of the desorption rate *vs.* the inverse of the temperature T for Ar and Xe on Pt(111), calculated by Tully (1981). Activation energies and prefactors for exponential fits are given for intermediate and high temperatures. Note the deviation from Arrhenius-type behavior at high T.

ics. Among these are articles by Cole and Toigo (1982), Barker and Auerbach (1985), Poelsema and Comsa (1989), Gomer (1990), and Tully (1994). Most comprehensive of all is a monograph by Kreuzer and Gortel (1986).

There are a few general statements that we believe to be consistent with experiments on regular surfaces. First, the Arrhenius form of desorption rate is found quite generally over an extended range of temperature. At low coverage, the activation energy is convincingly identified with the binding energy of single adatoms, as in the preceding Langmuir model. At higher coverage, there are contributions from the mutual interactions; there is ambiguity in the interpretation of these energies in terms of thermodynamic properties. Second, kinetics is not just thermodynamics! This is evident in discussions of the *compensation effect* (Cardillo and Tully 1984; Estrup *et al.* 1986), a term that reflects the fact that desorption rates are nearly continuous as systems pass through first-order transitions. The activation energies and prefactors on the two sides of the transition 'conspire' in a special way to achieve the near continuity. As discussed by Estrup *et al.* (1986), this may be seen as a statement about the near continuity of the sticking coefficient, since the incident gas flux gas is constant across the transition, and thus the desorption rate exhibits this compensation. There is a cautionary point to be made. An experimental study of the variation of the desorption rate may be done as a function of either coverage N or pressure P. In the former case, a first-order transition necessarily involves a domain of two-phase coexistence. Then the desorption rate must evolve continuously with N; this might be deemed a spurious compensation effect because in such circumstances every physical property varies continuously. Figure 4.2 shows that there is a temperature range where the desorption rate has an Arrhenius form, although there are departures at high T (Tully 1981).

Without delving deeply into the subject, we comment on the problem of the sticking coefficient s. The description is simplest at low coverage, which involves the energy loss mechanisms due to the dynamical interaction of the surface with a single approaching atom. Lennard-Jones and Devonshire (1936) showed that particularly simple behavior is expected in the limit of zero incident energy and a cold surface. Classically, s must equal one in this limit because a very cold surface cannot provide energy but must absorb energy during the virtually infinite time of the atom's approach to the surface. In contrast, the quantum limit of s is zero, because a long wavelength atom approaching a surface experiences an abrupt change of index of refraction associated with the adsorption potential. The abruptness leads to reflection of the atom at a large distance, before it can excite any phonons in the solid. Only recently have experiments demonstrated that this very distinctive quantum prediction is accurate (Yu et al. 1993; Carraro and Cole 1992, 1995; Clougherty and Kohn, 1992; Kohn, 1994). Figure 4.3 shows a comparison of theoretical and experimental results for the sticking of atomic hydrogen at a liquid helium surface.

Finally, and even more briefly, we address the issue of diffusion. Macroscopic measurements of this quantity are particularly susceptible to surface heterogeneity. In Chapter 1, we enumerated a set of microscopic probes of diffusion, including NMR, quasi-elastic atom and neutron scattering, Mössbauer, STM, and current fluctuations in field emission. The last of these has demonstrated for hydrogen the interesting transition from the quantum tunneling regime to the thermal hopping regime (Gomer 1990). Of the two regimes, the hopping phenomenon has a more straightforward theory to evaluate, typically in terms of the transition state model of thermal activation.

4.2 Continuum models

Many monolayers are approximate realizations of systems in mathematical two dimensions. This occurs when the width of the well in the laterally averaged holding potential $V_0(z)$ is narrow enough that the adsorbed atoms move mostly in a plane.* Analogy to three-dimensional phases suggests that there will be gas, liquid, and solid phases. The adsorption isotherms for Xe/graphite shown in Fig. 4.4 have features corresponding to these phases and a triple point and a critical

---

*The degrees of freedom perpendicular to the plane are significant for weakly bound layers and for layers under large lateral stress and then must be included in the models.

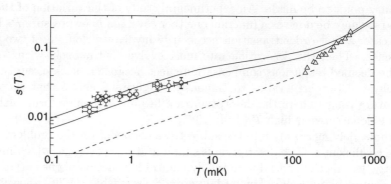

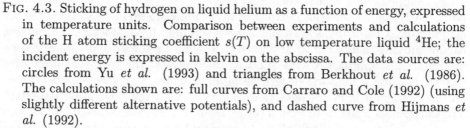

FIG. 4.3. Sticking of hydrogen on liquid helium as a function of energy, expressed in temperature units. Comparison between experiments and calculations of the H atom sticking coefficient $s(T)$ on low temperature liquid ^{4}He; the incident energy is expressed in kelvin on the abscissa. The data sources are: circles from Yu *et al.* (1993) and triangles from Berkhout *et al.* (1986). The calculations shown are: full curves from Carraro and Cole (1992) (using slightly different alternative potentials), and dashed curve from Hijmans *et al.* (1992).

point. We treat the dilute gas phase and the harmonic solid in this section and obtain rather explicit results. As for the analysis of the 3D phases, we use the leading terms in an expansion in terms of a small parameter which is (density × force-volume) for the gas and (root-mean-square displacement/lattice constant) for the solid.

4.2.1   *2D ideal gases*

We develop in succession the ideal gas limit and the virial expansion and use the second virial coefficient to construct an estimate of the gas-liquid critical point. First we treat the case of a noninteracting gas on a smooth planar surface.

An extreme alternative to the localized site models of Section 4.1 is a model in which the atoms wander freely across the surface. This means that the potential energy of adsorption is assumed to be a function of the normal coordinate, z, alone. In some cases, we further specialize to a model in which the potential varies quadratically,

$$V(z) = -D + \mathcal{K}z^2/2 \,, \tag{4.13}$$

where $\mathcal{K}$ is the force constant and the equilibrium position is $z = 0$. Such a description is an accurate representation of many adsorption potentials, sufficient

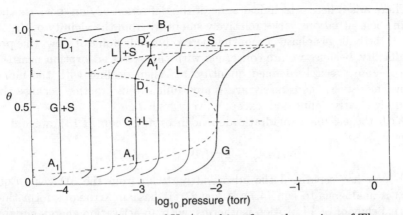

FIG. 4.4. Adsorption isotherms of Xe/graphite, from the review of Thomy *et al.* (1981). The solid lines trace fractional coverage as a function of the pressure of coexisting 3D gas at constant temperature. Vertical portions indicate first-order transitions. The dashed lines demarcating phase coexistence regions are thereby deduced. S, L, and G refer to 2D solid, liquid, and gaseous phases, respectively.

to describe the energetics and dynamics over an extended range of T.[*]

As in Section 4.1.1, we first omit completely the effects of mutual interactions among the adsorbed atoms. We consider in succession the classical and quantum mechanical approaches to this same problem. While the quantum description is always valid in principle, many systems are accurately and more conveniently described by the classical theory, in which case it is usually employed. Falling in the latter category, as seen in Chapter 5, are all adsorbates of interest apart from helium, hydrogen, and perhaps neon.

A classical gas of noninteracting particles has a simple solution for its density function in this external field:

$$n(z) = n \exp[-\beta V(z)] \, , \qquad (4.14)$$

normalized to the vapor density n as $z \to \infty$, where V vanishes. Then the film coverage on a surface of area A is given by the expression

$$N = A \int n(z) \, dz = nA \int \exp[-\beta V(z)] \, dz \, . \qquad (4.15)$$

This includes the particles in what is usually considered to be the 3D gas.

---

[*]For instance, Unguris *et al.* (1979) showed that the Debye–Waller factor of Xe/Ag(111) measured in diffraction exhibits a proportionality between the temperature and the mean-square vibration amplitude (in the harmonic approximation), $k_B T = \mathcal{K} \langle z^2 \rangle$. The harmonic approximation is valid typically when the well depth D is large compared to $k_B T$, in the classical case, and to the zero-point localization energy in the well, in the quantum case.

There arises here a technical issue of the definition of *film*, because the density in some situations varies relatively smoothly from the vicinity of the surface toward the bulk gas phase. From the thermodynamic point of view, this presents no difficulty, because we are concerned with the excess adsorption near the surface, a more precisely defined quantity.[15] When dealing with the monolayer regime, moreover, there rarely arises any difficulty in practice because the film is usually clearly delineated, except at very high T.

With the assumed quadratic dependence of V, eqn (4.13), we obtain the isotherm

$$N/(nA) = [2\pi/(\beta\mathcal{K})]^{1/2}\exp(\beta D) ,\qquad(4.16)$$

which now has separated the adsorbed particles from the 3D gas. Equation (4.16) is analogous to eqn (4.7) at low P. It has an Arrhenius form that is a generic behavior of adsorption isotherms, with an activation energy equal to the well depth. This property explains the utility of the thermodynamic technique as a probe of the well depth.[16]

We turn next to the analogous quantum treatment of the noninteracting gas on a smooth surface. In the very simplest model, an atom's energy has the form

$$E_k = \epsilon_o + \hbar^2 k^2/(2m) ,\qquad(4.17)$$

where $\mathbf{k}$ is the 2D wave vector parallel to the surface. This means that the atomic motion perpendicular to the surface corresponds to a unique energy level, ϵ_o, while motion along the surface is free. (At eqn (4.24) we treat a more general model in which other levels are populated.) Because the particles do not interact, the system's statistical mechanical properties are additively those of the groups of particles occupying independent states in mutual equilibrium; each such state k is characterized by an occupation number distributed in a statistical way according to the thermodynamic conditions. The grand partition function for the particles in state k is

$$\Xi_k = \sum_{j=0}^{j_U} \exp[j\beta(\mu - E_k)] ,\qquad(4.18)$$

where the upper limit j_U for the occupation number j in the sum is one for fermions and infinity for bosons. The sums give

$$\Xi_k = [1 \pm \exp[\beta(\mu - E_k)]]^{\pm 1} ,\qquad(4.19)$$

where the $\pm$ sign is $+$ for fermions and $-$ for bosons. Then the thermal average of the state's occupation number is

---

[15] Even then, there is a technical difficulty because one must define a nominal boundary for the region of adsorption (Griffiths 1980).

[16] On a reconstructed semiconductor surface, there may be sites of markedly different energy for adatoms. Inert gas atoms are used advantageously to probe, nonperturbatively, the binding energies of the various sites (Conrad and Webb 1983; Packard and Webb 1988).

$$n_k = \beta^{-1}(\partial \ln \Xi_k/\partial \mu)_\beta = 1/[\exp[\beta(E_k - \mu)] \pm 1] . \qquad (4.20)$$

The total number of adsorbed atoms is obtained by summation over all states, including the spin degeneracy factor g:

$$
\begin{aligned}
N &= g\sum n_k = [gA/(2\pi)^2] \int d^2k\, n_k \\
&= [gA/(2\pi)] \int_0^\infty dk\, k/[\exp(\beta(E_k - \mu)) \pm 1] .
\end{aligned}
\qquad (4.21)
$$

Here we have replaced the sum over the quasi-continuous variable k with an integral in the usual way. Then, for eqn (4.17), the integral can be done analytically, and we get

$$N\lambda^2/(gA) = \pm \ln[1 \pm \exp[\beta(\mu - \epsilon_o)]] . \qquad (4.22)$$

Equation (4.22) is the isotherm for the quantum *one-state model*. The connection between the quantum and classical results can be obtained by recalling that the classical regime is the limiting case of small occupation of every quantum state. The limit $\mu \to -\infty$ in eqn (4.22) gives

$$\beta(\mu - \epsilon_o) \simeq \ln[N\lambda^2/(gA)] . \qquad (4.23)$$

Equation (4.23) is the 2D analogue of eqn (4.1) for the ideal 3D vapor. In order to completely recover the quasi-2D classical isotherm eqn (4.16), we must improve upon our starting assumption about the omission of excited states of motion perpendicular to the surface. This involves a straightforward extension of the preceding mathematics, and the resulting generalization of eqn (4.23) is:

$$N\lambda^2/(gA) = \sum_j \exp[\beta(\mu - \epsilon_j)] . \qquad (4.24)$$

Here the sum is over all states of perpendicular motion, with a spectrum $\{ \epsilon_j \}$, and the relative occupation of levels i and j scales in this limit as the Gibbs–Boltzmann factor, $\exp[-\beta(\epsilon_i - \epsilon_j)]$. Finally, we specialize to the harmonic approximation, for which the level spectrum is

$$\epsilon_j = -D + (j + \frac{1}{2})\hbar\omega , \qquad (4.25)$$

where $\omega = (\mathcal{K}/m)^{1/2}$ is the classical angular frequency of vibration perpendicular to the surface. The result is an isotherm

$$N/(n\lambda A) = [2\sinh(\beta\hbar\omega/2)]^{-1} \exp(\beta D) . \qquad (4.26)$$

Equation (4.26) agrees with the classical result, eqn (4.16), in the correspondence principle limit, $\beta\hbar\omega \ll 1$, where the level spacing is small compared to the temperature.

We summarize the results of this section as follows: The noninteracting gas limit provides explicit expressions for the adsorption isotherm that are both convenient and accurate at low coverage. Analytic solutions require simplified forms for the dependence of the adsorption potential on z. Next, in Section 4.2.2, we show that the general virial expansion can be combined with band structure calculations to yield the first two terms in the expansion of a series expressing coverage in terms of gas pressure. The results of Section 4.2.1 follow as special cases.

The preceding discussion of the noninteracting quantum gas is appealing because it is so straightforward to calculate properties exactly. However, in practice, adatom–adatom interaction effects (*diffraction*) usually predominate over effects of quantum statistics except at the very lowest temperatures (Larsen *et al.* 1965; Siddon and Schick 1974a). This conclusion emerges in studies using the virial expansion, *inter alia*.

4.2.2   *2D gas: virial expansion*

In the case of 3D, the virial expansion provides a basis for deriving information about the interatomic potential from thermodynamic data. Even more information can be gleaned from such an expansion in the case of adsorption. The gas–substrate potential is reflected in the coefficient of leading term for the coverage as power series in the 3D pressure (Sams *et al.* 1960; Steele 1974; Patrykiejew and Sokolowski 1989). This fact is manifested in the preceding section's results, e.g., eqn (4.16); we now show it is a general property of the adsorption by developing what is termed *gas–solid virial theory*.* The amount adsorbed is identified as the difference between the total number of molecules in the system and the corresponding number that would be present in a comparison case where the system has the same volume V, temperature T, and fugacity z, but the gas–solid potential energy is simply the hard wall of the container.

As in 3D, the virial expansion for the 2D gas may be conveniently derived from the grand canonical partition function:

$$\Xi = \sum_N Q_N \exp(N\beta\mu) = \sum_N Q_N z^N \ , \qquad (4.27)$$

where Q_N is the canonical partition function for N atoms. We shall use the notation q_N for the comparison problem with no gas–solid attraction. The fugacity $z = e^{\beta\mu}$ is that of the coexisting dilute 3D gas. For a monatomic gas, it is related to the 3D pressure P by

$$z = (\beta P \lambda^3 / g) + (B_{3D} \lambda^3 / g)(\beta P)^2 + \dots \qquad (4.28)$$

---

*This approach has been criticized by Sokolowski and Stecki (1981), who argue that the preferred approach should be based on a justification of the choice of the Gibbs dividing surface. The differences in the results of the two approaches are minuscule for the case of a dilute 3D gas and a thin adsorbed layer.

where g is the spin degeneracy factor and $B_{3D}(T)$ is the 3D second virial coefficient

$$\begin{aligned} B_{3D}(T) &= -V(2q_2 - q_1^2)/(2q_1^2) \; ; \\ q_1 &= gV/\lambda^3 \; . \end{aligned} \tag{4.29}$$

Although we are discussing conditions where the 3D gas is so dilute that the second term on the right-hand side of eqn (4.28) is negligible, we retain it for the formal role it plays in the theory of the adsorption virial coefficients.

Since eqn (4.27) is effectively an expansion in the pressure, it provides a logical starting point for the virial expansion with the gas–solid interaction present. For this to be useful, successive terms in the series should become small sufficiently rapidly. We shall find the condition for this to be the case. Using the expansion of the logarithm, the average total number of atoms is

$$\begin{aligned} N = \partial \ln \Xi / \partial(\beta\mu)|_\beta &= Q_1 z + (2Q_2 - Q_1^2)z^2 \\ &+ (3Q_3 - 3Q_1 Q_2 + Q_1^3)z^3 + ... \end{aligned} \tag{4.30}$$

We determine the surface excess coverage, N_s, by subtracting from N the particle number N_0 that would hypothetically be present in the volume in the absence of an adsorption potential. The corresponding series is

$$N_0 = q_1 z + (2q_2 - q_1^2)z^2 + ... \tag{4.31}$$

Hence, the leading terms in the series for the coverage are

$$N_s = N - N_0 = [Q_1 - q_1]z + w_2 z^2 + ... \; , \tag{4.32}$$

where

$$w_2 = (2Q_2 - Q_1^2) - (2q_2 - q_1^2) \; . \tag{4.33}$$

This (and other thermodynamic properties) may be written in terms of virial coefficients and the pressure:

$$N_s = B_{AS}(\beta P) + C_{AAS}(\beta P)^2 + ... \tag{4.34}$$

The low P region exhibits a Henry's law proportionality between N_s and P, as in the models of Sections 4.1.1 and 4.2.1. The first two coefficients are

$$\begin{aligned} B_{AS} &= [Q_1 - q_1](\lambda^3/g) \; ; \\ C_{AAS} &= w_2(\lambda^3/g)^2 + B_{AS}B_{3D} \; . \end{aligned} \tag{4.35}$$

4.2.2.1 *Classical virial theory* In the classical case, if we take $g = 1$, the first coefficients become

$$B_{3D}(T) = -\frac{1}{2} \int d^3r [\exp(-\beta u(r)) - 1]$$

$$B_{AS} = \int d^3r [\exp(-\beta V(r)) - 1] \, ,$$

$$C_{AAS} - B_{AS}B_{3D} = \int d^3r_1 \int d^3r_2 f_{12}[\exp(-\beta[V(r_1) + V(r_2)]) - 1] \, ,$$

$$f_{12} = \exp[-\beta U(r_1, r_2)] - 1 \, . \tag{4.36}$$

The two-body potential for the 3D gas has been assumed to depend on the separation, $u(r)$. However, in the expression for f of the adsorbed particles, we allow the adatom–adatom potential U to depend quite generally on both particles' coordinates, as may arise from the effects of the surface on their interaction discussed in Section 2.3. The appearance of the Mayer f function in the expression for C_{AAS} reflects a similarity to the second virial coefficient B_{3D} for the 3D gas. Here, however, there are Boltzmann weighting factors due to the adsorption potential.

The 2D approximation One version of the *2D approximation* applies to the case when the adsorption potential is extremely deep and narrowly confined near the substrate. If we may employ the smooth surface approximation, then C_{AAS} takes on a particularly convenient form:

$$C_{AAS} = ([A2\pi \exp(2\beta D)]/[\beta \mathcal{K}]) \int d^2r \exp([-\beta U(r)] - 1) \, , \tag{4.37}$$

where term $B_{AS}B_2$ has been dropped because it has only a factor $\exp(\beta D)$. We may then form the ratio of the first two terms in eqn (4.34):

$$(\beta P)d^3 \exp(\beta D) \, , \tag{4.38}$$

where d is a microscopic length related to the integral in C_{AAS}. It is evident from eqn (4.38) that the surface virial expansion fails when T becomes much smaller than the adsorption potential well depth D divided by Boltzmann's constant.* This qualitative conclusion is analogous to the situation in 3D, for which the virial expansion fails due to condensation when T becomes comparable to the well depth of the pair potential.

The most famous mathematical description of condensation is the van der Waals theory. We now obtain its 2D analogue from the virial expansion. To do so, we make a set of assumptions that have qualitative validity, at best.

---

*The specific criterion evidently involves the vapor density prefactor. Not surprisingly, this condition is correlated with the requirement that the first virial term in the expansion of N_s be large.

The conclusion is a useful guide in many common circumstances, as for the 3D derivation.

We begin by citing (Steele 1974) an expression for the spreading pressure ϕ:

$$\beta\phi A = \ln\Xi - \beta PV = \ln(\Xi/\Xi_0) \ . \tag{4.39}$$

Rearrangement in terms of the leading virial coefficients leads to an expansion that bears a formal similarity to the 3D virial expansion:

$$\beta\phi A/N_s = 1 + B_{2D}(N_s/A) + \cdots \ , \tag{4.40}$$

where

$$B_{2D}/A = -w_2(\lambda^3/g)^2/(2B_{AS}^2) \ . \tag{4.41}$$

In the smooth surface, strong attraction, approximation that was used to obtain eqn (4.37), the two-body term is

$$B_{2D} = -\frac{1}{2}\int d^2r\,(\exp[-\beta U(r)] - 1) \ . \tag{4.42}$$

Equation (4.42) differs from the expression for the 3D second virial coefficient in eqn (4.36) only in the lower dimensionality of integration.

Van der Waals approximation for condensation The van der Waals theory is obtained by truncating the virial expansion and evaluating B_{2D} approximately.[*] If the potential diverges within a hard-core domain $r < r_h$ and the attraction is weak for $r > r_h$, the second virial coefficient becomes

$$B_{2D} = b - \beta a \ ; \quad b = \pi r_h^2/2 \ ; \quad a = -\frac{1}{2}\int_{r>r_h} d^2r\,U(r) \ . \tag{4.43}$$

The equation of state then is written in a form analogous to the 3D van der Waals equation,

$$\phi + a(N_s/A)^2 = k_B T[N_s/A]/[1 - bN_s/A] \ , \tag{4.44}$$

where it is assumed that $bN_s/A \ll 1$. The van der Waals theory of condensation is based on using eqn (4.44) for the pressure ϕ. Using the fact that the first and second derivatives of pressure with respect to coverage at fixed T vanish at the critical point leads to the following results for the critical parameters:[†]

$$k_B T_c = 8a/(27b) \ ; \quad (N_s/A)_c = 1/(3b) \ ; \quad \phi_c = a/(27b^2) \ . \tag{4.45}$$

Since formally identical equations correspond to the critical parameters in 3D, we may express the ratio of the respective critical temperatures as

$$T_c(2D)/T_c(3D) = (a/b)_{2D}/(a/b)_{3D} \ . \tag{4.46}$$

---

[*]See, for example, K. Huang, *Statistical Mechanics* (Wiley, New York, 1987).

[†]Note that the low density assumption of the derivation is violated at the critical point, as usual.

Table 4.1   *2D critical and triple point temperatures, in kelvin, for monolayer adsorption on the basal plane surface of graphite.*[†]

System	$T_c(2D)$	$T_t(2D)$	$T_c(2D)/T_c(3D)$	$T_t(2D)/T_c(3D)$
^{4}He	3	–	0.58	–
H_2	18	–	0.55	–
D_2	21	–	0.55	–
Ne	15.8	13.6	0.36	0.31
Ar	55	49	0.36	0.33
Kr	85	–	0.41	–
Xe	117	99	0.40	0.34
CH_4	69	56	0.36	0.29
N_2	48	–	0.38	–
CO	50	–	0.37	–
O_2	64	24	0.41	0.16

The ratio on the right-hand side of eqn (4.46) depends on the form of the pair potential. For example, in the case of the Lennard-Jones (12,6) potential

$$U(r) = 4\epsilon[(\sigma/r)^{12} - (\sigma/r)^6] \,, \qquad (4.47)$$

the hard core volume may be taken as the region $r < \sigma$. Then $b = 2\pi\sigma^3/3$ in 3D. Evaluating the a-integrals, we find $a_{2D}/a_{3D} = 27/(80\sigma)$, so $T_c(2D)/T_c(3D) = 0.45$. While the numerical value 0.45 depends on the interaction model used, the result that $T_c(2D)$ is roughly half of $T_c(3D)$ is typical of this theory. The explanation of the reduction factor is necessarily geometrical, since it comes from the integrations of identical functions in different dimensions; it can be interpreted as arising from the additional r-factor in 3D phase space, 'favoring' larger distances and hence the attraction relative to the hard core repulsion. This can be rephrased in terms of the higher coordination number favoring condensation in 3D. As seen in the Table 4.1, experimental values of the critical temperature for adsorption on graphite yield even lower relative temperatures in 2D; a typical ratio is about 0.4. The explanation of the further reduction is that the van der Waals theory is a mean-field theory. The neglect of the relatively large fluctuations in 2D causes the theory to overestimate T_c in 2D by more than it

[†]The two right-hand columns contain ratios of these temperatures to the 3D critical temperatures assembled in Table A.1; for Lennard-Jonesium the values are $T_c(2D)/T_c(3D) = 0.39$ and $T_t(2D)/T_c(3D) = 0.30$. Systems without an entry for the triple point temperature have commensurate lattices and the critical temperature that is listed is for the maximum of the gas–solid coexistence curve (^{4}He, H_2, and D_2) or for the melting temperature of a half-monolayer of solid (Kr, N_2, and CO). This compilation is based on a review by Steele (1996). The $T_c(2D)$ for ^{4}He is from a review by Greywall (1994) while the entries for H_2 and D_2 are from a review by Wiechert (1991a).

overestimates the value in 3D. This will be seen in the next chapter, where so-lutions of the lattice-gas model and computer simulations of the Lennard-Jones fluid are discussed. Table 4.1 also contains entries for the 2D triple point tem-perature and entries that indicate the additional phenomena which may arise from the corrugation of the substrate.

4.2.2.2 *Quantum virial theory* Light atoms and H_2 molecules require a quan-tum mechanical treatment. The evaluations of the partition functions in the quantum and classical cases involve disparate procedures, but many qualitative conclusions about the systems' behavior are similar. In the quantum case, the $N = 1$ canonical partition function is

$$Q_1 = \int dE \, \mathcal{N}(E) \exp(-\beta E) \,, \tag{4.48}$$

where $\mathcal{N}(E)$ is the density of states for one atom on a surface. To understand the physical content of this function, consider for simplicity the case of a smooth surface, so that the adatom energy spectrum has a form similar to eqn (4.17):

$$E_j(k) = \epsilon_j + \hbar^2 k^2/(2m) \,; \tag{4.49}$$

ϵ_j is an eigenvalue of the Schrödinger equation for motion perpendicular to the surface. For a given value of j, the energy range dE corresponding to a range dk of wavenumber is $\hbar^2 k dk/m$. For 2D area A, the number of states per unit area in $\mathbf{k}$-space is $A/(2\pi)^2$. The area of $\mathbf{k}$-space corresponding to a ring of width dk is $2\pi k dk$, so that

$$\mathcal{N}(E) = [mAg/(2\pi\hbar^2)] \sum_j H(E - \epsilon_j) \,, \tag{4.50}$$

where the Heaviside function $H(x)$ is equal to unity for positive argument and zero for negative argument, i.e., it is the unit step function. The function $\mathcal{N}(E)$ increases in a sequence of steps, such as is shown in Fig. 4.5 Also shown there is the density of states for a more realistic form of the potential, relevant to the case of helium on graphite. The principal differences between the results occur at low energy and at intermediate energy values. The former is associated with an enhancement of the mass, the so-called band mass, due to the perturbing potential. In the case shown in the figure, this is about a 5% increase. The higher energy anomalous behavior is associated with the presence of a gap in the spectrum of states and is quite analogous to behavior found in electronic bands in solids.

Upon assuming the smooth-surface form of the density of states, eqn (4.50), we obtain

$$Q_1 = (gA/\lambda^2) \sum_j \exp(-\beta\epsilon_j) \,. \tag{4.51}$$

If only this first term in the virial expansion is retained, the isotherm is identical to eqn (4.24) and hence leads to eqn (4.16) in the classical regime. The reason is

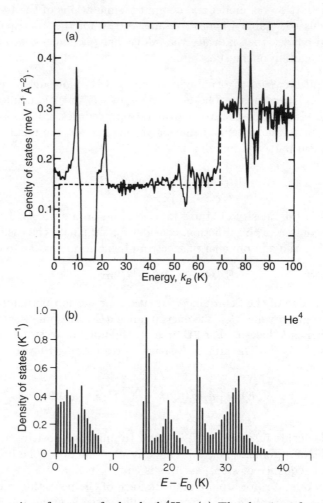

FIG. 4.5. Density of states of adsorbed ^{4}He. (a) The density of states is shown for ^{4}He on graphite, from Carlos and Cole (1980b). Comparison is made between the step-like dashed curve computed for the case of a hypothetical smooth surface and the more ragged result computed with the energy bands relevant to the corrugated potential actually present. (b) Density of states for ^{4}He adsorbed on Xe-plated graphite, from Novaco and Milford (1972). The difference between (a) and (b) is a result of the much more corrugated potential in (b).

that quantum statistics play no role in the one-particle problem. Band-structure corrections modify even this result, and Fig. 4.6 shows the consequences for the low-coverage specific heat C_1 in the case of the helium on graphite. The effects of both the gap in the spectrum and the excitation of motion perpendicular to the surface are seen rather clearly.

A substantial effort has been expended to treat the second term in the quantum virial expansion (Siddon and Schick 1974a,b,c; Rehr and Tejwani 1979; Guo and Bruch 1982). To second order, the heat capacity* satisfies an equation

$$(C - C_1)/Nk_B = -(N/A)\beta^2(d^2B_{2D}/d\beta^2) , \qquad (4.52)$$

where

$$C_1/(Nk_B) = \beta^2(d^2 \ln Q_1/d\beta^2) \qquad (4.53)$$

is the heat capacity of a noninteracting gas on the given surface, including the effects of surface corrugation, i.e., band structure. The quantum form of B_{2D} is based on eqns (4.33), (4.35), and (4.41), Since the right-hand side of eqn (4.53) is linear in coverage, the adequacy of this truncation of the series can be tested directly by comparison with experiments. Figure 4.7 shows one example that is noteworthy in several respects.

The calculations displayed in Fig. 4.7 were performed by evaluating the quantum-mechanical second virial coefficient in the 2D approximation (Siddon and Schick 1974a,b,c). To do so, the usual 3D scattering theory of the second virial coefficient was adapted to 2D. Figure 4.7 shows good overall agreement between the data and the theory. Some of the questions raised subsequently should be noted. One issue is the role of substrate heterogeneity, which is significant in the low density regime being probed by the virial theory. Another is the use in the theory of the free 3D interaction between atoms. As discussed in Chapter 2, there is theoretical reason to believe that the substrate weakens the attraction to some extent. There is, in fact, some indication (Vidali and Cole 1980) that including this effect would improve the agreement with experiment. Yet another issue is the role of the substrate corrugation (Guo and Bruch 1982). There is also the question of the adequacy of the two-term virial series used in analyzing the data; this is checked to some extent by the scaling used to plot the data.

Figure 4.7 shows that the magnitude of the second virial coefficient becomes large at low temperature. This suggests that higher order coefficients may also become important in this regime. Indeed, we believe that 2D ^{4}He, but probably not ^{3}He, condenses at sufficiently low temperatures (Novaco and Campbell 1975;

---

*Again, other properties are expressed with similar series. An interesting application is the magnetic susceptibility of ^{3}He gas. An ideal degenerate Fermi–Dirac gas has the Pauli paramagnetic susceptibility, which is temperature-independent in contrast to the Curie law result. The Pauli susceptibility is attained for ^{3}He gas only at temperatures much below 1 K. However, the data near 1 K (Hegde and Daunt 1978; Owers-Bradley et al. 1978) can be understood quantitatively with the second virial coefficient. There are off-setting effects of the quantum statistics and the mutual interactions in the relevant *exchange* second virial coefficient (Siddon and Schick 1974a).

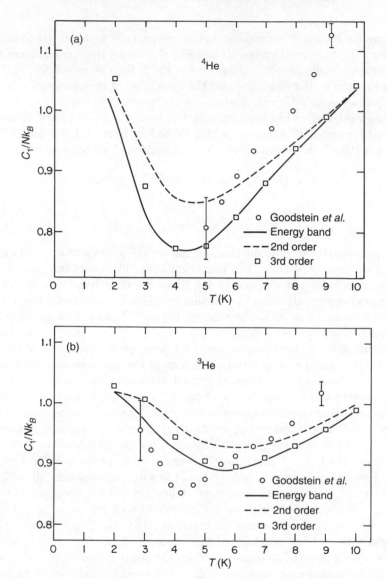

FIG. 4.6. Specific heat of low-coverage He/graphite, illustrating band structure effects. Note the suppressed zero and the expanded scale of the ordinate. The circles are obtained by extrapolating the experimental data to low coverage so that adsorbate-adsorbate interactions play no role (D.L.Goodstein, private communication; Silva-Moreira et al. 1980). The solid curve is the theory (Carlos and Cole 1980b) based on their calculated density of states (shown in the previous figure, Fig. 4.5). Note the deviation from the mathematical 2D result of unity. The high-T rise of the data above one is due to excitation of motion perpendicular to the surface. The dashed curve and squares are second- and third-order perturbation-theory calculations, respectively, of the band structure effects, by Guo and Bruch (1982), from which this figure is taken.

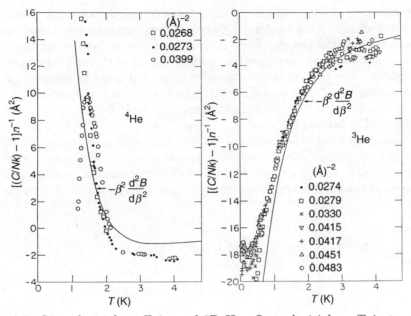

FIG. 4.7. Second virial coefficient of 2D He. Second virial coefficients of the helium isotopes on graphite, as calculated by Siddon and Schick (1974a) for a 2D quantum Lennard-Jones model and measured by Bretz *et al.* (1973). Figure from Dash *et al.* (1994).

Whitlock *et al.* 1988; Gottlieb and Bruch 1993). Thus, the low-T virial expansion will eventually fail.

In closing this discussion of virial expansion calculations, we emphasize that the evaluation of the quantum second virial coefficient is particularly difficult in the case of a periodic substrate field, as occurs in general.* The reason is simply that the two-particle problem no longer reduces to a one-particle problem involving the relative coordinate and an effective mass. Rehr and Tejwani (1979) have addressed this problem in terms of a representation involving localized states, but their formulation has not been implemented consistently in any case, to our knowledge.

4.2.3 Harmonic solid

We showed in Section 4.2.2 that a quantitative theory can be developed for the low-density gas by evaluating the leading virial coefficients, which become one-

---

*The quadratic term in a power series expansion in the corrugation amplitude, V_g, has been evaluated numerically using path integrals, but it is a far longer computation than for the zeroth-order term (Guo and Bruch 1982).

and two-body problems. Correspondingly, the harmonic solid has a concrete theory, based on a transcription to elementary excitations, the normal modes. We now develop the theory of the normal modes of a monatomic triangular lattice, and in doing so encounter the striking effects of thermal fluctuations in a 2D lattice. The potentials are expanded to second order in small-amplitude displacements from the sites, and the many-body problem is reduced to tractable form by transforming to normal coordinates. Several extensions of the analysis are described in Section 5.2.1.

Consider a monatomic triangular lattice of N_s sites and atoms of mass M interacting by central pair potentials ϕ and at a smooth surface, so that the holding potential is only a function $V_0(z)$ of the perpendicular coordinate z. An expansion of the total potential energy to second order in the displacements $\delta\mathbf{r}_j$ gives

$$\begin{aligned} \Phi &= \sum_j [V_0(z_o) + \frac{1}{2}V_0''(\delta z_j)^2] \\ &+ \frac{1}{2}\sum_{i\neq j}[\phi(r_{ij}) + \frac{1}{2}\nabla\nabla\phi : (\delta\mathbf{r}_i - \delta\mathbf{r}_j)(\delta\mathbf{r}_i - \delta\mathbf{r}_j)] \,, \end{aligned} \qquad (4.54)$$

where primes denote derivatives. The time-dependent displacements $\delta\mathbf{r}_j$ are expressed as a sum of complex running wave amplitudes $\mathbf{u}(\mathbf{k})$ over wavevectors in the first Brillouin zone

$$\delta\mathbf{r}_j = (1/\sqrt{N_s})\sum_{\mathbf{k}} \mathbf{u}(\mathbf{k})\exp[\imath(\mathbf{k}\cdot\mathbf{r}_j - t\omega(\mathbf{k}))] \,, \qquad (4.55)$$

with

$$\mathbf{u}(-\mathbf{k}) = \mathbf{u}(\mathbf{k})^* \qquad (4.56)$$

and

$$\omega(-\mathbf{k}) = -\omega(\mathbf{k}) \,. \qquad (4.57)$$

More complicated cases are handled by a generalization of the running wave amplitude $\mathbf{u}$ to have more than the three components that suffice for the monatomic triangular lattice.[*]

---

[*]When the lattice has a basis, there is a separate amplitude $\mathbf{u}_\alpha(\mathbf{k})$ for each atom α in the basis. The index j of eqn (4.55) becomes a composite with indices for the Bravais cell and for the component of the basis. For molecules with orientational degrees of freedom, the displacement vector and the amplitude $\mathbf{u}(\mathbf{k})$ have more than 3 components. These considerations lead to mostly straightforward extensions of the index notation used here for the simple monatomic lattice, but subtleties may arise in formulating the expansion with generalized coordinates (van Kampen and Lodder 1984; Bruch 1987).

The amplitudes $\mathbf{u}(\mathbf{k})$ are the eigenvectors of the dynamical matrix

$$\mathbf{D}(\mathbf{k}) = V_0'' \hat{z}\hat{z} + \sum_{\mathbf{r}_j \neq 0} \nabla\nabla\phi(r_j)[1 - \cos(\mathbf{k} \cdot \mathbf{r}_j)] \;, \qquad (4.58)$$

and satisfy

$$M\omega^2(\mathbf{k})\,\mathbf{u}(\mathbf{k}) = \mathbf{D}(\mathbf{k}) \cdot \mathbf{u}(\mathbf{k}) \;. \qquad (4.59)$$

The eigenvalues of this matrix $\mathbf{D}$ determine the frequencies of the normal modes of vibration. If the lattice is dynamically stable, the eigenvalues are positive. A dynamical matrix $\mathbf{D}$ of dimension d has d (possibly degenerate) frequencies for each $\mathbf{k}$. We take $\omega_\nu(\mathbf{k})$, $\nu = 1, ..., d$, to be positive numbers. A periodicity that follows from the definitions of the dynamical matrix eqn (4.58) and the adlayer reciprocal lattice vectors $\boldsymbol{\tau}$, eqn (3.8), is:

$$\omega_\nu(\mathbf{k}) = \omega_\nu(\mathbf{k} + \boldsymbol{\tau}) \;. \qquad (4.60)$$

In the present case, there is no coupling in the dynamical matrix between vibrations perpendicular and parallel to the substrate plane. The eigenvalue problem for the 3×3 matrix $\mathbf{D}$ then separates, and the frequency of perpendicular motion is given by

$$M\omega_z^2(\mathbf{k}) = V_0'' + \sum_{\mathbf{r}_j \neq 0} \partial^2\phi/\partial z^2 [1 - \cos(\mathbf{k} \cdot \mathbf{r}_j)] \;. \qquad (4.61)$$

Note that the gap at $\mathbf{k} = 0$ is just the frequency of an individual atom vibrating perpendicular to the surface. Expressions for the frequencies of vibrations polarized parallel to the surface are given for the triangular and square lattices in Appendix G.

The harmonic approximation to the Helmholtz free energy is:

$$F([a, z], T) = \Phi_0 + k_B T \sum_{\mathbf{k},\nu} \ln[2\sinh(\beta\hbar\omega_\nu(k)/2)] \;, \qquad (4.62)$$

where Φ_0 is the static lattice potential energy and $[a, z]$ denotes the explicit dependence on lattice constants a_i and interlayer spacings z_i of a more general lattice. The sum is over all polarizations ($\nu = 1, ..., d$) and over wavevectors in the first Brillouin zone. The equilibrium monolayer (or bilayer, ...) free energy has structural parameters that minimize F, subject to the constraint that the lattice occupies a specified area A:

$$\partial F/\partial a_i = \partial F/\partial z_j = 0 \;. \qquad (4.63)$$

This corresponds to the quasiharmonic approximation for the free energy (Hooton 1955), which is discussed in more detail in Section 5.2.1: the free energy is evaluated for a harmonic Hamiltonian with force constants given by second derivatives of the original interactions (Bruch and Wei 1980; Roosevelt and Bruch 1990).

The harmonic approximation that leads to the free energy eqn (4.62) can be expressed as a second quantized oscillator Hamiltonian

$$H_0 = \Phi_0 + \sum_{\mathbf{k},\nu} \hbar\omega_\nu(\mathbf{k})[a^+(\mathbf{k},\nu)a(\mathbf{k},\nu) + \frac{1}{2}] , \qquad (4.64)$$

with boson creation and annihilation operators. In terms of these operators, the time-dependent displacement is

$$
\begin{aligned}
\mathbf{u}_j = {}& \sum_{\mathbf{k},\nu} [\hbar/2N_s M\omega_\nu(\mathbf{k})]^{1/2} [a(\mathbf{k},\nu)\exp[i(\mathbf{k}\cdot\mathbf{R}_j - \omega_\nu(\mathbf{k})t)] \\
& + a^+(\mathbf{k},\nu)\exp[-i(\mathbf{k}\cdot\mathbf{R}_j - \omega_\nu(\mathbf{k})t)]]\mathbf{e}_\nu(\mathbf{k}) , \qquad (4.65)
\end{aligned}
$$

where the $\mathbf{e}_\nu(\mathbf{k})$ is a unit vector giving the polarization of the solution $\mathbf{u}(\mathbf{k})$ of eqn (4.59).

The thermal average $\langle ... \rangle$ of the mean-square displacement from a lattice site is

$$
\begin{aligned}
\langle \mathbf{u}_j^2 \rangle &= (\hbar/2MN_s)\sum_{\mathbf{k},\nu} [(2\langle a^+ a\rangle + 1)/\omega] \\
&= (\hbar/2MN_s)\sum_{\mathbf{k},\nu} \coth(\beta\hbar\omega/2)/\omega , \qquad (4.66)
\end{aligned}
$$

while the mean-square relative displacement is

$$
\begin{aligned}
W_{jj'} &= \langle (\mathbf{u}_j - \mathbf{u}_{j'})^2 \rangle \qquad\qquad\qquad\qquad\qquad (4.67) \\
&= (\hbar/MN_s)\sum_{\mathbf{k},\nu} [1 - \cos(\mathbf{k}\cdot(\mathbf{R}_j - \mathbf{R}_{j'}))]\coth(\beta\hbar\omega/2)/\omega .
\end{aligned}
$$

Equation (4.66) displays the much-discussed effect (Peierls 1934; Imry and Gunther 1971; Imry 1978) of the large thermal fluctuations at long wavelengths (small $k = |\mathbf{k}|$) in mathematical two dimensions. For a smooth substrate, the frequencies of the acoustic modes at small k are proportional to k. Since the phase-space factor in the wavevector integration has the form kdk and the summand varies as $1/\omega^2 \propto 1/k^2$ at small k, there is a logarithmic divergence in $\langle u_j^2 \rangle$ arising from modes near the zone center. The size of the effect is small even with the appropriate cutoff for a lattice of 10^{15} atoms (Abraham 1981a). More importantly, it is a feature of the center of mass of the 2D 'crystal', and there is no such divergence in the mean-square relative displacement, eqn (4.68). For large relative separations $|\mathbf{R}_j - \mathbf{R}_{j'}|$, the expectation value is dominated by the long wavelength excitations and an asymptotic evaluation gives

$$W_{jj'} \simeq (k_B T/M)[(1/c_l^2) + (1/c_t^2)](1/\pi n)\ln(|\mathbf{R}_j - \mathbf{R}_{j'}|/L) , \qquad (4.68)$$

where n is the number density and c_l and c_t are the speeds of longitudinal and transverse sound in the triangular lattice. By convention, several factors are combined into the Jancovici temperature T_q defined (Jancovici 1967) for a triangular lattice and a wave vector q by

$$1/(k_B T_q) = (q^2/\tau_0^2)(\pi/M\sqrt{3}) \times [(1/c_l^2) + (1/c_t^2)] , \qquad (4.69)$$

where the primitive reciprocal lattice vector is given in terms of the nearest-neighbor spacing by $\tau_0 = 4\pi/L\sqrt{3}$ and $n = 2/L^2\sqrt{3}$.

Other correlation functions for the harmonic triangular lattice for large separations are evaluated in similar fashion. The translational long-range order of the lattice is measured by the function

$$C_g(\mathbf{R}) = \langle \exp\{ig \cdot [(\mathbf{R} + \mathbf{u}(\mathbf{R}) - \mathbf{u}(0)]\} \rangle , \qquad (4.70)$$

where the average is over the thermally driven displacements $\mathbf{u}$. For a reciprocal lattice vector $\mathbf{g}$ and lattice vector $\mathbf{R}$, C_g becomes

$$\begin{aligned} C_g(\mathbf{R}) &= \exp[-(g^2/4)(\mathbf{u}(\mathbf{R}) - \mathbf{u}(0))^2] \\ &\sim (L/R)^{2T/T_g}, \; R \to \infty . \end{aligned} \qquad (4.71)$$

Thus, the translational order parameter decays to zero at large separations, but it has an algebraic, power-law, dependence rather than the exponential decay usual in cases with disorder. The thermal average of the structure factor eqn (3.20) becomes

$$S(\mathbf{k}) = N_s \sum_j \exp[i\mathbf{k} \cdot (\mathbf{R}_j - \mathbf{R}_o)] \exp[-(k^2/2)W_{jo}] . \qquad (4.72)$$

When $\mathbf{k}$ is a reciprocal lattice vector $\mathbf{g}$, the asymptotic evaluation of W_{jo}, eqn (4.68), leads to

$$S(g) \sim N_s^{2(1-T/T_g)} . \qquad (4.73)$$

The harmonic 2D solid has a *diffraction spot pattern* that is less sharp than for 3D solids (for which $S(\mathbf{g})$ is proportional to N_s^2), because the peaks have a power-law singularity for wavevectors $\mathbf{q}$ close to reciprocal lattice vectors $\mathbf{g}$:

$$S(\mathbf{q}) \sim |\mathbf{q} - \mathbf{g}|^{-2(1-T/T_g)} ; \qquad (4.74)$$

the finite size of an island of solid may change the effective line shape greatly (Weling and Griffin 1981, 1982). Equation (4.74) has important consequences for X-ray diffraction experiments on the solid near its melting transition. As discussed in Section 4.3, the Kosterlitz–Thouless theory predicts that the solid melts at a temperature T_m given by eqn (4.80). The Jancovici temperature T_{τ_o} for $q = \tau_o$, the magnitude of the leading reciprocal lattice vector, is $6.75T_m$, assuming the relations of a Cauchy solid (Stewart 1974, 1977). Therefore, the singularities for the *Bragg peaks* at larger reciprocal lattice vectors of magnitude $\sqrt{3}\tau_o$ and $2\tau_o$, with Jancovici temperatures $\frac{1}{3}$ and $\frac{1}{4}$ of T_{τ_o}, respectively, are so weak that they are scarcely observable in experiments performed near the melting temperature. This may be viewed as an extreme case of attenuation by a Debye–Waller factor. In fact, X-ray diffraction experiments from monolayer solids have concentrated thus far on the region of the primitive reciprocal lattice vector.

The characterization of the order in the 2D solid, and of the stages through which it evolves with increasing temperature, can be developed further. An analysis (Mermin 1968) that does not assume the harmonic approximation shows that the low-temperature solid has a well-defined orientational order that reflects the crystal anisotropy. At higher temperatures the orientational order correlation function has algebraic (power law) and exponential decay with increasing distance between two lattice sites (Halperin and Nelson 1978). However the nature of the transition(s) by which the solid melts as the temperature is increased is not firmly established and apparently depends on details of the interactions. We summarize features of the continuous melting proposals of Kosterlitz and Thouless (1973) and of Halperin and Nelson (1978) in Section 4.3.

4.3 Topological defects

Much of this book emphasizes discrete atomic character and the role of interactions in determining properties of physically adsorbed layers. Over a large range of conditions, the holding potential constrains the monolayer solid to such an extent that it is effectively a quasi-two-dimensional system. This creates situations where the long-range order is characteristic of continuous systems of reduced dimensionality and where strong fluctuation effects lead to subtly singular thermodynamic behavior. An introduction to the lessened correlations for the harmonic 2D lattice was given in Section 4.2.3. We now discuss the 2D solid in a broader context that relates the fluctuation effects to symmetry and dimensionality.

The discussion was initiated by Kosterlitz and Thouless (1973), who emphasized universal aspects of the fluctuation effects in a quite broad domain of systems with reduced dimensionality. They identified several 2D systems (the crystal, superfluid, and planar magnet) in which algebraic long-range order might be lost by thermal activation of topological (point) defects in the order parameter. Their proposals have been much extended (Kosterlitz and Thouless 1978; Halperin and Nelson 1978; Nelson and Halperin 1979; Young 1979) and much debated (Frenkel and McTague 1979; McTague *et al.* 1980; Toxvaerd 1980; Abraham 1980a, 1981a, 1982, 1986; Novaco and Shea 1982; Naidoo *et al.* 1993; Bagchi *et al.* 1996). The review literature is already voluminous (Kosterlitz 1980; Nelson 1980, 1983; Morf 1984; Strandburg 1988; Glaser and Clarke 1993).

Here we review the treatment of thermally activated edge dislocations in an isotropic elastic solid and their role in a proposed continuous 2D melting. Then we summarize the status of superfluidity at the monolayer level.

4.3.1 *Melting of a 2D isotropic elastic solid*

For a monolayer solid, an edge dislocation is a point defect that has a role analogous to vortex singularities in a planar magnet and in a superfluid film. The energy of an isolated edge dislocation in the monolayer depends logarithmically on the total area of the monolayer. Therefore, the corresponding entropy of the number of sites where it can be centered, also logarithmic on the total area, leads to a threshold temperature at which it can be thermally excited. It is a characteristic of the reduced dimensionality that such topological defects may have significant thermal populations and contribute in a dominant way to the thermodynamics and dynamics of the monolayer.

The continuum elasticity description of the monolayer solid (Nelson and Halperin 1979) is used here to formulate an estimate of the temperatures where thermally activated dislocations enter. In addition, several of the issues that arise in discussions about how this mechanism for melting may be realized are introduced. Calculations of monolayer melting for atomic models and experiments that show apparently continuous monolayer melting are reviewed in Chapters 5 and 6.

The deformation tensor u_{ij} for Cartesian components of a continuum distortion of positions x to x' is derived from the displacement vector $\mathbf{u}$ by

$$
\begin{aligned}
r'_i &= r_i + u_i(\mathbf{r}) \\
u_{ij} &= du_i/dr_j \ .
\end{aligned} \tag{4.75}
$$

While the strain may be 'smooth', the elasticity theory also admits singular, point defect, solutions for which the closed line integral about the site of a dislocation does not vanish

$$
\oint d\mathbf{u} = \mathbf{b}_0 \ . \tag{4.76}
$$

In eqn (4.76), $\mathbf{b}_0$ is the Burgers vector of the dislocation; for the lowest energy dislocations its length is a_0, the lattice constant of the underlying discrete lattice. This is the value used for the following discussion.

The elastic deformation energy of a 2D solid that is initially in a triangular lattice of total area A and spreading pressure π is

$$
\Delta E = \int_A d^2r[-\pi \sum_i \eta_{ii} + \frac{1}{2}\lambda(\sum_i \eta_{ii})^2 + \mu \sum_{ij} \eta_{ij}^2] \ , \tag{4.77}
$$

where λ and μ are the Lamé constants of an isotropic elastic medium and the Lagrangian strain tensor η_{ij} is defined in terms of the strains u_{ij} by

$$
\eta_{ij} = \frac{1}{2}[u_{ij} + u_{ji} + \sum_k u_{ki}u_{kj}] \ . \tag{4.78}
$$

Equation (4.77) arises from a reduction of a general quadratic form for the elastic energy in terms of η that uses the inversion and 6-fold-rotation symmetries of the triangular lattice.

Nelson and Halperin (1979) present an explicit expression for the asymptotic behavior of the strain field of an isolated dislocation based on eqn (4.77) and satisfying eqn (4.76). The increment in energy at zero spreading pressure is

$$\Delta E_1 = (a_0^2/4\pi)[\mu(\lambda + \mu)/(\lambda + 2\mu)] \ln(A/A_0) . \qquad (4.79)$$

The area A_0 is set by the small-distance *core* behavior of the dislocation, the form of which lies outside the elastic approximation (Joos *et al.* 1994). What is crucial in the result is the logarithmic dependence on the total area A. Since the entropy of the dislocation is $S \simeq k_B \ln(A/A_0)$, the free energy is lowered by spontaneous creation of such dislocations above a temperature T_m given by

$$a_0^2/[4\pi k_B T_m] = 1/\mu_R + 1/(\mu_R + \lambda_R) . \qquad (4.80)$$

Kosterlitz and Thouless (1973) gave a related argument that notes that below T_m there are bound pairs of dislocations with equal and opposite Burgers vectors, so that the $\ln(A)$ dependence in eqn (4.79) is canceled and the energy of a pair depends logarithmically on its separation. The separation of the pair increases with increasing temperature and diverges at T_m. To make the discussion more accurate requires including the effects of other thermally excited dislocation pairs on the energy of this pair; then there is a renormalization ('R') of the elastic constants that enter in eqn (4.80). The qualitative relation of thermally excited dislocations or unbound dislocation pairs to melting is that they provide a mechanism for the solid to release a static shear stress. That is, a solid is able to withstand a static shear and a fluid is not, so that the unbound dislocations are in a medium that 'has melted'.

The elastic constants that enter into this discussion are closely related to macroscopic properties of the elastic solid.* The Lamé constant μ is the shear modulus, and the speeds of sound for the medium of mass density ρ are given by

$$c_l^2 = (\tilde{\lambda} + 2\tilde{\mu})/\rho , \quad c_t^2 = \tilde{\mu}/\rho , \qquad (4.81)$$

where the pressure-renormalized Lamé constants are (Stewart 1977)

$$\tilde{\lambda} = \lambda + \pi , \quad \tilde{\mu} = \mu - \pi . \qquad (4.82)$$

The bulk modulus of the solid is

$$B = \mu + \lambda = \tilde{\mu} + \tilde{\lambda} . \qquad (4.83)$$

The results for the harmonic lattice, Section 4.2.3, may be expressed in terms of the continuum theory by restating the speeds of sound in eqn (4.68) in terms of the Lamé constants by eqn (4.81).

---

*The lattice dynamics of a solid under static stress was formulated by Barron and Klein (1965) and pressure-renormalized Lamé constants were defined by Wallace (1972) and by Stewart (1974, 1977). For a proof that the triangular lattice of a quantum solid has an isotropic elasticity theory see Bruch (1992).

For an incommensurate monolayer solid on a periodic substrate there is an additional term:

$$\Delta E' = \Delta E - (h/2) \int d^2r \cos(6\theta) , \qquad (4.84)$$

including the energy of orientational epitaxy, Section 3.6. The elastic constant h is expressed in terms of the curvature of the Novaco–McTague energy, Section 5.2.1, at the optimal orientation angle χ (Greif and Goodstein 1981)

$$(36/8)h = \gamma = (N/4A) \, d^2e_{NM}/d\chi^2 . \qquad (4.85)$$

The speeds of sound are then given by eqn (4.81) with redefined Lamé constants

$$\mu' = \tilde{\mu} + \gamma , \quad \lambda' = \tilde{\lambda} - 2\gamma . \qquad (4.86)$$

There also are consequences for correlation functions that are accessible in precise diffraction experiments. The long-range order in the 2D solid is characterized with correlation functions for translational order, eqn (4.70), and orientational order

$$c_6(\mathbf{R}) = \langle \exp[i6(\theta(\mathbf{R}) - \theta(0))]\rangle. \qquad (4.87)$$

An approximation to eqn (4.70) for the harmonic lattice was given in eqn (4.71) for reciprocal lattice vectors g of the average lattice. The angle $\theta(\mathbf{R})$ in eqn (4.87) is the orientation of the bond between two neighboring atoms relative to a fixed reference axis. It is expressed in terms of the displacement field $\mathbf{u(r)}$ by

$$\theta(\mathbf{r}) = \frac{1}{2}(du_y/dx - du_x/dy) . \qquad (4.88)$$

The Kosterlitz–Thouless theory characterizes the low temperature solid by long-range bond-orientational order in c_6 and an algebraic decay, power-law dependence on the inverse of the separation, of the translational order in c_g. Lattice defects such as dislocations are defined in terms of the local short-range order.

Kosterlitz and Thouless described a scenario for melting in terms of the evolution of these correlation functions with increasing temperature. An isolated dislocation has an energy, given in terms of elastic constants, that increases logarithmically with the area of the solid; thus it has very low thermodynamic probability at low temperatures. However, the strain field of pairs of dislocations can be localized enough that the pairs are thermally excited. With increasing temperature the pair is excited to larger relative separation and at the temperature T_m, given by balancing an entropy term, also proportional to the logarithm of the area, against the logarithmic energy, the dislocations unbind and move 'freely' relative to each other. Such a system is not able to support a static shear stress and is characterized as a liquid. How strong a damping this implies for transverse waves in the liquid near the melting condition is not known. An algebraic orientational long-range order may remain for a temperature interval after the static shear modulus has vanished. This is called a *hexatic* phase.

In terms of the correlation functions for the triangular lattice on a smooth surface, there is full long-range orientational order and algebraic decay of translational order up to T_m given by eqn (4.80) using renormalized Lamé constants. As T_m is approached from below, the solid has some dislocation pairs. At T_m, the shear modulus drops to zero, the correlation function $c_g(\mathbf{R})$ has exponential decay with increasing distance R, and the correlation function $c_6(\mathbf{R})$ has algebraic decay. Beyond a second higher temperature T_i, the algebraic orientational order is lost, and $c_6(\mathbf{R})$ has exponential decay for large R. The transitions at T_m and T_i are continuous with no conspicuous signatures in the specific heat.

The 2D lattice which most resembles the idealization for this discussion is the Wigner lattice of electrons on a liquid helium surface (Williams 1993). The lattice is observed to melt at a density-dependent temperature that is in agreement with theoretical calculations. The calculations also show the shear modulus vanishes at the transition. Observations, though not yet conclusive, are consistent with this (Grimes and Adams 1979; Morf 1979, 1984; Gallet et al. 1982); excitations of the liquid helium surface complicate the analysis (Marty and Poitrenaud 1984). There is no experimental evidence for the second transition (Deville et al. 1984), and there has been no latent heat detected at the melting transition (Glattli et al. 1988a,b).

The substrates for physically adsorbed inert gases all provide some lateral periodicity. Thus, the orientational alignment energy gives an effective field that drives some order even in the fluid. It has a role analogous to that of a uniform magnetic field in the paramagnetic phase of a ferromagnet, and the transition to an isotropic liquid is broadened. The phase transition at T_i is no longer present (Nelson 1983).

The signature of the transition at T_m in a diffraction experiment involves both the radial and transverse line shapes of a Bragg peak. The loss of long-range translational order is manifested by a broadening in the radial direction, while the orientational fluctuations involving h are manifested in the azimuthal, i.e., transverse, broadening. High-resolution X-ray diffraction from inert gases on single-crystal surfaces has measured both broadenings. The diffraction pattern consists of six-fold arrays of 'broadened spots'. The state of the monolayer above the melting temperature has been labelled the *hexatic liquid* and the *lattice liquid* for this reason. There have not yet been inelastic scattering experiments to characterize the excitations in the disordered phase. For physically adsorbed systems, there are only narrow temperature ranges, of the order of a few kelvin, over which the azimuthally broadened spots are observed (Zerrouk et al. 1994; Nuttall et al. 1994; D'Amico et al. 1990). By contrast, azimuthally broadened spots are observed over a range of about 50 K for a lattice liquid phase of K/Ni(111) (Chandavarkar and Diehl 1989). A search for the corresponding broadening for K/Ni(100) shows that it is limited to a range of at most 10 K (Fisher and Diehl 1992). The latter observations show the significance of the symmetry of the substrate field h. The two-stage melting apparently is found in a colloidal suspension of a 2D dipolar system (Kusner et al. 1995).

The theory of dislocation-mediated melting presents it as a scenario, a mechanism for the melting in the absence of other mechanisms that would preempt it at lower temperatures. It would be a continuous melting transition rather than the discontinuous first-order melting with a latent heat that occurs for 3D solids of the same adsorbate. Some calculations for 2D models purport to show that the continuous melting is indeed preempted (Toxvaerd 1978, 1980; Barker *et al.* 1981; Phillips *et al.* 1981; Kalia and Vashishta 1981). However, interpretation of the model calculations is not straightforward (Strandburg 1988) and there continue to be reports of a hexatic phase in some computer simulations (K. Chen *et al.* 1995). Equilibrium for the processes of the dislocation-mediated melting is not easily achieved on the time scale of computer experiments (Nelson 1983) and correlation lengths may be large but finite in samples of limited extent (Bakker *et al.* 1984). Adding to the delicacy of the interpretation is the result that the melting entropy per atom is much smaller in 2D than in 3D (van Swol *et al.* 1980; Kalia and Vashishta 1982) for model potentials for which computer experimenters concluded that the melting is first-order (Ramakrishnan 1982). Even the first-order melting of the classical hard disk system (Hoover and Alder 1967) has been called into question (Zollweg and Chester 1992; Lee and Strandburg 1992).

4.3.2 Monolayer superfluidity

Superfluidity was first seen in helium films in the 1930s. The transition temperature was observed to decrease from the bulk value 2.17 K as the film became thinner, eventually disappearing at a threshold thickness, often called the *inert layer*. Typically, this 'layer' corresponds to one or two atomic layers of helium. One interpretation of these data is based on the concept of the superfluid *order parameter* ψ (Ginzburg and Pitaevskii 1958; Ginzburg and Sobyanin 1982). This function, it was argued, should satisfy specific boundary conditions, e.g., vanishing at the substrate and having zero slope at the free surface, that could only be satisfied if the thickness exceeded a specific threshold value. A somewhat distinct point of view is based on the larger fluctuations in the restricted geometry of the film. If, as commonly believed, the superfluidity of the ^{4}He films is a consequence of Bose condensation, it should disappear in a thin film because of these fluctuations.* Kosterlitz and Thouless (1973) included thin-film superfluidity as a possible example of the effects of *topological* long-range order. In the superfluid case, this means that there may be superfluidity *without* Bose condensation. There exist long-range, but not infinitely long-range, correlations and these suffice to produce the remarkable superfluid properties. The theory can be expressed in terms of quantized vortices that play a role analogous to the dislocations in the theory of melting. The energy/entropy balance leads to a new onset criterion for superfluidity

---

*In fact, a theorem of Chester, Fisher, and Mermin (1969) proves that the Bose condensation is absent in **any** film of arbitrary thickness! It generalizes previous work of P. C. Hohenberg (1967).

$$\rho_s(T)/T = 2M^2 k_B/(\pi\hbar^2).$$ (4.89)

Here $\rho_s(T)$ is the superfluid mass per unit area at the onset temperature T and M is the atomic mass. This relation has proven to agree with both previous and subsequent experimental data. The contrast to the 3D case is remarkable, because here the superfluid density drops discontinuously to zero at the transition while for 3D it goes continuously to zero at T_λ.

Note that the Kosterlitz–Thouless relation is compatible with superfluidity within a single atomic layer. An estimate with $\rho_s(T)$ equal to the atomic mass times a typical monolayer helium number density ($\sim 0.1/\text{Å}^2$) yields a transition temperature $T \simeq 1$ K. Until quite recently, however, there has been no experimental evidence for superfluidity in the case of a monolayer of helium. Among the reasons offered for this absence is that the first layer is often compressed to the point of solidification. Three recent experiments have claimed to provide indications of monolayer superfluidity. Greywall (1993) found an anomalous heat capacity for low coverage ^{4}He/graphite that is consistent with predictions of Wagner and Ceperley (1992, 1994). The Wagner–Ceperley computer simulations indicate this behavior is coincident with superfluid onset. One particularly promising route to monolayer superfluidity is to explore weak-binding substrates, which ought not to solidify the first helium layer (E. Cheng et al. 1989). A pair of experiments provides examples of this behavior. Wyatt et al. (1995) found evidence of a submonolayer superfluid of He/Rb, another weak-binding surface. Mochel and Chen (1994), using *third sound* measurements (a normal mode unique to a superfluid film), found indications of superfluidity in the first layer of ^{4}He on a hydrogen substrate. All of these tentative observations are extremely interesting and controversial; they remain to be verified.

4.4 Correlations in the monolayer fluid

Simple approaches to the time dependence of correlations in a fluid, as, for instance, on the basis of linear kinetic theory, led to the expectation that at sufficiently long times the correlations decay exponentially. This was embedded in the language associated with transients and the approach to equilibrium. Therefore, molecular dynamics simulations of hard sphere and hard disk fluids that showed strongly persistent correlations that decayed only algebraically with increasing time (Alder and Wainwright 1967, 1970) drew intense interest. The correlation decayed as $1/t$ ($t =$ time) for the 2D fluid, and this eventually was understood to be a consequence of vortex motions that have a hydrodynamic analogue (Alder and Wainwright 1970; Pomeau and Résibois 1975). The *long time tails* are observed in computer simulations for 2D fluids on both flat and corrugated surfaces (Toxvaerd 1979) over periods of 10–20 mean collision times, which corresponds to time scales of *ca.* 10 ps for physically adsorbed layers of argon and nitrogen. The corresponding frequencies might be accessible in a future extension of capabilities in quasi-elastic scattering experiments. However, it is an open question what role the dynamics of the substrate will have in damping the hydrodynamic motions responsible for the persistence of the correlations.

In this section, we briefly define some of the quantities characterizing the diffusive motion in a monolayer fluid, as the issues of the persistence of correlations arise in molecular dynamics simulations carried to time scales of 100s of ps. Most succinctly, the (tracer) diffusion D is not defined in mathematical 2D.

Ideas of stochastic processes applied in two dimensions lead one to define a diffusion constant $D[\mathbf{r}(t)]$ in terms of the accumulated mean-square displacement by

$$D[\mathbf{r}(t)] = \langle [\mathbf{R}(t) - \mathbf{r}(0)]^2 \rangle / 4t \quad t \to \infty \, , \tag{4.90}$$

where the angular brackets denote an ensemble average. In practice, one seeks a range of times where the mean-square displacement is accumulating linearly with increasing time. Plausible magnitudes are obtained for $D[\mathbf{r}(t)]$, even though the procedure is somewhat subjective.

More fundamentally, the tracer diffusion constant may be formulated as an integral of the 2D velocity autocorrelation function:

$$D[\mathbf{v}(t)] = (1/2) \int_0^\infty dt \, \langle \mathbf{v}(t) \cdot \mathbf{v}(0) \rangle \, , \tag{4.91}$$

If the correlation function in eqn (4.91) decays exponentially at large times t, $D[\mathbf{v}(t)] = D[\mathbf{r}(t)]$.

However, the simulations (Alder and Wainwright 1970; Toxvaerd 1979) show that $\langle \mathbf{v}(t) \cdot \mathbf{v}(0) \rangle$ decays exponentially only for a few mean collision times and then falls only as $1/t$. The precise functional form of the very long time decay is still not firmly established (van der Hoef and Frenkel 1991).

Whether these simulations are presenting a phenomenon that occurs for monolayers is open to question. The assumption that the substrate may be viewed as a relatively inert support for the adlayer is used in defining the models. The only dynamic interactions of the adlayer and substrate manifested in experiments thus far are the hydbridization of monolayer-solid normal modes with the Rayleigh wave of the substrate and a radiative damping of modes near the Brillouin zone center (Gibson and Sibener 1985; Kern et al. 1987b; Hall et al. 1985, 1989). The substrate dynamics may provide a crucial damping mechanism for the monolayer hydrodynamic motions, but there have not been estimates of the relevant time scales. Analysis of the quasi-elastic scattering experiments thus far has been in terms of Lorentzian line shapes (Frenken et al. 1990) that would be the situation for a fluid with normal diffusion.

5

MANY-BODY THEORY OF MONOLAYERS

In this chapter we develop the theory of interacting monolayers, following a division into two classes: (1) those in which the substrate imposes such a strong lattice structure on the adsorbate that lattice gas approximations capture much of the essential physics and (2) those in which the continuous character of the configurations is dominant and for which an analogue of the theory of 3D rare gas solids is implemented. Section 5.1 treats the lattice gas models and results of the renormalization group theory of phase transitions (Domb and Green 1972, 1976). The concept of universality applies here and critical point exponents are expected to be insensitive to several parameters in the models. Section 5.2 treats the continuum systems, using methods applied to rare gas solids and liquids (Pollack 1964; Klein and Venables 1976, 1977). This is 'nonuniversal theory' and the goals are to account generically for the observations and to reach a quantitative understanding for a few systems. We emphasize the formulation of the theory and illustrate it with a few physical examples. Our aim is to highlight the considerations from which the theory begins and to show the relations among the different approximation methods. In Chapter 6 we give a more thorough account of a few series of adsorbates, covering the range from the initial monolayer condensation to the multilayer regime and contrasting the behavior to the 3D bulk phases of the adsorbate.

The monolayer domain extends from an initial low-density adsorbed gas to monolayer condensation and then to what is termed monolayer completion. The dilute 2D gas, already discussed in Section 4.2, is followed by a dense surface phase, which may be liquid or solid. There need not be a sharp condensation because a consequence of the gas–liquid critical point phenomenon is that at high enough temperatures there is no discontinuity as the fluid density increases with increasing pressure. The upper limit on the monolayer domain is set by further layer condensations: for physically adsorbed layers, this usually is the bilayer formation but in cases of extremely limited layer growth, it may terminate with the condensation of a dense 3D phase (Pandit et al. 1982; Gittes and Schick 1984; Muirhead et al. 1984; Schick 1990; Binder et al. 1994). The critical point phenomenon for higher layer condensations (De Oliveira and Griffiths 1978) again limits the mathematical sharpness of this characterization of the monolayer domain.

5.1 Lattice gases

The lattice gas idealization of monolayer phenomena is most extensively applied to chemisorbed systems, where there is strongly localized adsorption at preferred sites. However, there are also important applications to model order/disorder transitions of a commensurate physisorbed layer (He/graphite) and to generate schematic monolayer and multilayer phase diagrams. Two elementary lattice gas models of the monolayer, the Langmuir adsorption isotherm (Langmuir 1918) and the BET-isotherm (Brunauer *et al.* 1938), were presented in Section 4.1. Lattice gas models provide the main point of contact between physical adsorption and the renormalization group theory of critical point phenomena (Domb and Green 1976; Wilson 1983). Phase transitions in monolayers have been reviewed by Vilches (1980), Schick (1981), Weinberg (1983), Ohtani *et al.* (1986), Binder and Landau (1989), Binder (1992a), and Persson (1992a).

We present the principal lattice gas models which are used in describing physically adsorbed layers and outline several of the approximations with which they are treated (Section 5.1.1). Domany *et al.* (1978) displayed an underlying unity of order–disorder phenomena in adsorbed layers by constructing a classification scheme based on the symmetries found in the approximate solutions; this is summarized in Section 5.1.2. Applications of real space renormalization group transformations to He/graphite and to Kr/graphite are summarized in Section 5.1.3.

5.1.1 *Lattice gas models*

A lattice gas model consists of: (1) a lattice of adsorption sites, perhaps defined by the substrate, (2) interaction energies with the substrate, (3) interaction energies between filled sites, which depend on the distance between the sites, and (4) a site occupation index that enumerates the possible states at the given site. The entropy for such models reflects the discrete nature (1 atom/0 atom) of the occupation of the sites. When the interaction energy of nearest-neighbor pairs is strongly positive, so that the occupation of such pairs is energetically unfavorable, the low-temperature ordered state consists of a superlattice (subset) of the one-particle sites. The models may be generalized to include a distinction between dense-disordered and dense-ordered structures, so that there is an additional entropy contribution beyond the occupied/vacant combinatorial. We define the models in order of increasing complexity: the Ising lattice gas (Stanley 1971; McCoy and Wu 1973), the Potts model (Baxter 1982b), and the Potts lattice gas (Berker *et al.* 1978).

5.1.1.1 *Models* The Langmuir isotherm, Section 4.1, shows the qualitative relation of fractional surface coverage θ to the pressure of a coexisting 3D ideal monatomic gas. It was based on a lattice gas model with these ingredients: (1) each adsorbed atom has potential energy ϵ_0; (2) the adsorbing surface area A has N_s sites, each of which may have 0 or 1 atoms; and (3) the adsorbed atoms are distributed randomly over the sites. The grand partition function for the adsorbed phase at chemical potential μ was presented in Section 4.1.1, and the fractional coverage θ was given in eqn (4.7). The coverage is linear in the pressure at low coverages, with a proportionality constant that depends exponentially on the binding energy to the substrate. The simplicity of the derivation and of the result arises because adatom–adatom interactions have not been explicitly included.

Another historically important model based on lattice gas ideas is the Brunauer–Emmett–Teller isotherm, Section 4.1.1. As in the Langmuir model, the surface presents N_s adsorbing sites. Each surface site defines a stack of adsorption sites perpendicular to the surface plane, and the stacks are statistically independent. The result is again a closed form solution for the coverage as a function of 3D gas pressure, Section 4.1.1. However, lateral interactions do affect the shape of isotherms, and layer filling is not sharply sequential, effects that are included in the de Oliveira–Griffiths model (described below). We now turn to the more realistic models; a consequence of the increased complexity is that the solutions are not as explicit as in Section 4.1.

Ising lattice gas   The basic lattice gas Hamiltonian for a monolayer with two-body lateral interactions is (Stanley 1971; Yeomans 1992)

$$H_l = \frac{1}{2} \sum_{i \neq j} v_{ij} \sigma_i \sigma_j + \epsilon_0 \sum_j \sigma_j , \qquad (5.1)$$

where ϵ_0 is the binding energy to the substrate, the site occupation variables σ_j take on the values 0 and 1, and v_{ij} is the potential energy between two filled sites. The grand potential Ω is obtained from the grand partition function with chemical potential μ as

$$\Omega = -k_B T \ln Tr \exp[-\beta(H_l - \mu N)] , \qquad (5.2)$$

and the average fractional occupation of sites, θ, is

$$\theta = \langle N \rangle / N_s = \langle \sum_j \sigma_j \rangle / N_s . \qquad (5.3)$$

The substitution $\sigma_j = (1 + s_j)/2$ shows that the Hamiltonian H_l can be transformed to the Hamiltonian H_I of an Ising spin $\frac{1}{2}$ magnet in an external field:

$$H_l - \mu N = H_I + \frac{1}{8} \sum_{i \neq j} v_{ij} + (N_s/2)(\epsilon_0 - \mu) , \qquad (5.4)$$

with

$$H_I = \frac{1}{2} \sum_{i \neq j} J_{ij} s_i s_j - h \sum_j s_j$$

$$J_{ij} = v_{ij}/4$$

$$h = \frac{1}{2}(-\epsilon_0 + \mu) - \frac{1}{4}\sum_{j\neq 0} v_{j0} . \qquad (5.5)$$

Roughly, the lattice model with nearest-neighbor attractions corresponds to a ferromagnetic Ising model; with nearest-neighbor repulsions it corresponds to an antiferromagnet.

When attractive nearest-neighbor interactions in the lattice gas lead to a sharp condensation at low temperatures, the associated magnetic model has a ferromagnetic transition. Then the correspondence between the magnetic and lattice models is very informative: the magnetic transition occurs at zero field ($h = 0$), and this specifies the chemical potential of the condensation through eqn (5.5). Also, the fact that the coexisting spontaneous magnetic moments at zero field are of equal magnitude and opposite in sign implies a *particle–hole symmetry* in the lattice gas: the sum of the densities of the dilute and dense phases at coexistence is 1.

$$\theta(\text{dilute}) + \theta(\text{dense}) = 1 \; ;$$

$$\mu_0 = \epsilon_0 + \frac{1}{2}\sum_{j\neq 0} v_{j0} . \qquad (5.6)$$

There is some artistry in building a lattice gas Hamiltonian which has more general properties. For instance, it may be known from observations that a system does not have particle–hole symmetry at the condensation. Then the corresponding lattice gas model is generalized to include three-body interactions (Ching *et al.* 1978).* Also, the Hamiltonian in eqn (5.1) has no intrinsic dynamics; i.e., the site occupations are constants of the motion and the interactions necessary to reach thermal equilibrium are not explicitly included in the model. There is, in the computer simulations, what is called a 'Monte Carlo dynamics' for the model. The configurations may be inspected after an initial series of equilibration steps to get a measure of the relative occurrence of clusterings and fluctuation phenomena.

De Oliveira–Griffiths model De Oliveira and Griffiths (1978) extended eqn (5.1) to model multilayer adsorption. Their Hamiltonian, including the chemical potential term, is

$$H_{DG} = -\epsilon \sum_{\langle jk,j'k'\rangle} \sigma_{jk}\sigma_{j'k'} - \sum_j (\mu + v_j) \sum_k \sigma_{jk} . \qquad (5.7)$$

The layer index is $j(j')$ and sites within a layer are labelled $k(k')$. The $\langle ... \rangle$ notation denotes a restriction to distinct nearest-neighbor interactions: there are

---

*There is a related argument for 3D fluids where the shape of the coexistence curve is reported to demonstrate the presence of many-body interactions (Pestak *et al.* 1987).

'a' nearest neighbors within the layer and 'b' nearest neighbors in the adjoining layers with $a = 6$ and $b = 3$ for an fcc stack of triangular lattices. The potential energy of adatom–substrate interaction for an adatom in the j-th layer is $-v_j$ with

$$v_j = \epsilon D, \qquad j = 1$$
$$v_j = \epsilon CD/j^3, \quad j > 1, \tag{5.8}$$

in units of ϵ. The model stated in eqn (5.7) shows a succession of layer condensations with increasing chemical potential; the condensations end at critical points. A similar model was used by Pandit, Schick, and Wortis (1982) in their discussion of limited layer growth and wetting phenomena.

Potts model The Potts model has been used for the analysis of the order–disorder transitions of commensurate superlattices, such as the $\sqrt{3}R30°$ lattice of He/graphite (Alexander 1975; Schick 1980). The idea is that the short-range lateral interactions of adatoms are strongly repulsive and make occupation of nearest-neighbor sites unlikely, but that there is an attractive lateral energy when the atoms are at the sites of the superlattice. There are s equivalent superlattices. The model Hamiltonian is

$$H_P = -J \sum_{\langle i,j \rangle'} \sum_{\alpha=1}^{s} t_{i,\alpha} t_{j,\alpha} , \tag{5.9}$$

where the notation on the first sum specifies summation over distinct nearest-neighbor pairs of lattice sites. The site occupation index t takes on the values 0 or 1, with

$$\sum_{\alpha=1}^{s} t_{i,\alpha} = 1 . \tag{5.10}$$

For a 2D lattice of sites, there are exact analytic results for the ordering transition temperature: the transition is first order with a latent heat for large s values (Baxter 1982a,b; Baxter *et al.* 1979).

Potts lattice gas The Potts lattice gas has been used to model physisorbed layers such as Kr/graphite (Berker *et al.* 1978) and for the renormalization group treatment of the 2D Potts model (Nienhuis *et al.* 1979). Ebner (1983) cast it into a form that treats gas, liquid, and solid phases of a monolayer:

$$H_{PLG} = -J \sum_{\langle i,j \rangle'} \sum_{\alpha=2}^{s+1} t_{i,\alpha} t_{j,\alpha} - J\gamma \sum_{\langle i,j \rangle'} \sum_{\alpha=2}^{s+1} \sum_{\alpha'=2}^{s+1} t_{i,\alpha} t_{j,\alpha'} |_{\alpha \neq \alpha'} . \tag{5.11}$$

The Potts variable $t_{i,\alpha}$ has $s + 1$ components; the $\alpha = 1$ component corresponds to a vacant site. Both energy terms favor site occupation, but there are two qualitatively distinct dense phases: a phase with most of the sites having the same Potts component α_0 and one with a uniform distribution of the occupation over the Potts components 2 to $s+1$. The model has been generalized to include an external (substrate) potential that may depend on the Potts index α and to include farther neighbor interactions (Conner and Ebner 1987).

5.1.1.2 *Mean field solutions* Much of the emphasis in the statistical mechanical analysis of the lattice gas models is on continuous phase transformations, with a careful treatment of thermodynamic fluctuations. Here we summarize the mean field theory results. Although this approximation gives a poor account of critical fluctuations, it provides useful guides to the first order transitions and to the overall structure of the phase diagrams, Section 5.1.2.

Variational principle The basic step is a variational approximation for the Helmholtz free energy (canonical ensemble) or for the grand potential (grand canonical ensemble). It is based on inequalities first considered by J. Willard Gibbs in his formulation of the properties of the statistical mechanical entropy (Falk 1970). It corresponds to the Weiss approximation in ferromagnetism and the van der Waals approximation for the gas–liquid condensation.

The starting point is the inequality for nonnegative operators A and B

$$Tr(A \ln A - A \ln B - A + B) \geq 0 \ . \tag{5.12}$$

To obtain the Peierls–Bogoliubov inequality on the Helmholtz free energy, we choose

$$\begin{aligned}
A &= \exp(-\beta H_0)/Tr \exp(-\beta H_0) \ , \\
B &= \exp(-\beta H)/Tr \exp(-\beta H) \ ,
\end{aligned}$$

and define

$$\begin{aligned}
F(H') &= -k_B T \ln Tr \exp(-\beta H') \ , \\
\langle C \rangle_0 &= Tr\, C \exp(-\beta H_0)/Tr \exp(-\beta H_0) \ .
\end{aligned}$$

Then we have an upper bound in terms of a trial free energy F_{tr}

$$F(H) \leq F_{tr}(H) = F(H_0) + \langle H - H_0 \rangle_0 \ . \tag{5.13}$$

The corresponding bound on the grand potential Ω is obtained by including the chemical potential term $-\mu N$ in the definitions of A and B and $\langle C \rangle_0$. Then with definitions

$$\begin{aligned}
\Omega(H') &= -k_B T \ln Tr \exp[-\beta(H' - \mu N)] \ , \\
\rho &= \exp[-\beta(H_0 - \mu N)] \exp[\beta \Omega(H_0)] \ ,
\end{aligned} \tag{5.14}$$

we have

$$\begin{aligned}
\Omega(H) \leq \Omega_{tr} &= \langle H - \mu N \rangle_0 + k_B T\, Tr \rho \ln \rho \\
&= \Omega(H_0) + \langle H - H_0 \rangle_0 \ .
\end{aligned} \tag{5.15}$$

Ising lattice gas The mean-field approximation for the lattice gas models is obtained using a factored form for the trial phase space density operator

$$\rho = \prod_j \rho_j , \tag{5.16}$$

where the product is over the sites. The factorization expresses an approximation of statistically independent occupation of the lattice sites. Then the entropic term in the trial grand potential is

$$Tr\rho \ln \rho = \sum_j Tr\rho_j \ln \rho_j . \tag{5.17}$$

For the Ising lattice gas Hamiltonian H_l, eqn (5.1), we now make the further assumption that the lateral interactions v_{ij} give only nearest-neighbor attractions $-\epsilon$. Then the reference Hamiltonian H_0 is

$$H_0(l) = -\bar{\epsilon} \sum_j \sigma_j , \tag{5.18}$$

and the trial grand potential for a lattice of N_s sites with coordination number q is

$$
\begin{aligned}
\Omega_{tr}(H_l)/N_s &= -(\epsilon q/2)\langle \sigma \rangle^2 - (\mu - \epsilon_o)\langle \sigma \rangle \\
&+ k_B T \langle \sigma \rangle \ln \langle \sigma \rangle \\
&+ k_B T(1 - \langle \sigma \rangle) \ln(1 - \langle \sigma \rangle) ,
\end{aligned} \tag{5.19}
$$

with

$$\langle \sigma \rangle = 1/[1 + \exp(\beta[\bar{\epsilon} + \mu])] . \tag{5.20}$$

The parameter $\bar{\epsilon}$ in the comparison Hamiltonian H_0 is chosen to make the trial grand potential stationary and is

$$\bar{\epsilon} = -\epsilon_0 + q\epsilon\langle \sigma \rangle . \tag{5.21}$$

Using eqn (5.21) in eqn (5.20) gives the characteristic self-consistency equation of mean-field theory.

The quantity $\langle \sigma \rangle$ is the fractional occupation of a lattice site; at high temperatures it is a single-valued function of the chemical potential μ. For temperatures below

$$T_c(H_l) = q\epsilon/4k_B , \tag{5.22}$$

there are two coverages of equal grand potential for a chemical potential value

$$\mu_c = \epsilon_0 - (q\epsilon/2) . \tag{5.23}$$

This is the gas–liquid critical point phenomenon for the model. The isosteric heat in the single phase region is given in terms of the enthalpy of the 3D gas h_{3g} by

$$q_{st} = h_{3g} - \epsilon_0 + q\epsilon\langle \sigma \rangle . \tag{5.24}$$

The adsorption isotherm calculated under these approximations is known as the Fowler isotherm (Fowler 1966).

De Oliveira–Griffiths model In the corresponding treatment of the de Oliveira–Griffiths lattice gas Hamiltonian H_{DG}, eqn (5.7), the reference Hamiltonian H_0 has an effective one-body energy that depends on the layer index j:

$$H(DG)_0 = -\sum_j \epsilon_j \sum_k \sigma_{j,k} , \qquad (5.25)$$

and the fractional occupation of the N_s sites in layer j is f_j

$$\langle \sigma_{j,k} \rangle = f_j = 1/(1 + \exp[\beta(\epsilon_j + \mu)]) . \qquad (5.26)$$

The trial grand potential is made stationary by the set of f_j satisfying the coupled equations

$$k_B T \ln(f_j/(1 - f_j)) = \mu + v_j + \epsilon(af_j + bf_{j+1} + bf_{j-1}) \quad j \geq 1 , \qquad (5.27)$$

where $f_0 = 0$. De Oliveira and Griffiths (1978) showed that the mean field solution at low temperatures displays a succession of layer condensations with increasing chemical potential. Each condensation is primarily a discontinuity in the occupation of some layer j, but the condensation of layer j occurs before all vacancies are filled in the layers of index smaller than j. These features are quite plausible, and it is gratifying that the simple model has such a rich structure. Explicit forms can be given with this approximation for the vacancy populations and for the chemical potential values at the general layer condensation at low temperatures (Unguris *et al.* 1981). There is a critical point for each layer condensation. More accurate treatments of the model (Ebner 1981; I. M. Kim and Landau 1981) trace the limiting behavior of the layer critical points in the multilayer film and the relation to surface roughening of the 3D lattice.

Potts model The mean field treatment for the Potts model, eqn (5.9), has a site probability density ρ_j with weight x_α for the α-component of the Potts variable. The trial Helmholtz free energy of eqn (5.13) is

$$F_{tr} = -(qJ/2) \sum_{\alpha=1}^{s} x_\alpha^2 + k_B T \sum_{\alpha=1}^{s} x_\alpha \ln x_\alpha , \qquad (5.28)$$

with

$$\sum_{\alpha=1}^{s} x_\alpha = 1 .$$

Varying the x_α to make F_{tr} stationary with this pair of equations leads to an equation

$$x_\alpha = \tau \ln x_\alpha + \lambda , \qquad (5.29)$$

with a Lagrange multiplier λ and $\tau = k_B T/qJ$. For given τ and λ, this has at most two roots x. Although there are more general possibilities, numerical analysis for many examples shows that with two roots the lowest free energy is obtained when one of the x_α has one value and the other $s - 1$ have the second value. Thus the two phases to be considered are the disordered phase

$$x_\alpha = 1/s$$

$$F_{tr}/qJ = -(1/2s) - \tau \ln s \; ; \tag{5.30}$$

and the ordered phase

$$x_1 = (1/s)(1 + (s - 1)\sigma) \; ,$$

$$x_j = (1/s)(1 - \sigma) \quad j > 1 \; , \tag{5.31}$$

where σ is the root of

$$\sigma = \tau \ln[(1 + (s - 1)\sigma)/(1 - \sigma)] \; . \tag{5.32}$$

Kihara, Midzuno, and Shizume (1954) showed there is a first-order transition to the ordered phase at a temperature

$$\tau_c = (s - 2)/(2(s - 1)\ln(s - 1)) \; , \tag{5.33}$$

with an occupation splitting $\sigma_c = (s - 2)/(s - 1)$ and latent heat

$$Q = (qJ/2)(s - 2)^2/(s(s - 1)) \; .$$

However, this mean-field theory predicts a first-order transition for $s > 2$, while exact solutions on several 2D lattices have first-order transitions only for $s > 4$ (Baxter 1982a,b).

Potts lattice gas   The mean field theory for the Potts lattice gas Hamiltonian, eqn (5.11), is similar to that for the Potts model. The trial grand potential is formed with a factored density operator having probability x_α for the α-component of the Potts variable. The solution of the mean-field equations may show two kinds of transition. First, with increasing chemical potential μ there may be a transition from a dilute occupation (x_1 near 1, all other x_α near zero) to a dense occupation (x_1 near zero). Another transition may be from a dense disordered phase ($x_1 \simeq 0; x_\alpha \simeq 1/s, \alpha > 1$) to a dense ordered phase ($x_2 \simeq 1$). There could be a critical point for the dilute-to-dense disordered transition and a triple point where this transition line intersects with a low temperature condensation from dilute to dense-ordered. In fact, there is a parameter range for which this occurs. Also there is a range with an incipient triple point: there is a critical value of γ at which the triple point and critical point merge (Berker *et al.* 1978).

5.1.1.3 *Beyond mean field* We have dealt at some length with the mean-field theory of the lattice gas models, because the solutions provide a frame of reference for much of the theoretical discussion of site-wise adsorption and multilayer formation. However, there are some serious and gross failures of this level of approximation. The critical temperature (eqn (5.22)) for the lattice gas is much larger than an exactly known solution (Onsager 1944; Stanley 1971). For the square lattice the ratio of the mean-field value to the exact T_c is 1.76; for the triangular lattice it is 1.65. Similarly, exact solutions for the square and triangular Potts lattices show the ordering transition is first-order only for $s > 4$ but mean field theory gives first-order transitions for $s > 2$. A defect of the mean-field theory presented thus far is that there is no distinction between the long- and short-range order on the lattice. A second level of mean-field theory was devised to include this distinction.

The mean-field theory which has a parameter for the short-range order between nearest neighbors is variously referred to as the Bethe–Peierls, quasi-chemical, Fowler–Guggenheim, or Kikuchi cell cluster approximation (Fowler and Guggenheim 1939; Kikuchi 1951). It was first defined in terms of a self-consistent partition function for a cluster consisting of a central site and its q nearest neighbors, but it may also be stated as an approximation to the entropy in terms of a density operator with one-body and two-body weight factors.* The self-consistency equations become more complex, but some properties such as the critical temperature for the Hamiltonian H_l with nearest-neighbor attraction are still given algebraically:

$$k_B T_c(BP) = \epsilon/[2\ln(q/(q-2))] \ . \tag{5.34}$$

This approximation is quantitatively better than the Weiss approximation. For the square (triangular) lattices, it overestimates the critical temperature by a factor 1.27 (1.35). In one dimension, the Bethe–Peierls approximation yields the exact result of an ordering transition at zero temperature. The Bethe–Peierls approximation was applied to give a more quantitative account of the lattice gas model of adsorption by Campbell and Schick (1972). Asada (1990) evaluated the corresponding approximation for the de Oliveira–Griffiths model.

5.1.2 *Classification of order–disorder transitions*

Thus far we have discussed the lattice gas models in terms that distinguish only between first-order and continuous phase transitions. In fact, a great variety of transitions is observed (Weinberg 1983; Ohtani *et al.* 1986; Binder 1992a). Symmetry of the lattices and the nature of the ordered phase provide a basis for a qualitative understanding of the differences (Schick 1981). If the order–disorder transition is second-order, general arguments limit the possible variety of phenomena and provide a grouping of the thermodynamic singularities at the transition. Domany, Schick, Walker, and Griffiths (1978) created a classification

---

*In the case of the square lattice, it even provides a variational bound to the entropy (Schlijper 1985).

scheme that is based on work of Landau and Lifshitz and makes use of the Weiss mean-field approximation described in the preceding subsection.

5.1.2.1 *Critical exponents* For a continuous transition with critical temperature T_c, power law dependences of thermodynamic functions on the reduced temperature variable t

$$t = (T - T_c)/T_c \tag{5.35}$$

near T_c are expressed with critical exponents (Stanley 1971). For the specific heat the functional form

$$
\begin{aligned}
C &\simeq A_+ |t|^{-\alpha} \quad t > 0 \\
&\simeq A_- |t|^{-\alpha'} \quad t < 0 \,,
\end{aligned}
\tag{5.36}
$$

has critical exponents α and α'.* The magnetic susceptibility, fluid compressibility, or diffuse critical scattering is fitted to the form

$$\chi \simeq t^{-\gamma} \quad t > 0 \,, \tag{5.37}$$

and the intensity of the Bragg peak at the superlattice wavevector is

$$I(Q_s) \simeq |t|^{2\beta} \quad t < 0 \,. \tag{5.38}$$

A critical exponent ν for the power law increase of the correlation length is also defined.

The exponent β also governs the variation of the density difference between the coexisting gas and liquid near the fluid critical temperature. An example of such behavior near the critical point of a 2D fluid is shown in Fig. 5.1 for CH_4/graphite. The fitted exponent (H. K. Kim *et al.* 1986a) is close to that of the spontaneous magnetization in the 2D Ising model, $\beta = \frac{1}{8}$ (Yang 1952).

5.1.2.2 *Universality classes* Continuous transitions are grouped together in universality classes when they have the same values for the critical exponents. For adsorbed layers, Domany *et al.* (1978) found a great unification of order/disorder transitions because they can be assigned to a small number of universality classes. The differences between universality classes reflect symmetries, spatial dimensionality, and the number of components of the order parameter. The order parameter, which is zero in the disordered phase and nonzero in the ordered phase, is a quantitative measure of the amount of order; part of the theoretical formulation of these problems consists of identifying the order parameter (Stanley 1971; Yeomans 1992).

---

*An important special case is a logarithmic divergence, for which $\alpha = 0$.

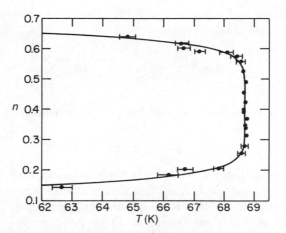

FIG. 5.1. Gas–liquid critical region of CH_4/graphite. The liquid–vapor coexistence curve of methane on graphite, from H. K. Kim and Chan (1984). The
solid line marks the boundary of the two-phase region in the temperature–
fractional coverage plane. Note the singular steepness of the curve near the
critical point. The order parameter measured in this experiment is the density difference between the two phases. It varies as reduced temperature t^β
with a fitted value 0.127 ± 0.01 for the critical exponent β, in good agreement
with the exact result $1/8$ for the 2D Ising model (Onsager 1944; Yang 1952).

To make the discussion more concrete, we treat the Hamiltonian H_l, eqn (5.1),
on a lattice of sites R_j, with local chemical potentials μ_j and grand potential

$$\Omega = -k_B T \ln Tr \exp[\beta \frac{1}{2} \sum_{ij} v_{ij} \sigma_i \sigma_j + \beta \sum_j \mu_j \sigma_j] . \qquad (5.39)$$

The local density is expressed as a sum over wavevectors in the first Brillouin
zone of the reciprocal lattice of $\mathbf{R}_j$ by

$$\rho_j = \langle \sigma_j \rangle = \rho + \sum_{\mathbf{q}} \exp[\imath \mathbf{q} \cdot \mathbf{R}_j] \rho(\mathbf{q}) , \qquad (5.40)$$

where ρ is the average occupation over all sites of the lattice. The generalization to spatially varying chemical potential (or field for the magnetic models) is
needed so as to be able to include phenomena such as superlattices and antifer-
romagnetism.

An explicit expression for ρ_j in terms of the temperature T and the set
of $v_{i'j'}$ and $\mu_{j'}$ is obtained with the Weiss mean-field approximation to the

grand potential, corresponding to eqn (5.19). The Fourier transform of the local chemical potential is obtained from

$$\mu(\mathbf{q}) = \frac{d}{d\rho^*(\mathbf{q})}[\Omega_{tr} + \sum_j \rho_j \mu_j]/N_s , \tag{5.41}$$

and a wavevector dependent compressibility $K(\mathbf{q})$ is defined by

$$1/K(\mathbf{q}) = \rho^2 d\mu(\mathbf{q})/d\rho(\mathbf{q}) . \tag{5.42}$$

At given average density, we identify the continuous ordering transition as occurring at that $\mathbf{Q}$ for which $K(\mathbf{q})$ first diverges with decreasing temperature. However, in general, there also are divergences at this temperature for the 'star' of $\mathbf{Q}$ formed of qs which are related to $\mathbf{Q}$ by point group operations of the space lattice. The ordering is specified by an order parameter whose number of components is equal to the number of independent terms in the star, i.e., those not related to each other by the addition of reciprocal lattice vectors. The only spatially varying components of the density in the ordered state have wavevectors in the star. Additionally, if the ordered state is commensurate with the lattice, the wavevectors must be at high symmetry points (at the center or on the boundary) of the Brillouin zone; this is the *Lifshitz rule*.

The mean-field free energy of a new model is expanded in powers of the hypothetical order parameter and is compared with the mean-field free energy of previously known models. If the first few terms of the expansions are identical, the model is assigned to the universality class of the known model and there is a prediction of its critical exponents. The logic is that either the model has a first-order transition, in which case the analysis leads to no conclusions, or the model has a second-order transition, in which case it must fall in one of the universality classes identified by this formal correspondence. It is by this novel argument that the mathematical structure of mean-field theory results is used to identify the behavior of a system in circumstances where fluctuation effects, omitted from the Weiss approximation, become dominant.

An immediate consequence of the analysis is to reveal that the order–disorder transitions are not restricted to the Ising class. The analogy between binary occupation and spin $\frac{1}{2}$ magnetism is incomplete. The square lattice with nearest-neighbor interactions does fall into the universality class of the 2D Ising model. However, the square lattice with first and second neighbor interactions admits other possibilities. Schick (1981) tabulated the commensurate ordered states which may be reached by continuous transitions for the five two-dimensional Bravais lattices and for the honeycomb lattice of the graphite basal plane surface.

Continuous transitions on the triangular lattice may fall into the universality class of the 3-state Potts model ($\alpha = 1/3$) or of the 4-state Potts model ($\alpha = 2/3$). The continuous transitions for a honeycomb lattice of adsorption sites include the possibility of a transition in the class of the ferromagnetic Ising model (logarithmic specific heat divergence). The distinction between the classes for

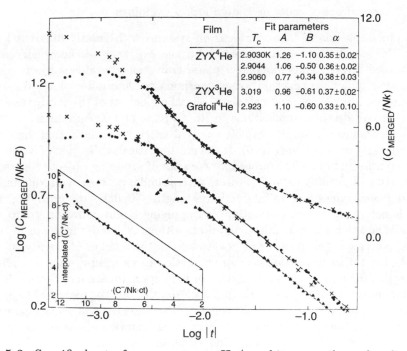

FIG. 5.2. Specific heat of commensurate He/graphite near the order–disorder transition. Log–log plot of the specific heat of He isotopes versus reduced temperature $(t - 1 - T/T_c)$ near the commensurate ordering transition to demonstrate power law behavior, from Bretz (1977). Data are shown for various forms of graphite (indicated in the inset with curve-fitting parameters). The exponent α characterizes the divergence of the specific heat and here is found to be $\alpha \simeq 0.35 - 0.37$, which is close to the value expected for the antiferromagnetic 3-state Potts model appropriate to this problem.

the triangular and honeycomb lattices was demonstrated with measurements of the specific heat divergence at the order–disorder transition of commensurate lattices of helium. The $\sqrt{3}R30°$ lattice of ^{4}He/graphite has $\alpha \simeq 0.36$ (Fig. 5.2) while ^{4}He on graphite preplated with krypton has a logarithmic divergence (Bretz 1977; Tejwani *et al.* 1980). The relatively strong specific heat divergence for the Potts model enhances the ability to distinguish the universality classes in 2D.*

---

*The 2×2 lattice of CF_4/graphite is a possible realization of the 4-state Potts model (Bak and Bohr 1983; Zhang *et al.* 1986).

5.1.3   *Applications of renormalization group mappings*

The implementation of Wilson's ideas on systems with greatly different length scales (Wilson 1983) was concurrent with the expansion of knowledge about physisorbed layers. Thus monolayer phase transitions became subjects for the application of renormalization group methods to determine phase diagrams, locate phase boundaries, and calculate critical exponents of thermodynamic singularities. We describe applications to He/graphite and Kr/graphite.

The monolayer examples exploit the real space renormalization method of Niemeijer and van Leeuwen (1976). It is a mathematical realization of Kadanoff's block transformations. The procedure consists of grouping objects together at one length scale leading to larger objects with a similar pattern of interactions but with different values for the interaction strengths. Usually the grouping, *mapping*, is not an exact transformation and applying renormalization group techniques is not as routine as following an algorithm. Niemeijer and van Leeuwen (1976) expressed their purpose as not 'to add to the number of exactly solvable models, but rather to provide for more realistic cases a computational scheme...' This approximation method and the mean-field approximation of Section 5.1.1.2 have an important difference in the point of application: the mean-field approximation is applied directly to the free energy, which has nonanalytic singularities; approximating the mapping does not carry an assumption about the analyticity of the free energy. The renormalization group method is most applicable in the vicinity of second-order phase transitions, where a correlation length becomes unboundedly large. The goal for the iterated approximate mapping is to retain the major physical processes while recursively eliminating other degrees of freedom.

5.1.3.1 $^4He/graphite   The monolayer phase diagram of ^{4}He adsorbed on the basal plane surface of graphite is reviewed in Section 6.1.1. Here we describe the modeling of the transitions from a fluid-like surface phase into ordered lattices of densities close to the density $n_c \simeq 0.0636/\text{Å}^2$ of a perfect commensurate $\sqrt{3}R30°$ lattice (Schick 1980).

The maximum temperature for the ordering occurs at a density 1% larger than n_c, and there is a line of second-order transitions in the density–temperature plane running through this point. The ordered state is a superlattice of the triangular lattice of adsorption sites in which next-nearest-neighbor sites are occupied. At lower temperatures, the order–disorder transition becomes a first-order condensation from low coverages; the nature of the low temperature transition at coverages greater than n_c is not settled.* The model for the line of second-order

---

*The suggestions include: a first-order commensurate–incommensurate transition (Schick 1980; Ecke and Dash 1983); a continuous transition from the commensurate lattice to a uniaxial striped lattice, in analogy to a phase diagram proposed for H_2/graphite (Motteler 1986; Motteler and Dash 1986); and a succession of continuous transitions from the commensurate lattice to a reentrant fluid to the uniaxial striped lattice (Halpin-Healy and Kardar 1985; Rabedeau 1989).

transitions allows for vacancies and interstitials in the commensurate lattice. We summarize the work of Schick, Walker, and Wortis (1977).

The model is a special case of the Hamiltonian H_l, eqn (5.1). The interactions are restricted to nearest-neighbor energies $v_{ij} = u(> 0)$ on a triangular lattice. The substrate (holding) potential ϵ_0 is included in the definition of the chemical potential.

The corresponding magnetic Ising model is a nearest-neighbor antiferromagnet on a triangular lattice in a uniform magnetic field. In zero field (or half coverage) this antiferromagnet has the pathology that there is no ordering at finite temperatures. For high enough magnetic field the ground state of the model has completely aligned spins; thus, at large enough chemical potential the lattice is filled. However, for a range of magnetic fields at low enough temperatures, there is antiferromagnetic ordering; this corresponds to one of three interpenetrating sublattices being occupied in the lattice gas.

A feature of antiferromagnets is that small uniform magnetic fields shift the transition temperature but do not destroy the ordering transition. The transition temperature and the critical coverage for the ordering of the lattice gas both vary with the chemical potential. The order of the transition remains continuous for a range of temperatures. The three-fold degeneracy of the ordered state, i.e., the three equivalent sublattices, suggested to the early workers (Alexander 1975) that the ordering transitions should belong to the universality class of the 3-state Potts model.

The real space renormalization group mapping which is constructed for this model is indicated in Fig. 5.3. Triads of spin on each sublattice, as shown, are mapped onto a single new spin; a 3-cell cluster with 9 spins is used to construct the recursion relations. The choice of this mapping is guided by the requirement that the scaled lattice preserve the symmetric three-sublattice (A, B, and C) structure of the original problem. The initial Hamiltonian, with pair interactions and one-body interaction with an external field, maps to a Hamiltonian which additionally has 3-spin (3-body) interactions and the further mappings are approximated by recursion relations for the three coupling constants H (field), K (2-spin), and P (3-spin). The results of the iterated mapping are surfaces in H–K–P space for the locations of the second-order transitions. The phase diagram for the original model is taken from the intersection of the surfaces with the P = 0 plane. An additional calculation determines the magnetization along the phase boundary and thus the critical temperature as a function of density $T_c(n)$ for the lattice gas.

The model has one parameter that is fitted to reproduce the maximum temperature of the order–disorder transition of ^{4}He/graphite. Figure 5.4 shows the region of the maximum; the experimental data were recalibrated to a density scale in which the maximum occurs at $1.01n_c$ (an offset arising from entropic effects). The approximate mapping used by Schick et al. (1977) failed to reproduce the specific heat critical exponent of the 3-state Potts model; it was later shown how to remedy this (Nienhuis et al. 1979).

The continuous order/disorder transition changes to a first-order dilute fluid

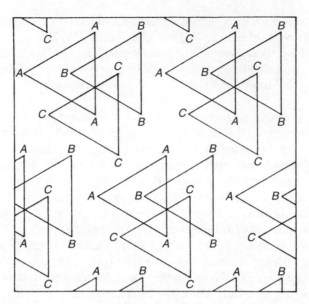

FIG. 5.3. Mapping for renormalization group theory of He/graphite. Grouping into interpenetrating clusters of three sites of the triangular lattice representing equivalent (A,B,C) adsorption sites on graphite, from Schick *et al.* (1977).

to commensurate solid transition at a multicritical point at lower temperatures (Ecke *et al.* 1985). This is reproduced with a lattice gas model which has attractive interactions between next-nearest-neighbor sites (Halpin-Healy and Kardar 1986). That model is similar to the Potts lattice gas model for Kr/graphite.

5.1.3.2 *Kr/graphite* Berker and coworkers (Berker *et al.* 1978; Caflish *et al.* 1985) used lattice gas models designed to mimic various aspects of the monolayer phase diagram of Kr/graphite. The phase diagram for their Potts lattice gas, eqn (5.11), already showed a variety of phase boundaries and multicritical points.* It was devised to model the submonolayer regime, especially the existence of a multicritical point where a high-temperature line of continuous fluid–commensurate solid transitions changes to a low-temperature line of first-order dilute fluid—commensurate solid transitions. Caflish *et al.* (1985) later devised a model for the transitions from the commensurate solid to higher density phases. We now discuss the work for the submonolayer regime.

Berker *et al.*(1978) make two approximate transformations of the atomic-scale description of Kr/graphite. The first, termed a prefacing transformation,

*The correspondence of notations between the Berker–Ostlund–Putnam (BOP) Hamiltonian and H_{PLG}, eqn (5.11), is: $s = 3$, $3\,J_{BOP} = \beta J(1 - \gamma)$, and $3\,K_{BOP} = \beta J(1 + 2\gamma)$, where β is the inverse temperature.

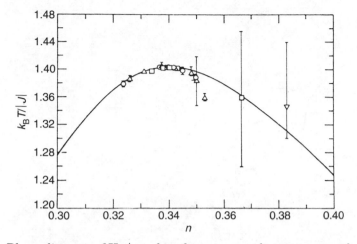

FIG. 5.4. Phase diagram of He/graphite from renormalization group theory, from Schick *et al.* (1977). The order–disorder temperature of the commensurate phase is shown as a function of fractional coverage, where $n = \frac{1}{3}$ would be a perfect commensurate monolayer. The data are for ^{4}He on Grafoil, from Bretz *et al.* (1973), Bretz (1977), and Hering *et al.* (1976).

is a set of arguments to construct the model Hamiltonian. The strong short-range krypton-krypton repulsions are incorporated as an exclusion of nearest-neighbor occupations of graphite adsorption sites and there are Kr–Kr attractive interactions from second-neighbor to fourth-neighbor sites of the graphite lattice.

The commensurate solid preferentially occupies one of the three sublattices A, B, or C shown in Fig. 5.3; the disordered fluid has equal average occupation of each sublattice. By grouping the sites into triads and defining a 3-Potts variable to label which site within the triad is occupied, the nearest-neighbor exclusion within the triad is incorporated into the model. An attractive energy is assigned to occupied nearest-neighbor triads and the concentration of vacant triads is controlled by a chemical potential energy term. The model is thus a 4-state Potts lattice gas with one projection of the Potts variable denoting a vacancy and three projections denoting occupancy of sites of the triad; the lattice of Potts sites is the $\sqrt{3}R30°$ superlattice of graphite adsorption sites. The model reduces to an Ising lattice gas and to the Potts model for special choices of the parameters in the Hamiltonian.

The actual mapping used by Berker *et al.* to recursively eliminate small length scales consisted of a reassignment of the bond strengths of some nearest-neighbor interactions to other bonds followed by a decimation in which the states of some of the sites were summed explicitly in the partition function. Each application of the mapping doubled the length scale of the triangular lattice of

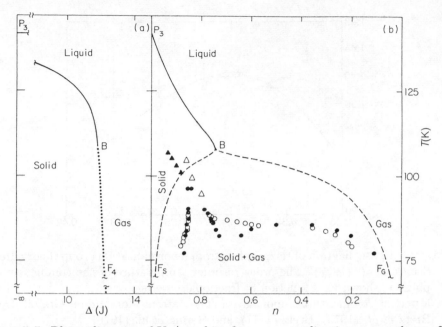

FIG. 5.5. Phase diagram of Kr/graphite from renormalization group theory for a Potts lattice gas, from Berker *et al.* (1978). The left panel presents the chemical potential–temperature plane; the dotted line indicates a first-order transition. The right panel shows the density–temperature plane, with a dashed line denoting the corresponding coexistence curve. P_3B is a line of Potts critical transitions in this model. Circles (triangles) denote experimental values for the first-order (continuous) transition line.

Potts sites. The recursion relations for the coupling constants were then analyzed for the evolution under iterations.

The resulting phase diagram (Fig. 5.5) does indeed have regions that fit ideas of commensurate solid, liquid/dense fluid, and gas/dilute fluid for the model. However, there is a qualitative difference in the connections of these regions for different values of γ in H_{PLG}, eqn (5.11). The solid is separated from the liquid by a line of Potts tricritical points and from the gas by a line of first-order transitions. For $\gamma < 0.5$, the liquid and gas phases join smoothly with no intervening phase transition, while for $\gamma > 0.5$ they have an intervening line of first-order phase transitions ending in an Ising critical point. Berker *et al.* (1978) assigned a value $\gamma = 0.27$ for the Kr–Kr interactions, so that their model Kr/graphite monolayer has no gas–liquid–solid triple point. This is a novel result that agrees with interpretations of the experimental data, Section 6.1.5. The submonolayer density–temperature phase diagram for this Potts lattice gas has a striking similarity to the experimental data.*

The Potts lattice gas model has a maximum occupation when all the Potts sites are filled. To treat the case of coverage greater than the commensurate density n_c, the exclusion of nearest-neighbor site occupation is relaxed. The objective for this *helical Potts lattice gas* is to mimic the situation in which large commensurate domains meet at domain walls (Caflish *et al.* 1985). The *helicity* enters because of the need to distinguish the energies of heavy and superheavy domain walls. Among the results of the analysis are the maximum temperature of the commensurate melting transition and the phase boundaries between commensurate solid, reentrant fluid and incommensurate solid at lower temperatures.

5.2 Continuum systems

A characteristic of the physically adsorbed layers, in contrast to most chemically adsorbed systems, is that a continuous variation of the adlayer solid lattice can be achieved experimentally. That is, these layers are compressible, highly compressible in the cases of helium and molecular hydrogen. The lattice constant variations with temperature and pressure reflect the monolayer equation of state and may be used to infer information about adsorption-induced interactions. When carried to the bilayer and trilayer regime, such studies give structural information important for understanding the initial stages of epitaxial growth. Further, phenomena which depend on the competition between the intrinsic ordering of the 2D solid and the periodicity set by the substrate may change greatly as the adlayer lattice constant varies. We present the theoretical formulation of the monolayer equation of state in Section 5.2.1, using the terminology for the lattices developed in Chapter 3, and give a summary of several approximation techniques. Two examples are presented in Section 5.2.2.

5.2.1 *Approximation methods for monolayers*

We outline several methods to calculate the properties of monolayers. Most of the analysis here is for solids in which the atomic displacements from the average lattice sites are small. The relative sizes of typical displacements and the lattice constant may be used to arrange a hierarchy of approximations. Quantum solids (e.g., helium and molecular hydrogen) have such large relative displacements that quite different methods are needed. To present that theory would take us some distance from the main development of the monolayer theory; we make contact with it only to validate some of the approximation methods used for

---

*Later work recovered qualitative features of the phase diagram with a Kikuchi cluster approximation (Osório and Koiller 1985) and with a density functional approximation (Sander and Hautman 1984).

semiclassical systems such as neon.*

After a discussion of the classical ground state, the state of minimum potential energy, we introduce the quasi-harmonic and cell approximations for the solid. Then follows a summary of methods, including computer simulations, applied to highly anharmonic or disordered states. Quantization of the normal modes of oscillation, the quasi-harmonic theory, incorporates most of the quantum effects for monolayer solids at low temperatures, while the Wigner–Kirkwood semiclassical corrections suffice at high temperatures. The Wigner–Kirkwood terms also give quantum corrections to the virial coefficients of the monolayer gas and to the liquid (Hirschfelder *et al.* 1964; Phillips and Bruch 1979; Glandt 1981).

5.2.1.1 *Static lattice sums* The most primitive approximation to the energy of a monolayer solid is to obtain the classical zero-temperature energy by minimizing the static potential energy of the solid in the presence of the substrate

$$\Phi = \sum_i V(\mathbf{R}_i) + \sum_{i<j} \phi(\mathbf{R}_i, \mathbf{R}_j) \ , \tag{5.43}$$

with an assumption of pair potential interactions among the adsorbed atoms or molecules. As a computational problem, this involves the optimization of many parameters in a nonlinear search. It may be a major task to minimize Φ as a function of structural parameters for a solid with many sublattices, i.e., a Bravais lattice with many components to the basis (Gooding *et al.* 1983; Gordon and Lançon 1985). If the lateral periodicity of V is neglected (i.e., retain only V_0 in eqn (3.9)), the minimization of Φ for an inert gas monolayer solid leads to a triangular lattice with a nearest-neighbor spacing $\sim 2\%$ larger than that which minimizes the potential energy of the 3D solid (Bruch *et al.* 1976a).

The corresponding analysis for the energy of a monolayer solid of a linear molecule must consider lattices with basis sets of several molecular orientations (Belak *et al.* 1985; Kuchta and Etters 1987a,b; Etters *et al.* 1980; Pan *et al.* 1982; Flurchick and Etters 1986). The zero temperature enthalpy can be constructed after evaluating the derivative of this energy with respect to area, with optimized values for the internal degrees of freedom of the layer. Solid-to-solid transitions between such lattices are investigated by comparing static approximations to the enthalpy (A is the area per molecule)

$$H_s = \Phi + \pi_s A \tag{5.44}$$

as a function of the static spreading pressure

$$\pi_s = -d\Phi/dA \ , \tag{5.45}$$

---

*The most complete treatment of zero-temperature 2D helium is that of Whitlock *et al.* (1988). Calculations for a commensurate helium monolayer on graphite at finite temperatures were initiated by Abraham and Broughton (1987) and have been extended by Broughton and Abraham (1988) and by Abraham *et al.* (1990).

for different structures. Symmetry relations on the sublattices may reduce the number of parameters in the search. For example, there are three internal degrees of freedom (two angles and a perpendicular height z of the molecule above the substrate) for the rectangular herringbone lattice of linear molecules. Another set of examples involves the determination of monolayer completion by a transition from monolayer solid to bilayer solid (Bruch *et al.* 1979, 1985; Kuchta and Etters 1987a); a related issue for the compressed monolayer solid is elastic stability (Phillips 1995).

The optimization becomes a truly massive task when the periodicity of V is included and the monolayer solid has a small misfit relative to a commensurate lattice. The solid then is composed of both extended regions, domains, where the atoms/molecules are in the commensurate sites and limited regions, known as discommensurations, walls, or misfit dislocations, where they are not, as discussed in Section 3.4. The configuration of such a solid is specified by a Bravais lattice with a very large unit cell; one must determine the positions of the many atoms within the basis. In practice, a relaxation search locates the positions of atoms in the 'walls' between the domains. The resulting atomic structure of these walls is used to calculate the form factor $F(\mathbf{k})$ entering in the interference function $S(\mathbf{k})$ for the diffracted intensity

$$S(\mathbf{k}) = N_s^2 |F(\mathbf{k})|^2 \ , \tag{5.46}$$

at a reciprocal lattice vector $\mathbf{k} = \tau$ of the Bravais lattice of N_s sites with N_b basis atoms

$$F(\mathbf{k}) = \sum_{j=1}^{N_b} \exp(i\mathbf{k} \cdot \mathbf{R}_j) \ . \tag{5.47}$$

Model solutions for inert gases on the basal plane surface of graphite are discussed in Section 5.2.2.

5.2.1.2 *Small amplitude distortions* The harmonic approximation to a triangular monatomic lattice was developed in Section 4.2.3 using a transcription to normal modes. Here, we discuss generalizations to more complex monolayer solids and to cases where the displacements increase beyond the scope of harmonic theory. Finally, we use the normal mode coordinates in a perturbation theory of the response of a homogeneous monolayer solid to the periodic components of the holding potential.

The normal mode problem for a complex solid of atoms or molecules interacting by central pair potentials and under the influence of the laterally averaged portion $V_0(z)$ of the holding potential differs from the development described in Section 4.2.3 primarily in the book-keeping requirements for the additional centers in the molecule or in the basis of the lattice (Luty *et al.* 1980; Cardini and O'Shea 1985; Roosevelt and Bruch 1990; Janssen *et al.* 1991). We describe now the considerations that arise in using these methods to give a quantitative account of the harmonic lattice dynamics.

Sum rules for the frequency spectrum provide a cross-check on detailed work. In fact, even-power moments of the frequency spectrum can be constructed without explicitly solving the eigenvalue problem (Wheeler and Gordon 1969; Blumstein and Wheeler 1973). An important special case is the second moment for the d-dimensional dynamical matrix, working from eqn (4.58):

$$\overline{\omega^2} \equiv \sum_{\mathbf{k},\nu} \omega_\nu(\mathbf{k})^2/(dN_s) = [V_0'' + \sum_{R_j \neq 0} \nabla^2 \phi(R_j)]/(md) \ . \tag{5.48}$$

The separation of variables perpendicular (z) and parallel ($\|$) to the plane for a monolayer Bravais lattice with a 1-atom basis was discussed for eqn (4.61). Then the second moment of the frequency spectrum for vibrations parallel to the plane is used to define an Einstein frequency

$$\omega_E^2 = \sum_{\mathbf{r}_j \neq 0} \nabla_\|^2 \phi/(2m) \ , \tag{5.49}$$

where the notation indicates that the 2D-Laplacian is used. For the planar monolayer the substrate term V_0'' usually is the dominant term on the right-hand side of eqn (4.61) and the second term is small, because first-order variations in the z-coordinate lead to only second-order variations in the interparticle spacing. Thus even if the structure of the monolayer is determined largely by the adatom–adatom potentials ϕ, the frequencies ω_z have only a small amount of dispersion (Gibson and Sibener 1988a).* The Einstein frequency enters also in the cell model approximation.

The small-amplitude vibration theory is most extensively used for inert gas solids. It has been applied also to low temperature orientationally ordered molecular solids. Then there are additional vibrational degrees of freedom, librations, associated with the molecular orientation (Cardini and O'Shea 1985). For molecules such as N_2 and CH_4 the restoring forces for the librations are rather small and the molecular solid becomes a plastic solid, an orientationally disordered phase, at low levels of thermal excitation. The harmonic approximation then has limited utility. However, orientationally ordered 3D solids of CO_2 and CF_4 persist almost to the melting temperature.

Special points summations   The harmonic approximation to the Helmholtz free energy was stated in eqn (4.62). Although it may be evaluated by a direct summation over the first Brillouin zone, there is an elegant and computationally efficient alternative that is based on the periodicity expressed in eqn (4.60). A function $f(\mathbf{k})$ with that periodicity can be expanded in a Fourier series

$$f(\mathbf{k}) = f_0 + \sum_{m=1}^{\infty} f_m A_m(\mathbf{k}) \ , \tag{5.50}$$

---

*See Fig. 1.2. Evidently this becomes less true in the case of a 'soft' holding potential that has a very wide potential minimum (E. Cheng *et al.* 1993a). Also, the molecular solids show more dispersion than the inert gas solids for which this argument is most pertinent.

where the function $A_m(\mathbf{k})$ is a sum over equivalent vectors of the direct lattice, i.e., vectors of equal length that are related to each other by symmetry operations of the direct lattice:

$$A_m(\mathbf{k}) = \sum_{|\mathbf{R}|=R'_m} \exp(i\mathbf{k} \cdot \mathbf{R}) . \tag{5.51}$$

The index m labels successive shells of neighbors in the Bravais lattice. The utility of the expansion in eqn (5.50) is based on the property that the amplitudes f_m for a smooth function $f(\mathbf{k})$ decrease rapidly as m increases. To apply this to the sum in eqn (4.62), note that the sum of $A_m(\mathbf{k})$ over the Brillouin zone vanishes for all $m \neq 0$, and so

$$\sum_{\mathbf{k}} f(\mathbf{k}) = f_0 N_s , \tag{5.52}$$

where N_s is the number of Bravais cells in the lattice. The art of the method of special points summation (Baldereschi 1973) is to identify small sets of $\mathbf{k}_i$ and weights p_i for which

$$\sum_i p_i A_m(\mathbf{k}_i) = 0 \tag{5.53}$$

for many shells m. Chadi and Cohen (1973) and Cunningham (1974) present recipes for constructing larger sets $(\mathbf{k}_i, p_i)$ once a set has been found that zeroes the sum in eqn (5.53) for some m-values. For several Bravais lattices there are tables of the wavevectors and weights that zero the sum to at least the tenth shell of neighbors. The Brillouin zone sum for the free energy can be performed with good accuracy with fewer than 50 special points in place of a quadrature with 1000 mesh points. This may produce a great reduction in labor if the dimension of the dynamical matrix is large. Additional care is needed for evaluating of a property, such as the low temperature specific heat, which primarily samples the Brillouin zone center (Hilliker and Hakim 1990).

Self-consistent phonons   The expansion of the potential energy to second order in the displacements is inadequate when the displacement amplitudes arising under thermal excitation or other driving forces are not small. This may occur at high temperatures or if the average lattice is so dilated by large quantum zero point energies that the harmonic force constants are small. In both circumstances, a vibrational frequency is determined more by force constants that are averaged over the atomic positions than by the force constants evaluated at the average positions. Such a formulation leads to self-consistent calculations (Werthamer 1976; Klein and Koehler 1976; Glyde 1995) for the frequencies and to generalizations of the expression for the zero-point energy. A further refinement is to include short-range correlations in the calculations; this is the *t-matrix* formulation used by Glyde and Khanna (1971) and by Glyde and Goldman (1976). The applications to monolayer solids are for high levels of thermal excitation by Novaco (1979), Moleko *et al.* (1986), and Hakim *et al.* (1988) and for quantum solids with large zero-point motions by Hakim and Glyde (1988) and Novaco (1992). Iterative solution of the self-consistency equations appears to suffice even for monolayer solids of molecular hydrogen (Novaco 1992).

Phonons by molecular dynamics There are a few molecular dynamics calculations of the spectral density at a given wavevector for thermally excited classical monolayers (Marchese *et al.* 1984; Cardini *et al.* 1985). This is the nominally exact solution of the main problem addressed in the self-consistent phonon theory. A closely related type of calculation is to evaluate the Van Hove dynamic response function $S(\vec{q}, \omega)$ (see, e.g., Pines and Nozières 1966) that enters in the analysis of energy losses for inelastic neutron scattering by a dense target. For a nearly harmonic solid, the location and width of the energy loss peak at given $\vec{q}$ give the frequency and lifetime of the 'normal mode'. For the solid at high thermal excitation, this provides a generalization of the concept of elementary excitations of the many-body system. The dynamic response function has been evaluated in molecular dynamics simulations of adsorbed nitrogen (Cardini and O'Shea 1985; Hansen and Bruch 1995) and of adsorbed krypton (Shrimpton and Steele 1991). While it appears to be straightforward to extract information about the frequency from the results, there are subtleties in the analysis of the intensities, especially for modes that are long-lived on the time scale of the simulation (Shrimpton and Steele 1991).

Hybridization It is assumed in most model calculations of the vibrational spectra of physisorbed layers that the role of the substrate is to provide a static external potential for the adatoms and thus to provide a support for the adlayer. The reasoning is that the substrate is so much stiffer, with larger force constants and higher frequencies, than the adlayer that the respective normal modes do not mix. However, mixing does occur for wavevectors near the Brillouin zone center; there the frequency of perpendicular vibrations of the adlayer becomes comparable to the frequency of the substrate Rayleigh wave. This is manifested both by an anomalous linewidth for inelastic atomic scattering and by distortion of the frequency-wavevector dispersion relation (Gibson and Sibener 1985; Kern *et al.* 1987b; Toennies and Vollmer 1989; Cui *et al.* 1992). The effect has been demonstrated in calculations of physically adsorbed layers on graphite (De Rouffignac *et al.* 1981), on metals (Hall *et al.* 1985, 1989), and on ionic crystals (Hoang and Girardet 1991). There is also radiative damping of adlayer normal modes near the Brillouin zone center, which become surface resonances in a more complete theory (Hall *et al.* 1985) when the substrate dynamics are included.

The Novaco–McTague rotation The solutions of the eigenvalue problem eqn (4.59) can be used in a perturbation theory calculation of the deformation of the adlayer driven by the substrate corrugation, i.e., by the periodic components V_g of the adatom–substrate potential.

Let Φ_0 be the ground state potential energy found for eqn (5.43) with the amplitudes V_g omitted from V; this coincides with the model treated in Section 4.2.3. We now calculate a shift $\Delta\Phi_0$ that is proportional to V_g^2. It arises from a modulation of the adlayer by the substrate, which is sometimes called a *mass density wave* to emphasize the analogy to charge density waves in 3D solids.

We treat the case of a triangular monatomic lattice on a triangular surface lattice. This is the case originally examined by Novaco and McTague (1977a,b) in their treatment of an inert gas solid on the basal plane surface of graphite. Generalizations have been made for a molecular adsorbate, for an adlayer lattice with a basis, and for adlayer and substrate with different Bravais lattices. The additional features which then arise are sketched at the end of this subsection.

Write the potential energy corresponding to eqn (5.43) as

$$\Phi = \Phi_0 + N_s V_0' v_z(0) \; + \; \frac{1}{2} \sum_k [\mathbf{v(k)}^* \cdot \mathbf{D(k)} \cdot \mathbf{v(k)}]$$
$$+ \; \sum_{j,\mathbf{g}} V_g(z_j) \exp(i\mathbf{g} \cdot \mathbf{r}_j) \, , \tag{5.54}$$

using the potential energy minimum Φ_0 for the adlayer on a smooth surface and the corresponding dynamical matrix $\mathbf{D}$, eqn (4.58). That potential energy has been expanded to second order in displacements from the equilibrium positions $\mathbf{R}_j$ in the undistorted lattice using

$$\mathbf{r}_j = \mathbf{R}_j + \mathbf{v(0)} + (1/\sqrt{N_s}) \sum_k \mathbf{v(k)} \exp(i\mathbf{k} \cdot \mathbf{R}_j) \, . \tag{5.55}$$

We retain only the first shell of substrate reciprocal lattice vectors in the V_g series. With the choice of origin made in eqn (3.2) the V_g are real numbers[*] that depend on the magnitude of $\mathbf{g}$. The series in eqn (5.55) corresponds to the running wave amplitudes of eqn (4.55) and provision is made for a uniform translation $\mathbf{v(0)}$ of the adlayer relative to the substrate. For an incommensurate adlayer, the expansion to first-order in $\mathbf{v(k)}$ of the last (V_g) term on the right-hand side of eqn (5.54) is

$$\delta\Phi_g = \sqrt{N_s} \sum_\mathbf{g} \exp(i\mathbf{g} \cdot \mathbf{v(0)})[V_{g_o}(z_0)(i\mathbf{g} \cdot \mathbf{v(-g)}) + V_{g_o}' v_z(-\mathbf{g})] \, . \tag{5.56}$$

The argument of the function $\mathbf{v(g)}$ in eqn (5.56) is taken to be in the first adlayer Brillouin zone, using the same periodicity as in eqn (4.60).

The set of $\mathbf{v(k)}$ values that minimize the resulting quadratic form for Φ is

$$v_z(0) \;=\; -V_0'/V_0''$$
$$v_z(\mathbf{g}) \;=\; -\sqrt{N_s} V_{g_o}' \exp[i\mathbf{g} \cdot \mathbf{v(0)}]/D_{zz}(\mathbf{g})$$

[*]More carefully, this is valid if the V_g are determined by the surface layer of the substrate; using a complex value for the amplitude is a way to distinguish the energies for fcc and hcp stacking sites of an fcc(111) surface (Bruch 1993).

$$v_\parallel(\mathbf{g}) = -i\sqrt{N_s}V_{g_o}\,\mathbf{D}(\mathbf{g})^{-1}\cdot\mathbf{g}\,\exp[i\mathbf{g}\cdot\mathbf{v}(0)]\,, \tag{5.57}$$

using the same separation of z- and $\parallel$-motion as in eqn (4.61). The final energy is then

$$\Phi_f = \Phi_0 + N_s(e_z + e_\parallel),$$

with

$$e_z = -[V_0'^2/2V_0''] - \frac{1}{2}\sum_{\mathbf{g}}V_g'^2/D_{zz}(\mathbf{g})\,,$$

and

$$e_\parallel = -\frac{1}{2}\sum_g V_g^2\mathbf{g}\cdot\mathbf{D}(\mathbf{g})^{-1}\cdot\mathbf{g}\,. \tag{5.58}$$

McTague and Novaco (1979) included thermal averaging and showed that the energies e_z and $e_\parallel$ are independent of temperature in the harmonic lattice approximation.* To complete the reduction of $e_\parallel$, the inverse of the dynamical matrix is written as a spectral decomposition

$$\mathbf{D}(\mathbf{k})^{-1} = \sum_\nu [1/M\omega_\nu(\mathbf{k})^2]\,\hat{\mathbf{u}}_\nu(\mathbf{k})\,\hat{\mathbf{u}}_\nu(\mathbf{k})\,, \tag{5.59}$$

with the normal mode frequencies $\omega_\nu(\mathbf{k})$ and polarization unit vectors $\hat{\mathbf{u}}_\nu(\mathbf{k})$ from eqn (4.59). The scalar product $\mathbf{g}\cdot\hat{\mathbf{u}}_\nu(\mathbf{g})$ vanishes for transverse polarizations in aligned lattices with small misfit. Since the frequency of long wavelength transverse vibrations is less than that of the longitudinal vibrations at a given misfit, the minimum of $e_\parallel$ arises when the lattices are rotated with respect to each other. Thus there is a significant shearing component to the modulation of the adlayer. The energy e_z has only a weak dependence on the orientation because of the small dispersion in the frequency spectrum of perpendicular vibrations of the monolayer.

The total energy Φ_f depends on the orientation of the adlayer relative to the substrate through the dependence on the misfit wavevector $\mathbf{q}$ formed from $\mathbf{g}$, as in eqn (3.16), by translation to the first adlayer Brillouin zone. This dependence, exhibited in Fig. 3.11) is responsible for *orientational epitaxy*, a locking of the orientation of the adsorbate lattice to the substrate. The misfit wavevector varies in magnitude and in direction as the orientation of the adlayer varies relative to the surface lattice. The optimal orientation θ is primarily determined by geometric factors in $e_\parallel$ and is insensitive to the value of V_g in the prefactor. Thus measurements of θ for a physical system have not yielded experimental values of V_g. Greif and Goodstein (1981) calculated the curvature of Φ_f with respect to θ and thereby evaluated a Frank elastic constant that enters in a continuum account of the response of the modulated monolayer, Section 4.3.1.

---

*Abernathy *et al.* (1992) developed a mean-field theory to describe the temperature dependence of the orientational epitaxy of the reconstructed (001) surfaces of gold and platinum.

The solution eqn (5.57) shows that the monolayer has a modulation, i.e., a *mass density wave*. The associated diffraction signature has been observed at the misfit wavevector (Kern *et al.* 1986d, 1987a) and as satellites to the diffraction peaks of the average monolayer lattice (Stephens *et al.* 1979, 1984; D'Amico *et al.* 1984; Faisal *et al.* 1986; Hong *et al.* 1989).

The calculation of Φ_f has been extended to bilayer and trilayer films of Ne/graphite (Bruch and Ni 1985). The energy extremes as the angle θ varies are broader than for the monolayer but are centered at nearly the same angle as the monolayer for bilayers and trilayers of the same lattice constant.

Clearly, the approximation fails when **g** is in the reciprocal lattice of the unperturbed adlayer and frequencies in eqn (5.59) vanish. Another way of expressing the limitation of the theory is that this is a perturbation treatment of the two-dimensional domain wall problem which retains only modulations with the basic misfit wavevector. At very small misfits the modulations have important Fourier components with multiples of the basic misfit vectors; domain wall structures become sharper (Shiba 1979, 1980; Bruch 1985).

Another prominent limitation, that arises in applications to coincidence superlattices, is that Φ_f has no dependence on the uniform component of the displacement, $v_{\parallel}(0)$. The formulation in such a case, a Bravais lattice with a basis, uses perturbation theory to locate the atoms in the basis. At second order, the energy then depends on the displacement of the Bravais lattice by $v_{\parallel}(0)$ relative to the substrate. Such *phase locking* of superlattices was treated by Theodorou and Rice (1978) and has been illustrated for Ne/graphite (Bruch 1988). On the other hand, there is a first-order term from higher amplitudes V_g for which **g** is in the adlayer reciprocal lattice. Thus there may be several significant contributions to the energies that stabilize superlattices with large bases: second-order energies from the distortion of the basis by the leading amplitude V_{g_o} and first-order energies from a higher amplitude V_g that is much smaller than V_{g_o} in magnitude.

The angular coordinates in an orientationally ordered molecular adsorbate are also modulated by the substrate corrugation. It was conjectured that the modulations of the additional degrees of freedom might enhance the observability of the satellites in a diffraction experiment, but model calculations so far do not show this to be a large effect (Bruch 1987).

The formalism has been generalized (Bruch and Venables 1984) to the case of different Bravais lattices for the adlayer and substrate. Examples arise in the adsorption of metals on metals. The approach based on the Fourier decomposition of the holding potential emphasizes coincidences in the reciprocal lattice rather than a coincidence of the rows of one space lattice with those of the other. As with the Novaco–McTague theory, this is a mechanism for an alignment that is not along an elementary symmetry direction and which might otherwise be attributed to some high-order coincidence superlattice. A complication that arises for a triangular adlayer on the (110) face of a body-centered cubic substrate is that more distinct amplitudes V_g must be retained in the series although there is little information on their relative magnitudes.

Novaco extended the formalism beyond the harmonic approximation using self-consistent phonon theory. While the additional effects are small for Ar/graphite (Novaco 1979), they are large for H_2/graphite (Novaco 1992, 1994). Only in a few cases has one gone beyond the level of perturbation theory in determining the modulated structure at small misfits and the orientation relative to the substrate; there is little systematic knowledge derived from the solutions (Shrimpton *et al.* 1988). For instance, the stability of different modulation structures depends on qualitative features such as the sign of V_{g_o} of a triangular lattice (Gottlieb 1990a). The treatment of orientational alignment for systems with well-developed domain walls remains an active area of research (Vives and Lindgård 1993).

As already noted, the Novaco–McTague alignment energy is independent of temperature in second-order perturbation theory. Therefore the observation of a temperature-dependent orientation angle, after correction for thermal expansion of the adlayer lattice, demonstrates that anharmonic processes are making significant contributions.

To summarize, small-amplitude vibration theory and the corresponding approximation for monolayer response to a spatially varying external potential are relatively straightforward theoretical formulations that apply to a domain of experimental data. The extension to self-consistent lattice dynamics, including anharmonic response, is analogous to work on 3D inert gas solids, but because of computational resources now available, it may have reached a higher level of sophistication than the 3D work (Glyde and Goldman 1976; Hakim *et al.* 1988). The harmonic dynamics of the incommensurate layer of small misfit can be cast in the form of lattice dynamics for a very large basis. While cumbersome, it has been used to identify the low-frequency spectra and motions associated with domain wall vibrations (Shrimpton *et al.* 1986; Shrimpton and Joos 1990).

5.2.1.3 *Cell model approximations* The small-amplitude vibration theory presented in Section 5.2.1.2 has the merit that it includes correlations among particle displacements. However, it is unwieldy to extend it to treat the anharmonic corrections. Cell models provide another starting point in which the one-particle anharmonicity is easily included, but at the price of neglecting correlations. They are an elaboration of the Einstein independent oscillator model of a solid. The leading Lennard-Jones and Devonshire cell model is a simply stated approximation that has fair accuracy at an intermediate stage of thermal excitation of a solid (Barker 1963; Hirschfelder *et al.* 1964; Price and Venables 1976; Phillips *et al.* 1981; Phillips 1986). The self-consistent cell theory of Cowley and Barker (1980) is an extension of this approximation that has been applied to the melting of the 2D Lennard-Jones solid (Barker *et al.* 1981).

The formulation of the potential energy for the cell model approximation to a lattice with atoms at average positions $\mathbf{R}_j$ and instantaneous configuration $\mathbf{r}_j$ is

$$\Phi = \Phi_0 + \sum_i \psi(\mathbf{r}_i) + \sum_{i>j} \Delta_{ij} \ . \tag{5.60}$$

The static lattice energy for particles in the holding potential V and with pair potentials ϕ is

$$\Phi_0 = \sum_i V(\mathbf{R}_i) + \sum_{i>j} \phi(\mathbf{R}_i - \mathbf{R}_j) \ . \tag{5.61}$$

The cell potential is defined as

$$\psi(\mathbf{r}_i) = V(\mathbf{r}_i) - V(\mathbf{R}_i) + \sum_{j(\neq i)} [\phi(\mathbf{r}_i - \mathbf{R}_j) - \phi(\mathbf{R}_i - \mathbf{R}_j)] \ , \tag{5.62}$$

and the remainder is

$$\Delta_{ij} = \phi(\mathbf{r}_i - \mathbf{r}_j) - \phi(\mathbf{r}_i - \mathbf{R}_j) - \phi(\mathbf{R}_i - \mathbf{r}_j) + \phi(\mathbf{R}_i - \mathbf{R}_j) \ . \tag{5.63}$$

The Hamiltonian for the elementary cell model of Lennard-Jones and Devonshire is obtained by omitting the remainder terms:

$$H_{LJD} = \Phi_0 + \sum_j [\mathbf{p}_j^2/2M + \psi(\mathbf{r}_j)] \ . \tag{5.64}$$

Since the Hamiltonian H_{LJD} is a sum of one-particle Hamiltonians, the corresponding Helmholtz free energy is

$$F_{LJD} = \Phi_0 - N_s k_B T \ln z, \tag{5.65}$$

with the one-body partition function

$$z = Tr \exp(-\beta[\mathbf{p}^2/2M + \psi(\mathbf{r})]) \ . \tag{5.66}$$

This enables calculations from very low temperatures, where the quantization of motions in the cell potential ψ is dominant, to high temperatures, where the motions are nearly classical. Usually one bypasses the problem of determining the energy spectrum of a 3D nonseparable potential by the additional approximation of making a circular (or cylindrical) average of the cell potential (Phillips and Bruch 1979; Tsang 1979; Phillips 1986).

The semiclassical approximation to the partition function z is stated in terms of the cell free volume v_i, which is an integral of the Boltzmann factor for the cell potential over the volume of the cage formed by the neighbors,

$$\begin{aligned} z &= (1/\lambda^3)\, v_i \ , \\ v_i &= \int_{cell} w_i \exp[-\beta\psi(r_i)] \, dr_i \ , \end{aligned} \tag{5.67}$$

with $\lambda = \sqrt{2\pi\beta\hbar^2/M}$ and the quantum correction factor

$$w_i = \exp\{(\beta\lambda^2/24\pi)[\beta/2(\nabla\psi)^2 - \nabla^2\psi]\} \ . \tag{5.68}$$

The quantum cell model was developed for 3D solids by Levelt and Hurst (1960; Hurst and Levelt 1961). For 2D, Phillips and Bruch (1983) did a full quantum calculation of z by numerically integrating a Schrödinger equation to derive the eigenvalues. They applied it to approximate the ground state of monolayer solids of neon and of methane. Thus they were able to establish temperature ranges where the semiclassical calculation is an accurate approximation to the full quantum theory and where the classical calculation suffices.

An advantage of the cell model approximation is that it is easily adapted to estimate effects of lattice vacancies. This has been done for monolayer and bilayer solids at intermediate temperatures (Wei and Bruch 1981). The most serious defect of the approximation is that it omits the correlations of displacements in neighboring cells. Consequently, the low T specific heat* has an exponential temperature dependence

$$C \propto (\beta\hbar\omega_E)^2 \exp(-\beta\hbar\omega_E) , \qquad (5.69)$$

as in the Einstein model of a solid rather than the T^2 dependence of Debye theory. The Einstein frequency in eqn (5.69) enters in the coefficient of the small amplitude expansion of the cell potential:

$$\psi(r) \simeq (M/2)\omega_E^2 (\mathbf{r} - \mathbf{R}_i)^2 . \qquad (5.70)$$

In practice the cell model approximation for monolayer solids is remarkably effective, just as for 3D solids (Holt *et al.* 1970; Holian 1980). Harmonic lattice dynamics arises as a systematic expansion; thus, it possesses some elegance and also produces the correct power law for the specific heat at very low temperatures. However, the thermal expansion calculated from the quasi-harmonic free energy increases too rapidly with increasing temperature in the intermediate temperature range, as judged by computer simulations. Figure 5.6, based on calculations with parameters for 2D xenon, shows there is a smooth transition from the low-temperature quasi-harmonic theory to the quantum-corrected cell theory, which in turn has a good overlap with the results of a classical Monte Carlo simulation. Thus the cell model gives more realistic thermal expansion at intermediate temperatures, judging by comparisons to experiment and to computer simulations. Further, the cell model internal energy is quite close to the result in simulations. The fact that the uncorrelated cell approximation does this well suggests that terms crucial for determining the phase transitions are a small part of the total energy.

The remainder Δ_{ij} is of fourth order when expanded in powers of the displacements; it is omitted in many applications of the Lennard-Jones and Devonshire cell model. When it is included, the formalism corresponds to the finite temperature generalization of the Hartree variational approximation to the ground state of a quantum solid (Nosanow and Shaw 1962; Ni and Bruch 1986; Bruch 1987). Neon solids have large quantum correction terms in both 2D and 3D.

---

*Hunzicker and Phillips (1986) give a careful study of the temperature dependence of the specific heat in a cell model approximation for CH_4/graphite.

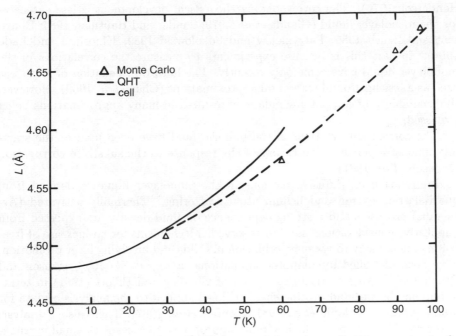

FIG. 5.6. Thermal expansion of 2D Lennard-Jonesium. Lattice constant of a zero spreading pressure (nearly unconstrained) 2D Lennard-Jones triangular lattice as a function of T, computed using parameters for Xe, from Phillips *et al.* (1981). Results obtained with the quasi-harmonic solid theory (full curve), quantum-corrected cell model (dashes), and classical Monte Carlo (triangles) methods are shown.

5.2.1.4   *Monolayer fluids* As for 3D phases, thermal properties of a moderately dense gas can be analyzed with a cluster series in terms of virial coefficients. The second virial coefficient was discussed in Section 4.2.2, where both the classical and quantum calculations were reviewed. The series in mathematical 2D has been evaluated through the fifth virial coefficient for the Lennard-Jones (12,6) potential (Barker 1981) and through the eighth virial coefficient for hard disks (van Rensberg 1993). There are few experimental data to be analyzed in terms of the virial series (He: Goellner *et al.* 1975; Novaco 1975; Schick 1980; CH_4: Elliott *et al.* 1993; N_2, O_2, CO:Bojan and Steele 1987a,b), in part because the low coverages where the series is most applicable are also the coverages where effects of substrate heterogeneity are most troublesome. The virial coefficients for the LJ (12,6) model are used to estimate the 2D gas–liquid critical temperature, see Section 5.2.2.

There is a long history of approximations to the equilibrium correlations of the dense fluid (liquid) phase in 3D systems (Hirschfelder *et al.* 1964; Barker and

Henderson 1976). The corresponding theoretical development is less extensive for the monolayer liquid (Glandt *et al.* 1979; Sander and Hautman 1984; Bonissent and Bruch 1985; Patrykiejew and Sokolowski 1989; Kinoshita and Lado 1994). In part this is because experiments to measure the correlations in the monolayer liquid have come only recently. The hard disk equation of state was used as a testing ground for several approximations (Chae *et al.* 1969). However, the availability of computer simulations resulted in many approximations being bypassed.

The correlation functions of the uniform fluid have been used as the starting point of a perturbation theory of the response to the substrate corrugation (Monson *et al.* 1981).

Information on diffusive motions in the monolayer liquid is derived from quasi-elastic neutron and helium atom scattering. Thermally activated (Arrhenius) processes and a strong dependence on fluid density are expected from qualitative considerations and are observed. Mechanisms for an increase of free-volume, in analogy to vacancy formation in solids and involving layer promotion, have been identified in computer simulations (Etters *et al.* 1993; Hansen and Bruch 1995). An empirical description of Chudley and Elliott (1961) in terms of hopping between adsorption sites has been applied to the analysis of data for fluid methane (Coulomb *et al.* 1981; Bienfait *et al.* 1987). There also is analysis in terms of longer atomic flights, more appropriate to cases with small corrugation (Ellis and Toennies 1993; Sholl and Skodje 1994; De Jong and Jansen 1994). The role of the substrate dynamics in such models is to provide a mechanism for dissipation (Chen and Ying 1993; Persson 1992b, 1993). Dynamic correlations of adlayer and substrate as an important mechanism for enhancing adatom motion have been identified in some experiments for chemisorbed species (Kellogg and Feibelman 1990; Feibelman 1990; Kellogg 1994) and in some simulations (Ehrlich 1994; Black and Tian 1993).

5.2.1.5 *Nonuniform fluids* Approximations have been developed to cover inhomogeneous fluids and their ordering transitions, such as freezing. This subject has been thoroughly reviewed recently by Evans *et al.* (1986); see also Evans (1990) and Henderson (1992). Among the most prominent of the calculational approaches are perturbation theories and density functional theories. Goals of the work include describing freezing of fluids and the layer-like structure in fluid films.

At first sight perturbation theories may not seem well suited for the description of a film, because the film–vapor interface is so extremely inhomogeneous. One way to overcome this obstacle is to assume at the outset that one is concerned only with phenomena within the monolayer that might be adequately described by a two-dimensional description. Sokolowski and Steele (1985a,b) followed this procedure and expanded the film's thermodynamic properties in deviations of the density from that of the uniform system. As an application, they were able to describe the solidification of krypton on graphite. Their methodology bears a similarity to that appearing in the 3D solidification theory of

Ramakrishnan and Yussouff (1979). An important difference is that the latter theory predicts spontaneous symmetry breaking associated with the pair correlation function of an isolated liquid, while the 2D theory incorporates the substrate field. As a consequence, Sokolowski and Steele were able to explore the competing effects of external and intrinsic forces and found a solidification transition that is affected by the substrate in several respects, including an orientational ordering in its field.

The zero-temperature density functional method has been described in Chapter 2 with a focus on electronic properties. The theorem stating that the ground state energy of a system of particles is a unique functional of the density has been extended by Mermin (1965) to cover the case of the free energy at finite temperature. As in the electronic case, the theorem provides a rationale for the density functional technique, but incomplete knowledge of the relevant functional limits the accuracy of the solution of the resulting equations. A key point in most applications to fluids is that a completely local version of the theory, 'local thermodynamics', is totally inadequate to describe the rapid variation of the density at interfaces and in solids. Saam and Ebner (1977) were able to express the free energy functional exactly in terms of the density-dependent direct correlation function of the fluid, i.e., a nonlocal theory. Approximate nonlocal density functionals have been used with some success to describe virtually every type of problem involving inhomogeneous fluids. Among these applications are the bulk liquid–vapor interface (Ebner et al. 1976), prewetting transitions in multilayer films (Ebner and Saam 1977), and capillary condensation (Evans et al. 1986).

While it is possible, in principle, to study monolayer films with these 3D functionals, there have been few such applications. Instead, there have been attempts to apply a 2D version of the theory (Fairobent et al. 1982; Sander and Hautman 1984). To generate a 3D density functional that has a sufficiently accurate compensation in the approximations that it will automatically apply in 2D is a daunting prospect. It might be hoped that the use of the 2D simplification would avoid that challenge and at the same time simplify the calculation because of the smaller 'phase space' in 2D than in 3D. By solving the free energy minimization conditions, Sander and Hautman were able to generate an array of alternative phase diagrams characterized by the individual values of the reduced length and energy parameters. These are simply the ratios of the adsorbate hard core and well depth parameters to the lattice constant and corrugation strength V_g, respectively. This approach is sensible and credible if the assumed density functional is reliable, but comparison with experiments suggests that their results are only qualitatively accurate.

Another subject of extensive investigation with density functional methods is helium at zero temperature; see, e.g., Dupont-Roc et al. (1990). While such approaches have achieved semiquantitative success in their predictions of the structure of the gas–liquid interface (Stringari and Treiner 1987) and the wetting behavior of films (Cole et al. 1994), there remains some uncertainty about their accuracy (Clements et al. 1994). Furthermore, to date these theories have focussed on the liquid near its equilibrium density and do not reliably predict solidification, which occurs at some 50% higher density.

We may summarize the situation in the following way. Density functional methods are potentially valuable in treating monolayers, but there remains the need to assess any hypothetical functional by comparing its predictions with those based on other methods, as well as experiments. We expect that the approximation can be developed to the point that the remaining uncertainty is mostly in the assumed holding potential, especially its corrugation.

5.2.1.6 *Computer simulations* Theorists are prone to regard computer simulations as 'computer experiments'. We give the same level of description here as in the survey of experimental techniques, Section 1.6: an enumeration of the techniques and their capabilities, but with little detail on the implementation. The most extensive applications are to the treatment of single phases which have substantial disorder, such as solids at high thermal excitation and dense fluids.

The simulations group roughly into treatments of equilibrium ensembles (phase space averages with Monte Carlo samplings) and dynamical solutions (molecular dynamics, phase-space trajectory averages). Monographs on the former have been written by Kalos and Whitlock (1986), Binder and Heermann (1992) and Nicholson and Parsonage (1982) and on the latter by Hoover (1986, 1991) and Allen and Tildesley (1987). Time-averaged properties from the two approaches must be equal, the *quasi-ergodic hypothesis*.

A series of concerns and their resolutions arise in the applications to monolayers and their phase transitions.

1. The shape of the simulation cell in a calculation that uses periodic boundary conditions may bias, or even determine, the outcome for crystallization or molecular orderings. Parrinello and Rahman (1980) devised a method to explore many cell shapes during the simulation.

2. The number of atoms/molecules in the simulation cell may be modest, on the scale of hundreds to thousands, compared to the number included in the correlation range for a fluid near a phase transition. Lee and Kosterlitz (1991) developed an extrapolation technique, *finite size scaling*, to estimate the results for very large samples. It has been applied to the hard disk system (Lee and Strandburg 1992). Other studies of size dependence in simulations of 2D solids include Toxvaerd (1983), Udink and van der Elskin (1987), Zollweg *et al.* (1989), Morales (1994), and Bagchi *et al.* (1996).

3. Especially near continuous transitions and near weakly first-order transitions, the running averages in a simulation display substantial 'slow fluctuations' and equilibrium averages are hard to determine. The *multiple histogram method* of Ferrenberg and Swendsen (1989) was devised to cope with this and even to exploit the information contained in the fluctuations. It has been applied to the hard disk system by Zollweg and Chester (1992) and to adsorbed molecular nitrogen by Kuchta and Etters (1993).

4. It frequently is difficult to distinguish first-order phase coexistence from a continuous transition in a simulation in the $(N\text{-}A\text{-}T)$-ensemble, a Monte Carlo calculation at temperature T for a specified number N of atoms in an area A. The analogous problem in 3D experiments is the phase coexistence in a 'closed cell' experiment for a solid at its melting curve. The pressure ensemble, $(N\text{-}p\text{-}T)$, is a variant of the Parrinello–Rahman procedure that has been used to select a single phase near the first-order coexistence (Abraham 1980a, 1986). Another choice is the grand ensemble, the $(\mu\text{-}A\text{-}T)$-ensemble, where the chemical potential μ is taken as one of the thermodynamic variables. This has been applied especially for fluid films by Nicholson and Parsonage (1982); see also Lane and Spurling (1976), Parsonage and Nicholson (1978), and Bottani and Bakaev (1994). A recent development for determining phase coexistence equilibria is the Gibbs Ensemble method (Panagiotopoulos 1995).

The $N\text{-}A\text{-}T$-ensemble ($N\text{-}V\text{-}T$ of 3D) is the traditional Metropolis ensemble (Metropolis *et al.* 1953) and it remains the most widely used of the Monte Carlo simulations. The number of particles in the periodically extended simulation cell depends on the complexity of the interaction models. Typical values for the LJ(12,6) pair potential are a few hundred atoms, but a calculation with $N = 161,604$ was performed for a model of Kr/graphite (Abraham *et al.* 1984). For contrast, some X-ray experiments on xenon monolayers have probed states with correlations lengths of 2000 Å, that have 5×10^5 atoms in the corresponding disk. There is a straightforward but tedious way to construct absolute free energies from the simulation, and hence to determine first-order phase boundaries. The ensemble averages of the pressure and the internal energy are used to integrate differentials for the Helmholtz free energy from the harmonic low-temperature solid and from the dilute high-temperature gas (approximated by virial series). For a model with pair potential interactions, the internal energy and the pressure (via the virial theorem) are evaluated as averages of two-particle correlations; however, the entropy is based on a many-body correlation and is not so easily obtained.

The implementation of molecular dynamics simulations is, in principle, a matter of integrating Newton's equations of motion. The scale of elapsed time now achieved for inert gas models is on the order of 100–1000 ps. Considerations which arise in the implementation are:

1. Discretization of the time derivatives. There is a trade-off of accuracy and computational efficiency (Gear 1971; Allen and Tildesley 1987).

2. Systems with widely different time scales. Simulations of dense phases of molecules are confronted with the fact that intramolecular vibrations have a far shorter period than the time scales of rotational and center-of-mass translational motions. Algorithms of molecular dynamics with constraints have been devised to cope with this (van Gunsteren and Berendsen 1977; Ryckaert *et al.* 1977; Edberg *et al.* 1986).

3. Comparison to Monte Carlo thermal averages. The traditional molecular

dynamics simulation is a microcanonical ensemble average, but there is a variant termed *isothermal molecular dynamics* (Evans *et al.* 1983).

A problem common to both the molecular dynamics and Monte Carlo simulations is how to test whether equilibrium has been achieved in the averagings and whether transients associated with the starting configuration have thoroughly decayed. One test has been to look for hysteresis as a function of temperature changes, although it is recognized that changes from a solid lattice to a fluid with increasing temperature occur more readily than the formation of a crystal upon cooling. This is the issue of superheating *vs.* supercooling. A related comment on monolayer simulations is that phase transitions involving the loss of order seem to occur without much metastability (Etters *et al.* 1990; Roth and Etters 1991; H.-Y. Kim and Steele 1992); that is, using a criterion of single-phase instability to locate a phase transition has had fewer problems than for 3D simulations (Meijer *et al.* 1990). This may well be correlated with the fact that many of the transitions being simulated for monolayers are second-order or weakly first-order phase transitions.

There are several reviews of the application of computer simulations to adsorbed layers (Abraham 1981a, 1982, 1986; Binder and Landau 1989; Persson 1992a). There is also relevant information in the proceedings of recent workshops (Binder 1992b; Strandburg 1992; Landau *et al.* 1993).

5.2.2 *Illustrative examples*

We present two examples to illustrate the application of the methods discussed in Section 5.2.1: Lennard-Jonesium, an ideal 2D system interacting by the Lennard-Jones (12,6) pair potential, and the 2D harmonically modulated solid. Model solutions for these idealizations of the physical monolayer have been good guides to understanding the phenomena observed in many adsorption systems.

5.2.2.1 *Lennard-Jonesium* We assume the monolayer system is represented by atoms moving in a plane and interacting by the Lennard-Jones 12-6, LJ(12,6), pair potential:

$$\phi(r) = 4\epsilon[(\sigma/r)^{12} - (\sigma/r)^6] = \epsilon[(r_0/r)^{12} - 2(r_0/r)^6] \,, \tag{5.71}$$

where the separation r_0 of the pair potential minimum is:

$$r_0 = 2^{1/6}\sigma \,. \tag{5.72}$$

The information on Lennard-Jonesium is presented in the following order: dilute gas, gas–liquid critical point phenomenon, low-temperature solid, and the solid at melting. The parameters ϵ and σ set energy and length scales for the dense phases. Figure 5.7 shows a composite phase diagram for the 2D Lennard-Jonesium, including the limit set by bilayer formation (using parameters for Xe/Ag(111)).

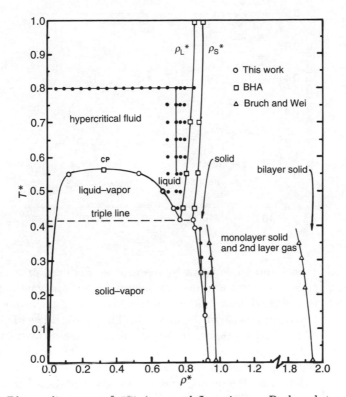

FIG. 5.7. Phase diagram of 2D Lennard-Jonesium. Reduced temperature–density phase diagram for the 2D Lennard-Jones system, from Phillips *et al.* (1981). Circles are results of their computations and squares are results of Barker *et al.* (1981). The triangles denote the coexistence of monolayer and bilayer solid, computed by Bruch and Wei (1980) for Xe/Ag(111) parameters.

Corresponding states It is conventional to express energy, area, density, spreading pressure, temperature, and compressibility for this model in scaled (star) form:

$$
\begin{aligned}
E^* &= E/\epsilon \\
A^* &= A/\sigma^2 \\
\rho^* &= \rho \times \sigma^2 \\
\pi^* &= \pi \times \sigma^2/\epsilon \\
T^* &= k_B T/\epsilon \\
K^* &= K\epsilon/\sigma^2
\end{aligned}
\tag{5.73}
$$

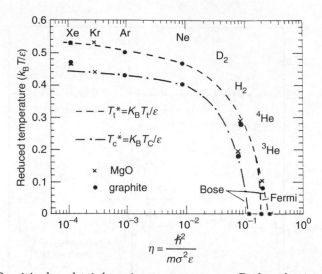

FIG. 5.8. 2D critical and triple point temperatures. Reduced critical temperatures and triple point temperatures measured for various gases on graphite and MgO, as indicated, as a function of the adsorbate's quantum parameter η; note the logarithmic scale of the abscissa. The classical law of corresponding states implies that there are universal values of these two temperatures; this holds for the higher mass cases at the left end of the scale. The data for the lighter, more quantum cases show systematic deviations from the classical results. The asterisks on the abscissa indicate predicted termini of the curves, calculated by Miller and Nosanow (1978) and Ni and Bruch (1986) for fermions and bosons, respectively. The figure is taken from E. Cheng *et al.* (1991a).

The law of corresponding states (Hirschfelder *et al.* 1964) is a systematic development along these lines: when data for several systems are organized in a common scaled form with 'appropriate' choices of σ and ϵ, functional values for another system may be estimated by interpolation in the scaled data. Quantum effects, including isotopic effects for light atoms and molecules, are ordered (De Boer 1948; De Boer and Bird 1964; Nosanow *et al.* 1975; Miller *et al.* 1977; Nosanow 1977) in terms of the de Boer parameter Λ^* defined by

$$\Lambda^* = h/(\sigma\sqrt{M\epsilon}) . \tag{5.74}$$

Figure 5.8 shows an example of such a graph using a quantum parameter $\eta = (\Lambda^*/2\pi)^2$: the 2D critical temperatures for several combinations of adsorbate and substrate.

Table 5.1 *2D gas–liquid critical point parameters for Lennard-Jonesium*[†]

truncation	T^*	ρ^*	π^*/ρ^*T^*	source
B_3	0.615	0.268	0.333	Glandt(1978)
B_4	0.492	0.119	0.377	Glandt(1978)
B_5	0.475	0.100	0.379	Barker(1981)
pert	0.56	0.32	0.25	Henderson(1977)
MC	0.537	0.365	0.189	Reddy and O'Shea(1986)
MC	0.515	0.355	–	Smit and Frenkel(1991)
GE	0.497	0.35	–	Panagiotopoulos(1994)

Virial series and critical point   The leading terms in the virial series for the equation of state of the dilute 2D gas are:

$$\pi^* = \rho^*T^*[1 + B_2^*(T^*)\rho^* + B_3^*(T^*)\rho^{*2} + ...] \tag{5.75}$$

The interaction terms in other thermodynamic functions are constructed from this expansion by using Maxwell relations. For example, the 2D analogue of the relation

$$\partial S/\partial V|_T = \partial P/\partial T|_V$$

leads to an expansion for the entropy of the monatomic gas

$$s = s_{id} + s_2 + ... ,$$

where

$$s_{id}/k_B = 2 - \ln(\rho\lambda_T^2)$$

and

$$s_2/k_B = -\rho \times \frac{d}{dT}(TB_2) . \tag{5.76}$$

As discussed in Section 4.2 there are systematic procedures to evaluate B_2 over the whole temperature range from extreme quantum (low T^*) to classical (high T^*) conditions. Phase shift analysis (Siddon and Schick 1974a) and path integrals (Guo and Bruch 1982) have been applied for helium in mathematical 2D. The temperature dependence of B_2 is closely related to the ordering effect of interactions that is reflected in the entropy. In classical statistical mechanics it is easy to prove that s_2 is negative; this corresponds to a basic theorem of information theory that the maximum entropy is achieved for the most uniform phase space distribution (Baierlein 1968). The coupling of the momentum and position phase space variables in quantum physics disrupts the argument. In

---

[†]Results for LJ(12,6) potential. 'Pert' denotes results of a perturbation theory. 'MC' denotes results derived from Monte Carlo simulations, while 'GE' denotes the result of a Gibbs Ensemble Monte Carlo determination for potential cutoff at $r_c^* = 5.0$.

fact, the second virial coefficient for 2D LJ(12,6) helium has a range of low temperatures where s_2 is positive (Bruch *et al.* 1976b; Lim 1985).

The tabulated virial coefficients (Morrison and Ross 1973; Kreimer *et al.* 1973; Glandt 1978; Barker 1981) can be used to estimate the gas–liquid critical point by constructing a truncated density series from eqn (5.75) and determining the onset temperature for a van der Waals loop. The equations to be satisfied are

$$\partial \pi^*/\partial \rho^* = \partial^2 \pi^*/\partial \rho^{*2} = 0 \ . \tag{5.77}$$

The first approximation uses the second and third virial coefficients.

Table 5.1 shows the evolution of the results with higher-order virial coefficients and includes results from a perturbation theory (Henderson 1977) and from two computer simulations (Reddy and O'Shea 1986; Smit and Frenkel 1991). Henderson applied the 2D version of Barker–Henderson liquid state perturbation theory and found that the apparent convergence of the perturbation series for the 2D liquid is slower than for the 3D liquid. Several other groups also tried to determine the gas–liquid critical point parameters with Monte Carlo and molecular dynamics simulations (Fehder 1969; Tsien and Valleau 1974; Knight and Monson 1986; Rovere *et al.* 1993). Smit and Frenkel (1991) reviewed the situation: their experience was that the task is much more delicate than for the 3D case and that truncation of the range of the pair potential, a step usually considered to be a convenience but not a crucial ingredient, has major effects on the results. This is evident in the subsequent calculations reviewed by Panagiotopoulos (1994, 1995).

The table shows that the truncated virial series gives a fair estimate of the critical temperature and a poor estimate of the critical density. For reference, the ratio $\pi^*/\rho^* T^*$ is 0.375 for the gas–liquid critical point of the van der Waals equation. The approximation to the second virial coefficient in Section 4.2.2.1, eqns (4.43) to (4.45), gives $T_c^* = 0.356$ and $\rho_c^* = 0.212$. The critical temperature T_c^* for 3D Lennard-Jonesium is estimated from Monte Carlo simulations to be in the range 1.32–1.36 (Smit 1992; Barker and Henderson 1976; Verlet 1967; Levesque and Verlet 1969).

Thermally excited capillary waves are expected to cause relatively large fluctuations at the liquid–vapor interface in 2D. This has been demonstrated in computer simulations (Abraham 1980b; Sikkink *et al.* 1985; L. J. C. Chen *et al.* 1991).

Ground state of the 2D solid   The 2D solid of Lennard-Jonesium has been extensively studied for the development of computational methods on monolayer solids and as a possible example of dislocation-mediated melting, Section 4.3. Strandburg (1988) reviewed the simulations of melting. We begin with a summary of information on the ground state of 2D Lennard-Jonesium as a function of the De Boer parameter Λ^*.

The nearest-neighbor spacing in the ground state of the classical LJ(12,6) lattice is (Bruch *et al.* 1976a)

$$L_0 \ = \ 0.99019...r_0 \quad \text{2D triangle}$$

Table 5.2   *Properties of the ground state
of 2D solids using Lennard-Jones models for
the inert gases.*[†]

System	Λ^*	L_0/r_0	E^*	K^*
	0.	0.9902	−3.382	0.0176
Xe	0.0628	0.9956	−3.25	0.0181
Ar	0.186	1.007	−3.01	0.0223
Ne	0.579	1.051	−2.25	0.0382
H_2	1.72	1.23	−0.608	0.126

$$L_0 = 0.97123...r_0 \quad \text{3D fcc} . \tag{5.78}$$

The separations are smaller than $L_0 = r_0$ because of the attractive energy of pairs of atoms which are not nearest neighbors, an effect that is larger in the 3D packing.

In the harmonic lattice approximation, the ground state lattice constant L_0 is the value that minimizes the sum of the potential and zero-point energy

$$E^* = \frac{N}{2}[A_{12}(r_0/L)^{12} - 2A_6(r_0/L)^6] + (\hbar/2\epsilon)\sum_{\mathbf{k},\nu}\omega_\nu(\mathbf{k}) , \tag{5.79}$$

with lattice sums $A_{12} = 6.009814$ and $A_6 = 6.375882$ for the triangular lattice (Zucker 1974). The values of L_0/r_0 and E_0^* for several values of Λ^* are given in Table 5.2. The scaled lattice constant and compressibility are increasing functions of the zero point energy. However, for $L/r_0 > 1.102$ some of the harmonic frequencies become imaginary because the atoms sit at local maxima in the potential field of their neighbors (De Wette and Nijboer 1965; Wei and Bruch 1981); this approximation to the ground state certainly fails for $\Lambda^* > 0.704$. The application of quasi-harmonic theory to the ground state of the neon lattice was validated by comparison to the results of Hartree and Jastrow variational calculations (Ni and Bruch 1986). The entry for H_2 in Table 5.2 is the result of a quantum mechanical calculation with a Jastrow correlated wave function.

At larger de Boer parameters the ground state is a liquid rather than a solid (Liu *et al.* 1976); for Bose statistics and the Jastrow trial function this occurs at $\Lambda^* > 2.16$ (Ni and Bruch 1986; Whitlock *et al.* 1988). For even larger Λ^*, the ground state is not self-bound; the threshold for self-binding of many bosons in 2D that is obtained from a Jastrow trial function is $\Lambda^* = 3.27$ (Miller and Nosanow 1978; Bruch 1978). This is also the threshold for a bound dimer in 2D. See Fig. 5.8, where these values provide zeroes for T_i^* as a function of Λ^*.

[†]The energy E^*, compressibility K^*, and nearest-neighbor spacing L_0 of the ground state of the triangular lattice are given for several values of the de Boer parameter Λ^*.

2D thermal expansion and sublimation curve   The monolayer thermal expansion is a goal of many calculations because the most precise data with single crystal metal surfaces are the structural properties derived from diffraction experiments rather than thermodynamic properties. The so-called *unconstrained solid* is the state of a submonolayer solid in equilibrium with 2D gas, i.e., at the sublimation curve. The thermal expansion of the near-classical solid along its sublimation curve has been approximated in several ways. The density and pressure of 2D gas in coexistence with the solid are quite low; π^* at the classical triple point is estimated (Barker *et al.* 1981) to be 0.006. Therefore, to a good approximation, the lattice constant $L_0(T)$ at the sublimation curve is the solution of

$$\partial F/\partial L|_T = 0 \;, \tag{5.80}$$

where F is the Helmholtz free energy. Equation (5.80) is a condition of zero spreading pressure and this is the reason for saying the solid is effectively unconstrained. The solution to eqn (5.80) with xenon parameters was shown in Fig. 5.6.

There is good agreement between values for the internal energy at the sublimation curve from Monte Carlo calculations and from the cell model (Phillips *et al.* 1981); discrepancies are the order of 1%. On the one hand this is encouraging for the utility of the simple approximations. On the other hand, it does not eliminate reservations having to do with thermal equilibration of the simulation: there are few thermally excited defect structures in the simulation and the concern has been that the time scale for them to equilibrate is longer than the effective duration of the simulation (Nelson 1983).

The entropy of the solid along the sublimation curve has a somewhat counterintuitive behavior. When the entropies at the sublimation curve of the quasi-2D solid of Xe/Ag(111) and of 3D–Xe are compared at the same temperatures, there is a considerable range where the monolayer has the higher entropy per atom (Phillips and Bruch 1985). This might be surprising because of the ordering effects anticipated from confinement to the monolayer. However, the temperature power laws for the entropy from Debye lattice theory ($\sim T^D$ where D is the dimension) confirm that it is the required ordering at low temperatures. At intermediate temperatures, the ordering involves the entropy associated with monolayer vibrations perpendicular to the plane and therefore depends on details of interaction models.*

The gas density along the sublimation curve also provides a surprise. For a nearly classical 3D inert gas at the sublimation curve and near the triple point temperature, the product of the second virial coefficient and the gas density, $-\rho B_2$ is approximately 0.025 and the ratio of the gas density to the solid density is about 0.0024. The corresponding product for the 2D Lennard-Jones gas is about 0.15 and the ratio of gas to solid density is 0.02. The prospect for

---

*For ideal gases at the same temperature and particle spacing, the entropy of the 3D gas is greater than that of the 2D gas, as expected from phase space arguments.

physically adsorbed gases is even more extreme, as discussed by Unguris *et al.* (1982): because the leading term in the gas chemical potential depends only logarithmically on density, 10% reductions in the effective interactions in the condensed phase can lead to drastically increased densities of the monolayer gas at the sublimation curve. The sensitivity of the gas density to the interaction model is illustrated in Fig. 5.9, which shows calculations for the Xe/Ag(111) system. The LJ(12,6) xenon gas is at a density for which the logarithmic chemical potential is still sharply rising; however, the multiparameter xenon model has an estimated gas coexistence with the solid in a density range where the 2D gas chemical potential increases only slowly with increasing density.

Melting Computer simulations on samples of more than 200 particles show defect structures, such as dislocations, in the solid near melting (Cotterill and Pederson 1972). In fact, in spite of the debates on the mechanism of the melting and differences as to whether the transition is first-order or continuous, the various workers are in good agreement on the temperature range where the transition(s) occurs (Toxvaerd 1978; Strandburg 1988).

A free energy construction for the liquid and solid, in analogy to the 3D constructions, leads (Barker *et al.* 1981) to values for the classical triple point of $T_t^* = 0.415$ and $\pi_t^* = 0.0056$ and an entropy of fusion $\Delta s = 0.8 k_B$. Another determination (Phillips *et al.* 1981) by extrapolation of the sublimation and vaporization curves to an intersection gave $T_t^* = 0.40 \pm 0.01$ and $\Delta s/k_B = 0.68 \pm 0.13$, with a solid density $\rho^* = 0.85$, i.e., $L/r_0 = 1.04$. Van Swol *et al.* (1980) determined the fusion entropy to be 0.49 at the melting curve at $T^* = 0.8$; this value is much smaller than the value 1.75 for the 3D solid at the same temperature (Hansen and Verlet 1969). For comparison the triple point temperature for 3D Lennard-Jonesium is (Hansen and Verlet 1969) $T_t(3D)^* = 0.68 \pm 0.02$ and the entropy of fusion is 1.8–1.9 k_B. Locating the melting transition by such constructions has an inherent bias toward first-order transitions because accumulated errors in the free energy constructions may differ in the two phases.

There is evidence that the order of the melting transition changes with increasing pressure (K. Chen *et al.* 1995). At low 2D pressures Barker *et al.* (1981) find the transition is first order. The density difference between the coexisting liquid and solid is expected to decrease with increasing pressure. Chen *et al.* find an apparently continuous transition at high pressure ($\pi^* = 20$) at a temperature that is roughly on the extrapolation of the Barker *et al.* results. However, the chemical potential is then so high that further layer condensations would be expected to intervene for a physisorbed monolayer.

The theories of dislocation-mediated melting focus attention on the correlation functions of translational and orientational order, Section 4.3. Computer simulations with several thousand atoms in the cell have followed the orientational order correlation function through the apparent phase transitions. In one (Bakker *et al.* 1984), a large but finite correlation length for exponential decay of order in the fluid phase was found. In another, a comparative study was made of 2D Lennard-Jonesium and of hard disks (Strandburg *et al.* 1984) and no qualitative difference between the two systems was found. Although the hard disks

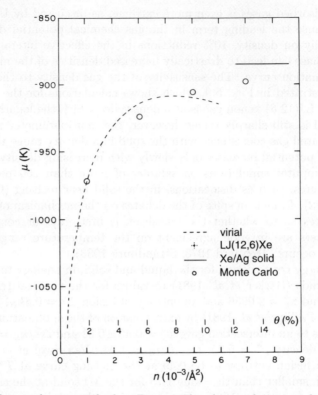

FIG. 5.9. Chemical potential of a 2D gas. The chemical potential as a function of density of a 2D gas of Xe on Ag(111), computed by Unguris *et al.* (1982). The horizontal line at the upper right indicates the chemical potential of the coexisting solid phase. The virial curve is based on an expansion truncated at the second virial coefficient. A realistic Xe/Ag(111) interaction model was used for it and the Monte Carlo simulation. The substantial deviation of the Monte Carlo data from the virial curve and that curve's nonmonotonic behavior reflect the high density of the gas at equilibrium.

were once thought (Hoover and Alder 1967) to be well established as a system with first-order melting, more extended simulations (Zollweg and Chester 1992; Lee and Strandburg 1992) leave the situation uncertain.

The shear elastic modulus of the solid has also been followed as the temperature increases (Abraham 1981b; Toxvaerd 1983; Morales 1994), as it is an important property in the Kosterlitz–Thouless theory.* The dilation of the lattice by

*For model calculations of the shear modulus of a quantum monolayer solid see Bruch and Gottlieb (1988) and Leung and Chester (1991).

thermal expansion is a mundane mechanism for a decrease in the shear modulus. Evaluating the elastic constants with static lattice sums at $L/r_0 = 1.04$ for the thermally expanded solid at $T^* = 0.4$ and using the Kosterlitz and Thouless estimate for the melting temperature leads (Phillips et $al.$ 1981) to $T_m^* = 0.7$. Abraham included thermal fluctuation effects in the elastic constants by Monte Carlo averaging and found a dislocation-induced instability temperature $T^* = 0.45$ for a solid of density $\rho^* = 0.83$ and a pressure $\pi^* = 0.05$, but his free energy constructions showed that the system had already melted at $T^* = 0.415$. In contrast, there does seem to be a close correlation between the shear modulus and the melting temperature for a lattice of electrons on the surface of liquid helium (Gallet et $al.$ 1982) and the latent heat at melting has been undetectable (Glattli et $al.$ 1988a,b).

Another question is whether there is a glass transition in the dense 2D fluid, involving localization and loss of atomic mobility. As noted in Section 4.4, the identification of an effective diffusion constant involves some subjective judgments, because of the long-time tail in the velocity autocorrelation function. The evidence from molecular dynamics simulations (Ranganathan and Dubey 1993; Ranganathan 1994) is that the development of localization at $T^* = 0.5$, just above the triple point, is qualitatively similar to that for 3D Lennard-Jonesium.

The 2D LJ solid has been studied in much more detail than the 3D solid has, and there is more information on atomic-scale correlations. Simplistic hopes that computer simulations would be easier than for the 3D solid, because linear dimensions of the cell increase as $\sqrt{N}$ rather than $N^{1/3}$ (N the number of particles in the simulation volume), were certainly not borne out. Fluctuation effects are indeed more prominent in the 2D system, as the theory of the harmonic solid indicated, Section 4.2.3. It seems now that the phase diagram remains similar to that for the 3D system, without the appearance of new phases.*

5.2.2.2 *Harmonically modulated 2D solid* A geometric description of structures that arise for a monolayer solid on a structured substrate was given in Section 3.4. Here we outline the theory of the frustrated structures that arise from the competing interactions in slightly incommensurate lattices on triangular substrates. Pokrovsky and Talapov (1984) reviewed this subject, especially quasi-one-dimensional cases, in the contexts of field theory and of nonlinear mechanics. Den Nijs (1988) has reviewed the domain-wall theory of the commensurate–incommensurate transition, especially for lattice models. Griffiths (1990) gave a modern introduction to the use of Frenkel–Kontorova models for commensurate–incommensurate phase transitions and Selke (1992) reviewed the relation of such models to magnetic models with frustrated interactions.

Qualitative aspects of the phenomena are present already in the Frenkel–Kontorova linear chain of $N + 1$ atoms with a potential energy

---

*For a treatment of the melting transition of a Lennard-Jones solid in the presence of the field of an fcc(100) surface, see Patrykiejew et $al.$ (1995).

$$\Phi = J \sum_{n=0}^{N-1} (x_{n+1} - x_n - a)^2 - A \sum_{n=0}^{N} \cos(2\pi x_n/b) , \qquad (5.81)$$

where the atom–atom interactions are Hooke's law springs with equilibrium separation a and the external harmonic potential has period b. When a is close to an integer multiple of b, solutions display the phenomena of misfit dislocations and extensive registered domains separated by *walls*. In the continuum approximation, conditions for the extended $(N \to \infty)$ line of springs to form a commensurate structure are known explicitly, as is the threshold chemical potential for the transition from the commensurate structure to an incommensurate structure at zero temperature. The literature is extensive (Frenkel and Kontorova 1938; Frank and van der Merwe 1949a,b; Ying 1971; Villain 1980a; Bak 1982; Aubry and LeDaeron 1983; Pokrovsky and Talapov 1984; Lyuksyutov et al. 1992).

We treat here the deformations of a monatomic triangular lattice by the periodic potential from a triangular substrate surface lattice, a 2D generalization of eqn (5.81). The model potential energy is a one-Fourier-component simplification of eqn (5.43)

$$\Phi = \frac{1}{2} \sum_{i \neq j} \phi(r_{ij}) + 2V_g \sum_{\mathbf{g}} \sum_{j} \cos(\mathbf{g} \cdot \mathbf{r}_j) . \qquad (5.82)$$

Again ϕ is the adatom–adatom potential energy; the sum on reciprocal lattice vectors $\mathbf{g}$ runs over three vectors in the first shell of the triangular lattice corresponding to $\boldsymbol{\tau}_1$, $\boldsymbol{\tau}_2$, and $-(\boldsymbol{\tau}_1 + \boldsymbol{\tau}_2)$ in Fig. 3.1. The atomic positions $\mathbf{r}_j$ are assumed to lie in a plane. Shiba (1980) was the first to give a detailed treatment of a physisorbed layer at zero temperature based on eqn (5.82). He replaced the adatom–adatom energy by a harmonic continuum approximation, an isotropic elasticity theory corresponding to eqn (4.77), and included the many multiples of the misfit wavevector in the adlayer response at small misfit. He compared the relative stability of monolayers with hexagonal and with uniaxial modulations and showed that, beyond a threshold misfit, there is rotation of the adlayer axes relative to the substrate axes. Later work, summarized in this subsection, extended the zero temperature work to anharmonic pair potentials and found necessary conditions on the potentials for the occurrence of observed structures. The role of thermal excitations has been discussed qualitatively and and has been demonstrated in a molecular dynamics simulation of Kr/graphite monolayer, Fig. 5.10. The Novaco–McTague perturbation theory, reviewed in Section 5.2.1.2, may be regarded as the leading small-distortion approximation to the detailed theories (Shiba 1980; Bak 1982; Villain and Gordon 1983; Gooding et al. 1983; Shrimpton et al. 1984, 1986, 1988).

Since published work differs in details such as the reference point of expansions and in the terminology used to describe structures and solutions, we begin with an internally consistent set of definitions. Then we explain which features the various authors have treated and the relative degrees of approximation in

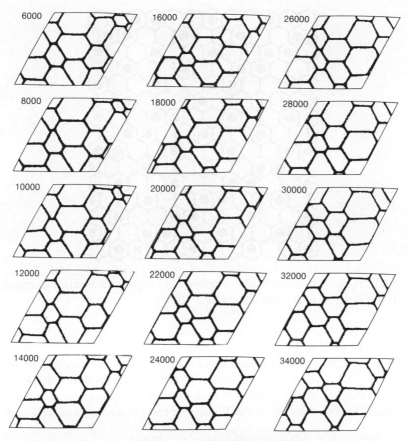

FIG. 5.10. Domain wall breathing modes. Breathing motion of the domain wall structure of Kr on graphite, from Abraham (1986). Numbers refer to time, expressed in units of hundreds of computer time steps.

their work. The review is limited to incommensurate layers on a substrate with the honeycomb symmetry of the basal plane graphite. There are other lattices that differ in the sign of V_g: while this makes no difference for the perturbation theory of Section 5.2.1.2, it has an important role in the relative stability of modulation structures in nearly commensurate phases. The model solutions for triangular incommensurate lattices have been for the case $V_g < 0$ which holds for inert gases/graphite.

Networks of domain walls The geometry of the modulation structures is subtle and is more easily presented pictorially than algebraically. Figures 5.11 and 5.12 are for the case of an adlayer close to the conditions for the $\sqrt{3}R30°$ commensurate lattice. At zero misfit only 1/3 of the honeycomb centers are filled and there are three equivalent, i.e., degenerate, ground states according to which set

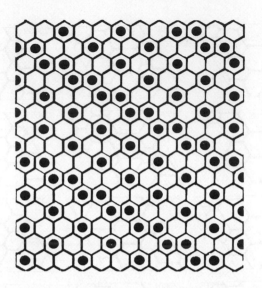

FIG. 5.11. Parallel domain walls. Compressed superheavy stripe domain walls (SHW), with $\mathcal{L} = 4$, from Joos *et al.* (1983).

of sites (A, B, or C in Fig. 5.3 or 3.8) is occupied.

Figure 5.11 illustrates the packing of a uniaxial striped configuration with a superheavy wall; two stripes are shown. For the corresponding array of heavy walls, see Gooding *et al.* (1983). In a domain of $\mathcal{L}$ walls ($\mathcal{L} = 4$ in Fig. 5.11), the average density relative to the density of the commensurate lattice is

$$n/n_c = 3\mathcal{L}/(3\mathcal{L} - 1) \quad \text{heavy}$$
$$n/n_c = 3\mathcal{L}/(3\mathcal{L} - 2) \quad \text{superheavy} . \tag{5.83}$$

The superheavy wall was treated in Section 3.4. The average lattice is the centered rectangular lattice and the misfit wavevectors $\mathbf{q}_1$ and $\mathbf{q}_3$ were defined in eqn (3.16). The positions in the adlayer can be expanded in a Fourier series in multiples of the misfit wavevector

$$\mathbf{r}_j = \mathbf{R}_j + \mathbf{u}\,(\mathbf{R}_j)$$
$$\mathbf{u}\,(\mathbf{R}_j) = \sum_{s=-\infty}^{\infty} \mathbf{A}_s \exp(\imath s \mathbf{q}_1 \cdot \mathbf{R}_j) . \tag{5.84}$$

We reserve the term *striped phase* for uniaxially incommensurate lattices with nonoverlapping walls.*

*The striped phase may have a larger-scale structure, termed a *chevron* pattern, in which the stripe "folds" to lie along successive 60° symmetry lines of the substrate. This is observed for the reconstructed surface of Au(111) (Barth *et al.* 1990) and is attributed to long-range elastic distortions of the underlying layers (Narasimhan and Vanderbilt 1992; Alerhand *et al.* 1988).

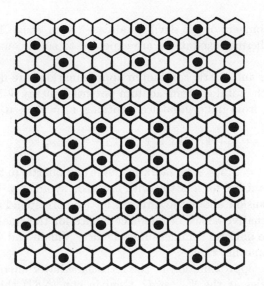

FIG. 5.12. Static honeycomb array of domain walls. Expanded hexagonal structure with superlight walls (SLW), with $\mathcal{L} = 3$, from Joos *et al.* (1983).

Figure 5.12 illustrates the hexagonal packing of domains with superlight walls. For the corresponding arrays of heavy and superheavy walls,[†] see Gooding *et al.* (1983). The hexagonal domain $\mathcal{L}$ atoms on a side contains $3\mathcal{L}^2 - 3\mathcal{L} + 1$ atoms and the coverages for the compressed structures relative to the commensurate value n_c are:

$$n/n_c = [9\mathcal{L}^2 - 9\mathcal{L} + 3]/[9\mathcal{L}^2 - 12\mathcal{L} + 4] \quad \text{heavy}$$

$$n/n_c = [9\mathcal{L}^2 - 9\mathcal{L} + 3]/[9\mathcal{L}^2 - 15\mathcal{L} + 7] \quad \text{superheavy} . \tag{5.85}$$

In these cases the expansion of eqn (5.84) is generalized to include misfit wavevectors $\mathbf{q}_i, i = 1, 2, 3$, for the hexagonal symmetry. Gordon and Lançon (1985) use the terms 'centered' and 'noncentered' to describe special symmetries of the superheavy walls.

The constructions with integer values of $\mathcal{L}$ correspond to a discrete set of densities with uniform sized domains. As $\mathcal{L}$ increases, the spacing of the densities becomes very close and the results of model calculations fall on smooth curves. However, an observation of a continuous variation at densities much different than n_c effectively means that a picture of sharply defined hexagonal domains has lost its validity. Hidden under these comments is a question concerning the existence of a *devil's staircase* of lock-in structures at commensurate superlattices (Bak 1982; Bak and Bruinsma 1982). For physisorption there has been no observation of such a succession. There are a few observed superlattices

[†]For treatments of the relative stability of heavy and superheavy walls, see Gordon (1987), Houlrik and Landau (1991), and Bruch (1993).

that may be harbingers of the devil's staircase (Kern *et al.* 1987a; Zeppenfeld *et al.* 1992a). Thermal excitations are expected to smear out much of the fine structure.

The harmonic and elastic continuum approximations are defined in terms of an expansion of the total adatom–adatom potential energy Ψ in powers of small displacements $\mathbf{u}_j$ from uniform lattice positions, $\mathbf{r}_j = \mathbf{R}_j + \mathbf{u}_j$, as

$$\Psi = \Psi_0 + \Psi_2 + \Psi_3 + \Psi_4 + \ldots \tag{5.86}$$

The terms Ψ_i for $i = 2, 3, 4$ are quadratic, cubic and quartic in the displacements; the linear term vanishes for a lattice with inversion symmetry. This Taylor series could be an expansion about the uniform lattice corresponding to the average density. An alternative choice of reference lattice for the small misfit case is to expand about the uniform commensurate lattice. Villain and Gordon (1983 eqn (2.25)) used yet another choice.

The harmonic approximation to the intrinsic adlayer energy is a truncation of the Taylor series at the Ψ_2 term. Used in eqn (5.82) this leads to a two-dimensional version of the Frenkel–Kontorova chain. Currat and Janssen (1988) review domain wall solutions for such models.

The harmonic continuum approximation* has the further step of replacing the discrete set of relative displacements $\mathbf{u}_i - \mathbf{u}_j$ by a continuous displacement function $\mathbf{u}(\mathbf{R})$ through

$$\mathbf{u}_i - \mathbf{u}_j \simeq \mathbf{R}_{ij} \cdot \nabla \mathbf{u}(\mathbf{R}) \, , \tag{5.87}$$

where $\mathbf{R}$ might be taken as the center of $\mathbf{R}_i$ and $\mathbf{R}_j$. Then retaining only terms second order in the gradients and using the six-fold rotation symmetry of the triangular lattice, the intrinsic adlayer energy takes the form of the elastic continuum expression, eqn (4.77). The Lamé constants are given by

$$\lambda = \mu = (n/16) \sum_{j \neq 0} [R_j^2 \phi''(R_j) - R_j \phi'(R_j)] \, , \tag{5.88}$$

where n is the number density, R_j are distances in the reference lattice and primes denote derivatives. A solid for which the Lamé constants are equal is termed a *Cauchy solid.*[†] We achieve the conditions of a Cauchy solid here by combining (1) a triangular lattice, (2) central pair potentials, (3) classical mechanics and (4) zero spreading pressure.

Shiba (1980) treated the adlayer with a combination of the elastic continuum energy eqn (4.77) and the substrate corrugation energy term of eqn (5.82). He

---

*The 1D harmonic continuum approximation is known as the Frank and van der Merwe (1949a,b) model. Results for the frequency spectrum of the 1D model are available in the work of Sutherland (1973), McMillan (1977), and Pokrovsky and Talapov (1978).

[†] For treatments of the pressure renormalization to the Lamé constants see Barron and Klein (1965), Wallace (1972), or Stewart (1974, 1977).

followed solutions as a function of misfit using a truncated Fourier series similar to eqn (5.84) for the adlayer modulations.[‡]

Numerical searches for the minimum energy configuration for the potential energy of eqn (5.82) have been made for models of Kr/graphite and Xe/graphite (Gooding *et al.* 1983; Joos *et al.* 1983; Gordon and Lançon 1985). The adatom positions are varied until the force on each atom vanishes. The searches are formulated as a relaxation of periodic arrays of commensurate domains separated by walls. The adlayer interactions are usually taken to be the realistic multiparameter interactions from the 3D phases supplemented by an estimate of adsorption-induced interactions. Shrimpton *et al.* (1984) developed a more efficient method that leads to the same result: it is based on the use of a series similar to eqn (5.84) for the displacements and an expansion for the forces based on eqn (5.86).

Such zero-temperature calculations, in the varying degrees of approximation, address the questions of whether the ground state is a commensurate structure and, if it is, what sequence of structures occurs under compression from the commensurate state. The pattern found in the calculations for Kr/graphite is that a commensurate structure is followed by a UIC lattice, then by an aligned triangular lattice and then by a rotated lattice with sheared domain walls. In terms of the energetics of domain wall structures, the relevant quantities are the energy per unit length of the wall (negative energy implies spontaneous formation of an incommensurate lattice) and the crossing or intersection energy of walls (negative energy implies the hexagonal array is favored over the striped one). The wall energy and wall-crossing energy have been extracted from calculations of honeycomb networks of domain walls for Kr/graphite models (Gordon and Lançon 1985; Schöbinger and Abraham 1985). The Peierls pinning energy, the energy barrier to sliding of the domain wall across the surface, is linked to the atomic discreteness of the adlayer (Pokrovskii 1981; Joos 1982). It is usually smaller than the numerical uncertainties in the calculations for the hexagonal arrays, but has been evaluated for the UIC lattice (Black and Mills 1990; Gottlieb and Bruch 1991a).

Calculations for Kr/graphite, with a large enough amplitude V_g that the commensurate lattice is the ground state, show that the first incommensurate structure is the UIC lattice (Gooding *et al.* 1983). Gordon and Lançon (1985) find, for similar models, that the signs of the energy/length of isolated walls and of wall intersection energies are correlated:[*] when the energy to create a wall is positive, so that the commensurate lattice is stable, the wall-crossing energy is positive and the UIC lattice is stable at small misfit.

The calculations of Shrimpton *et al.* (1988) illustrate the balance of en-

---

[‡]He also used $\tilde{\lambda}/\bar{\mu} = 1$, which is distinct from the Cauchy solid condition when the layer is under stress.

[*]Novaco (1994) has noted the need to make stability comparisons at equal grand potential and concluded that the Bak *et al.* (1979) argument must be amended to a statement that the UIC phase is stable when the wall-crossing energy is 'positive enough'.

ergy contributions that determines whether the adlayer is rotated relative to the substrate lattice. In the rotated adlayer, the domain walls have a staggered intersection and the compressional adlayer elastic energy of the intersections is reduced. At small misfits this effect is outweighed by the shearing energy for the walls themselves, but at larger misfits, and smaller hexagonal commensurate domains, there are many vertices and the reduction of the vertex energy in the rotated lattice becomes the larger effect. This interpretation is supported by analysis of the vibrational dynamics of the domain walls in the rotated and the nonrotated layer. The effect of anharmonicity in the Kr/graphite case is to reduce the threshold misfit for rotation from the value in Shiba's calculation; it enters particularly in the energy of the compressed vertices that correspond to the wall crossings.

Energy balances A primitive method of estimating the stability of the commensurate structure is to compare the adatom–adatom potential energy difference between the uniform lattice, with the 'intrinsic' 2D lattice constant, and the commensurate lattice to the registry energy, $6V_g$ in the commensurate lattice with $V_g < 0$. This leads to an underestimate of the magnitude of V_g required for the commensurate lattice to be stable, because modulation of the incommensurate lattice also lowers the energy. However, model calculations for Kr/graphite and for some uniaxial registry lattices (Bruch 1985) show that estimates of the threshold V_g with this elementary energy balance argument are within 25% of the results of detailed work.

Anharmonicity Calculations (Shrimpton *et al.* 1984; Bruch 1985) in which the adlayer potential energy is expanded in an anharmonic series such as eqn (5.86) help to demonstrate the role of the anharmonic effects in the modulation response to the substrate corrugation. Qualitatively, the cubic anharmonicity is more important in the cases where the intrinsic 2D lattice is dilated to reach the commensurate structure than in those where it is compressed. It has greater effect for Kr/graphite than for Xe/graphite. This is because of its role in determining effective elastic constants: as the intrinsic lattice is dilated, the harmonic elastic constants decrease and the harmonic theory predicts that the adlayer responds to a spatially periodic perturbation with sharp domain walls and a large energy shift. The cubic anharmonicity for the dilated Kr/graphite layer has the effect of increasing the effective elastic constant and limiting the sharpness of the response, so that the domain wall thickness is increased. In the Fourier expansion of the adlayer displacements, the effect of the cubic anharmonicity then is to reduce the number of terms that must retained for a given misfit.

Structure factor The zero temperature calculations with the large effective basis also yield the structure factor of the monolayer with small misfits, through eqns (5.46) and (5.47). The Novaco–McTague perturbation theory of the adlayer response, eqn (5.57), leads to a structure factor that peaks at the reciprocal lattice wavevector of the average space lattice and symmetric pairs of modulation satellites displaced from the reciprocal lattice wavevector τ by misfit wavevectors $\mathbf{q}_i$. For the most thoroughly studied case, Kr/graphite, the observed satellite pattern is not symmetric. The X-ray experiments of Stephens *et al.* (1979,

1984) have a powder-pattern average which washes out the 6-fold pattern of modulation satellites; there are two rather than three peaks. The calculations of Gooding *et al.* (1983) reproduce this asymmetry of one-sided satellites for a hexagonal array of domain walls. The relative intensities of the satellites of modulated layers are a sensitive function of the atomic positions in the domain walls (Cui *et al.* 1988; Cui and Fain 1989; Freimuth *et al.* 1990).

Thermal excitation Thermal excitation changes the relative free energies of the striped and hexagonal arrays of domain walls (Villain 1980b,c). In the striped array, meandering of the walls increases the total domain wall length and therefore also increases the potential energy. Villain noted a special property of the hexagonal array, illustrated in Fig. 5.13. Breathing modes of the walls do not change the total wall length or the total number of wall intersections, so that the potential energy variation is small. The breathing mode indeed has the lowest frequency in the calculations of domain wall dynamics (Shrimpton *et al.* 1988). It is observed to be the major thermal excitation in molecular dynamics simulations of the Kr/graphite layer at small misfit (Koch and Abraham 1986; Koch *et al.* 1984; Abraham *et al.* 1984). Villain remarked that the breathing modes have a large entropy and that this lowers the free energy of the hexagonal array relative to the striped array.

The thermally driven meandering of the walls of the UIC lattice at very small misfit leads to a novel mechanism of disorder (Coppersmith *et al.* 1982). In a striped array with very small misfit, the average spacing between the domain walls is much larger than both the width of the walls and the range of the mechanical interactions between walls. There is then only a very small variation in the potential energy of the walls as they meander; the effective interaction between the walls is determined mostly by the entropy of the thermally excited motions. Occasional crossings of the walls may have the effect of introducing free dislocations into the adlayer, so that it has zero shear modulus and some of the properties of a fluid. This *domain-wall fluid*, also termed a *reentrant fluid*, is a possible phase to succeed the commensurate solid at finite temperatures. It in turn may be succeeded by an incommensurate solid when the misfit corresponds to a large enough density of walls so that the direct mechanical interactions between walls are larger than the entropic effects. The density range of the reentrant fluid is expected to be very narrow at low temperatures, but at higher temperatures it might preempt the full range of the uniaxially striped solid. Symmetry considerations on the commensurate superlattice determine both which domain wall crossings are equivalent to free dislocations (Coppersmith *et al.* 1982) and the size of the Burgers vector of the dislocations. Then the Kosterlitz–Thouless analysis, Section 4.3, may be extended to treat the elastic instability of the incommensurate solid relative to the reentrant fluid. Just as for the theory of the melting of 2D solids by formation of free dislocations, the occurrence of the reentrant fluid may be preempted by a first-order phase transition. There is experimental evidence for the reentrant fluid in X-ray experiments on Kr/graphite (Moncton *et al.* 1981, 1982; Specht *et al.* 1987), but not for the Xe/Pt(111) system (Kern *et al.* 1988).

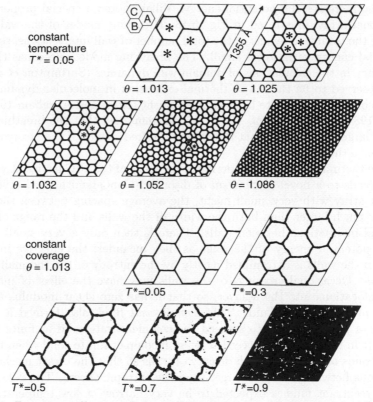

FIG. 5.13. Domain wall network for 103,041 Kr atoms on graphite as a function
of reduced temperature (in units of the Kr–Kr well depth) and coverage
(measured relative to commensurate), from Abraham *et al.* (1984).

Adsorption on fcc(111)   The calculations for an inert gas on graphite correspond to the case $V_g < 0$. It was anticipated that the case $V_g > 0$ would correspond to an inert gas on the (111) face of a face-centered cubic metal, fcc(111), with energy minima at the three-fold hollow sites for hcp and fcc stacking sites of the substrate lattice. The energetic topography of the holding potential surface then changes greatly. Roughly, it has valleys connected by rather low saddles in the presence of high peaks. The low saddles and the near equality of the energy at the hcp and fcc sites leads to a circumstance where the energy to create domain walls is much less than would be the case for the graphite models with V_g of the same magnitude.* The Xe/Pt(111) system with a triangular commensurate lattice and a UIC lattice is a candidate for modeling in these terms (Gottlieb and Bruch 1991b; Barker and Rettner 1992; Rejto and Andersen 1993). The structure of the walls that separate domains centered atop the two types of three-fold hollow sites is more complex than for the SHW that arises in the striped phase of inert gases on graphite. The SHW may be described by a one-component scalar displacement field while the HW and the modulations here require vector displacement fields. Still, rather explicit solutions are available in the continuum limit corresponding to the Frank and van der Merwe approximation for the SHW (Bruch 1991b, 1993).

---

*Such considerations arise in modeling the reconstructed Au(111) surface (Okwamoto and Bennemann 1987; Ravelo and El-Batanouny 1989, 1993).

6

MONOLAYER EXAMPLES

In this chapter we survey several examples of physically adsorbed monolayers chosen according to two criteria: (1) to demonstrate the interesting variety of physical phenomena present and (2) to illustrate the outstanding features of the general theory in Chapters 4 and 5. Fluid phases (dilute gases and liquids) have correlation lengths that usually are small except near the gas–liquid critical point, but in the monolayer the correlation length *may reach thousands of angstroms near a melting transition* (Specht *et al.* 1985; Nuttall *et al.* 1994). Further, the spatially periodic external potential arising from the substrate lattice may cause the fluid density to be quite nonuniform. The solid phases have sharp diffraction patterns and present a most diverse set of structures in the monolayer regime; we emphasize the types of lattices and their evolution under increased temperature and lateral stress.

Apart from the quantum solids of helium and hydrogen, the characteristic length, e.g., the nearest-neighbor spacing, varies by less than 10% from monolayer solidification to the onset of further layer condensations. The limited compressibility of van der Waals solids is a consequence of the fact that the width of the attractive potential well is narrow compared to the thickness of the hard-core overlap repulsions.[*] Therefore, a preliminary step to achieve conditions that enhance effects of the near coincidence of the adsorbate and substrate length scales is to scan across a series of adsorbates, such as the inert gases, to find an example sufficiently close to the desired match that the remaining difference can be manipulated with thermomechanical stresses. An example of such a scan is shown in Fig. 6.1. Physical phenomena are illustrated and our qualitative understanding is increased by pursuing systematic trends within a family which differs only in scale. A pattern in the experimental study of monolayer physics is to follow trends in a family of adsorbates on a given substrate.[†] Thus, in this chapter we emphasize inert gas adsorption on graphite, on two metals, and on

---

[*]A useful guide to the length scales for the inert gases is to use the Lennard-Jones (12,6) potential, eqn (5.71), with $\sigma = 1$. Then the core radius R_c is identified as 1, the separation at the minimum of the potential is $2^{1/6} \simeq 1.12$, and at 2.1 the potential is only 5% of its value at the minimum. By contrast, the nucleon–nucleon interaction has a thinner core relative to the range of its attraction. Brueckner (1966) argued that the many-fermion problem posed by liquid ^{3}He is more difficult than the nuclear matter problem because the ratio of the core diameter in the pair potential to the nearest-neighbor spacings in the fluid is much larger (0.8) for ^{3}He than for nuclear matter (0.4).

[†]This is done in modeling as well, for instance in a comparative study of the melting of argon, krypton, and xenon monolayers on graphite (Abraham 1983b).

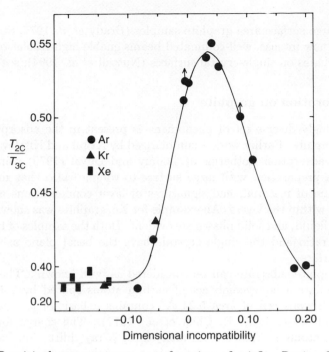

FIG. 6.1. 2D critical temperatures as a function of misfit. Ratio of 2D to 3D
critical temperatures for Ar, Kr, and Xe on various lamellar halide surfaces,
from Millot et al. (1982). The dimensional incompatibility is the excess above
one of the ratio of the spacing between neighboring anions of the substrate
and the interatomic spacing of the (111) plane of the bulk adsorbate.

an ionic crystal. By contrast, processes for molecular adsorbates, e.g., CO_2, are
illustrated by comparing results for the same adsorbate on various substrates
(Girardet et al. 1994).

The available examples reflect two general requirements on the experiments:
(1) to realize an ideal extended monolayer requires substrate homogeneity over
length scales of hundreds to thousands of angstroms; (2) the experiments must
distinguish the signals of the bulk substrate from those of atomically thin adsorbate layers. The first of these requirements imposes standards on the homogeneity and size of adsorbing powders and on the perfection of single-crystal adsorbing
surfaces. The second requirement imposes an interplay between sample and experimental method: methods which are intrinsically surface sensitive, such as
electron and atomic beam scattering, are applicable on single-crystal substrates
of modest surface area. Most of the thermodynamic and neutron scattering measurements require samples, such as powders, where the surface-to-volume ratio
is enhanced to the point that the adsorbate signal is a fair fraction of the total.
There has been a historic evolution: X-ray diffraction was first done for adsor-

bates on large surface area graphite samples (Brady *et al.* 1977; Stephens *et al.* 1979), but now intense, well-collimated beams enable high-resolution diffraction from adsorbates on single-crystal surfaces (Nuttall *et al.* 1994); see Fig. 1.3.

6.1   Adsorption on graphite

An exceedingly diverse set of phenomena is present in the adsorption of inert gases on graphite. Earlier work is summarized by Avgul and Kiselev (1970). The pioneering adsorption isotherms of Thomy and Duval (1970) using exfoliated graphite, a preparation with large surface-to-volume ratio that may reach 30 m^2 per gram of material, had signatures of layer condensations and of phase transitions within the layers. An example for Xe/graphite was shown in Fig. 4.4 where gas, liquid, and solid phases are evident. Both the samples of large surface-to-volume ratio and the single crystals have the basal plane as the primary exposed surface.

The graphite substrate can be considered as 'well known'. The basal plane surface exhibits a honeycomb net of carbon atoms spaced by 1.42 Å and its triangular Bravais cell of area 5.24 $Å^2$ contains a basis of two carbon atoms (Baskin and Meyer 1955; R. Chen *et al.* 1977). The charge form factor of the carbon atoms is known from analysis of X-ray diffraction intensities (R. Chen *et al.* 1977), but a recent determination of the quadrupolar distortion by Whitehouse and Buckingham (1993) differs from the result of the X-ray work. Results for the elastic constants have been summarized by Lee and Lindsay (1990) and for the phonon spectra by Benedek and Onida (1993). Godfrin and Lauter (1995) present an extensive review of the properties of graphite substrates.

A partial listing of the experimental techniques used to study the adsorbed layers includes: volumetric measurements (Thomy *et al.* 1981), specific heat measurements (Bretz *et al.* 1973), ellipsometry (Hess 1991; Quentel *et al.* 1975), Auger spectroscopy (Suzanne *et al.* 1974), low energy electron diffraction (LEED) (Suzanne *et al.* 1975; Fain 1982), helium atomic scattering (HAS) (Scoles 1988), elastic and inelastic neutron scattering (Taub *et al.* 1977; Nielsen *et al.* 1977, 1980a; Godfrin and Lauter 1995), quasi-elastic neutron scattering (Bienfait 1987b), nuclear magnetic resonance (Richards 1980), and X-ray diffraction (Specht *et al.* 1987). A culmination of the work for a specific adsorbate/substrate combination is the construction of a phase diagram in temperature and density or pressure variables. Because the experimental techniques tend to be complementary in their domains of utility, the phase diagram is a synthesis of contributions of many workers.

We survey the observations for the inert gas series from helium to xenon, including molecular hydrogen which is a closed-shell, nearly spherical object for our purposes (Silvera 1980; Novaco and Wroblewski 1989). This spans the range from quantum to classical layers as well as encompassing varying degrees of mismatch between overlayer and substrate. We conclude with a brief description of the additional features that arise for molecular adsorbates.

Table 6.1 *Survey of observed lattices for spherical adsorbates on graphite.*[†]

Adsorbate	Structure	Comments
He	$\sqrt{3}$	Potts model exponents
H_2	$\sqrt{3}$, UIC	UIC satellite
Ne	$\sqrt{7}$	superlattice
Ar	none	–
Kr	$\sqrt{3}$	no triple point – domain wall fluid
Xe	$\sqrt{3}$	by compression at monolayer completion

As an overview we present in Table 6.1 a survey of the commensurate structures observed for the simple gases on graphite. They are discussed in more detail in the following sections. We encounter three generic kinds of monolayer phase diagrams (Chan 1991). (1) The 2D analogue of the phase diagram of 3D inert gases. There are distinct gas, liquid, and solid phases. This includes Ar/graphite and Xe/graphite. (2) Phase diagrams in which the interactions that lead to commensurate lattices so dominate the physics that there is only one distinct monolayer fluid phase. These are called *incipient triple point systems*; examples are Kr/graphite (Butler *et al.* 1979) and N_2/graphite (Migone *el al.* 1983a). (3) Phase diagrams of quantum monolayers. These are similar to the previous class, with only fluid and solid phases. Examples are phase diagrams presented for He/graphite by Schick (1980) and for H_2/graphite by Wiechert (1991a).

6.1.1 *Helium*

The physics of condensed phases of helium is dominated by the combined effects of the small atomic mass and the weak van der Waals attractions. Qualitatively striking quantum effects are prominent in the properties of solids and liquids of helium in three dimensions (Wilks 1967; Woo 1976) and in the monolayer regime (Elgin and Goodstein 1974; Dash and Schick 1978; Dash 1978). The richness of the monolayer regime was first demonstrated by Bretz *et al.* (1973).

As for three dimensions (3D), there is no gas–liquid–solid triple point for helium in mathematical 2D (Miller and Nosanow 1978; Ni and Bruch 1986) and there are dramatic differences at low temperatures that reflect the spin/statistics of the ^{3}He and ^{4}He isotopes (Bretz *et al.* 1973; Dash and Schick 1978). Questions that arise for the monolayer phase diagram include whether there is a condensation to a quantum liquid at very low temperatures and which solid lattices form.

[†]Commensurate structures on the basal plane surface of graphite are listed. There are also triangular incommensurate lattices for all these adsorbates and for all, except helium, orientational epitaxy (Novaco–McTague rotation) is observed for those lattices.

A comparison of phase diagrams for H_2 and ^{4}He on graphite is shown in Fig.(6.2), from Motteler (1986). A later proposal for the phase diagram of ^{4}He/graphite and the corresponding diagram for ^{3}He/graphite are shown in Fig. 6.3. The phase diagrams are rather similar to those proposed for adsorbed ^{3}He and ^{4}He by Schick (1980). A discussion of the areas of agreement and divergence illustrates how much information is contained in such a diagram. The diagrams all include a commensurate $\sqrt{3} \times \sqrt{3} R\, 30°$ lattice as the initial solid phase at low temperatures and an incommensurate lattice as the final solid phase before second layer condensation. Neutron diffraction shows (Carneiro *et al.* 1981; Lauter *et al.* 1987) that the final solid phase is a triangular lattice. For ^{3}He, the identification of the phase boundary from incommensurate solid to fluid as a melting transition is substantiated by nuclear magnetic resonance experiments (Rollefson 1972; Richards 1980) that show marked differences in atomic motions in the two phases. It is believed that there is no self-bound liquid phase of 2D ^{3}He, and Fig. 6.3(b) indicates that the fluid range extends, without abrupt condensation, to about $0.04/Å^2$ at very low temperatures (Greywall and Busch 1990a,b). We now discuss the phases in more detail.

At low coverage, where effects of helium–helium interactions are negligible, the specific heat departs from the equipartition value of 1 k_B/mol (Silva-Moreira *et al.* 1980). This arises from the band structure of helium atoms moving in the periodic potential of the graphite (Cole *et al.* 1981). Similar effects arise in the selective adsorption scatterings discussed in Section 2.4.2. There still is discussion (Greywall 1993; Gottlieb and Bruch 1993) of whether there is a transition from dilute 2D gas to a self-bound 2D liquid of ^{4}He at low temperatures before a condensation to the commensurate solid. Theoretical modeling shows the occurrence of the self-bound liquid to be sensitive to details of the substrate-mediated interactions, while experiments have a signal that may be the specific heat of the liquid mixed with that of the commensurate solid. Also, there are calculations that suggest there may be a self-bound monolayer of ^{3}He/graphite (Brami *et al.* 1994).

The commensurate $\sqrt{3} \times \sqrt{3} R\, 30°$ lattice is a dominant feature of the phase diagrams in Figs. 6.2 and 6.3. The fact that this lattice is present over a range of densities is attributed to effects of vacancies in the first layer and second layer gas; neutron diffraction experiments confirm the existence of the lattice. For both ^{3}He and ^{4}He, the phase boundary between the commensurate lattice and the fluid has a maximum temperature of about 3 K. This region, which is determined largely by features of the substrate potential, has been investigated in some detail as a realization of the order/disorder critical point of the three-state Potts model, Section 5.1.2 (Alexander 1975; Bretz 1977; Schick 1980). There is an impressive degree of agreement between the experiment and the theory for the specific heat critical exponent.

The transition from commensurate to incommensurate solid near 1 K is manifested experimentally by very narrow specific heat peaks (Hering *et al.* 1976), Fig. 6.4. The identification of the phase(s) that succeeds the commensurate solid under isothermal compression remains undecided; analogy to the mono-

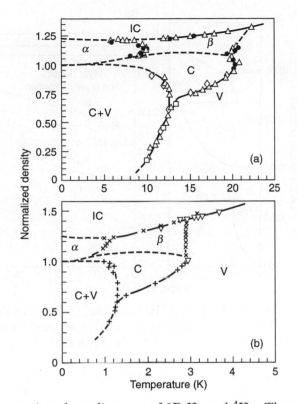

FIG. 6.2. Comparative phase diagrams of 2D H_2 and ^{4}He. The phase diagrams
of (a) parahydrogen and (b) ^{4}He on graphite are shown, from Motteler (1986).
This is constructed from the following data sources: circles, Motteler (1986);
triangles, Wiechert and Freimuth (1984), Freimuth and Wiechert (1985);
diamonds, Chaves et al. (1985); squares, Bretz and Chung (1974); crosses,
Hering et al. (1976); inverted triangles, Ecke and Dash (1983); plusses, Ecke
et al. (1985). The solid lines denote well-determined phase boundaries; the
dashed lines denote suspected phase boundaries. V, C and IC denote vapor
(or fluid), commensurate solid and incommensurate solid, respectively. The
α and β phases for H_2 have been discussed in terms of domain wall structures,
Section 6.1.2. For ^{4}He, they may reflect a two-phase coexistence (Greywall
1993).

layer isotopic hydrogen series (Wiechert 1991a) suggests there may be a continuously compressible striped uniaxially incommensurate solid (Rabedeau 1989).
However, both the illustrated phase diagrams show a coexistence between commensurate and incommensurate solids at low temperature, characteristic of a

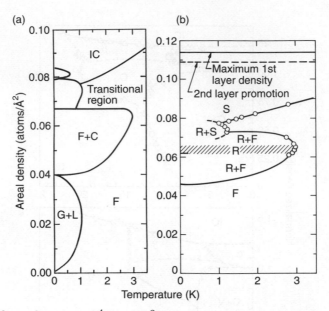

FIG. 6.3. Phase diagram of ^{4}He and ^{3}He/graphite, from Greywall (1990, 1993). (a) ^{4}He. G, L, F, C, and IC denote gas, liquid, fluid, and commensurate and incommensurate solid phases, respectively. Note that this diagram presents a 2D liquid phase as the ground state of a submonolayer patch at very low temperatures. (b) ^{3}He. F, R, and S denote the fluid and commensurate and incommensurate solid, respectively. Note the absence of a boundary denoting a transition to a self-bound liquid.

first-order transition (Ecke and Dash 1983). Not shown in the ^{3}He phase diagrams is a boundary in the fluid at millikelvin temperatures that may mark a superfluid transition or a magnetic ordering transition. The specific heat of the ^{3}He fluid has a linear dependence on temperature (*Sommerfeld*) over a wide temperature range and the specific heat of the incommensurate solid has a quadratic dependence on temperature (*Debye*) below 0.5 K. However, at millikelvin temperatures, the nuclear spins dominate the specific heat of the solid. Godfrin and Lauter (1995) have reviewed the phase diagram of the ^{3}He film on graphite.

The lattice constants of quantum solids are dilated by the 'soft' zero-point energy, so that such solids generically are highly compressible. To set a scale for comparisons, the nearest-neighbor distances L_{nn} in the 3D solids at the melting curve at zero temperature are 3.75 Å for ^{3}He and 3.65 Å for ^{4}He (Wilks 1967). On this scale, the $\sqrt{3}$ lattice with $L_{nn} = 4.26$ Å is quite dilated. For ^{3}He, neutron diffraction shows that L_{nn} in the triangular lattice at monolayer completion is 3.21 Å and for ^{4}He it is 3.15 Å (Carneiro *et al.* 1981; Lauter *et al.* 1987). Thus the ratio of initial and final density of the solids is 1.7; to achieve nearest-neighbor spacings of 3.2–3.3 Å in the 3D solids requires pressures of about 0.4

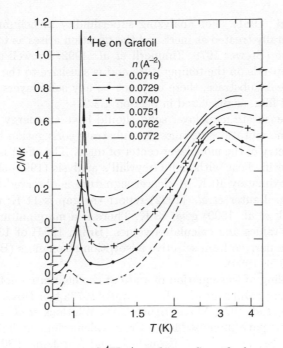

FIG. 6.4. Ordering transition of ^{4}He/graphite. Specific heat as a function of T showing the very sharp, almost T-independent, ordering transition of ^{4}He on graphite, from Hering *et al.* (1976). The ordered phase lies in a density region between the commensurate and incommensurate solid phases.

kbar. A further reflection of the large zero-point motions in the monolayer solids is that bilayer solids form with mutually incommensurate layers. For ^{4}He neutron diffraction shows L_{nn} within the two layers to be 3.15 Å and 3.57 Å, a 14% difference.[*] The interpretation is that the periodic potential exerted by the first layer lattice on the second layer atoms is sharply reduced by the zero-point motions, which reduces the energy cost arising from the mismatch of the layers.

A different way of expressing the large range of compression of the monolayer solid is that the holding potential (*ca.* 16 meV) of a helium atom on graphite provides a large thermomechanical force to draw atoms from the 3D gas to the surface and thus to compress the monolayer solid. Neutron diffraction shows that two solid layers form on graphite and further adsorption is as a fluid film

---

[*]Another possibility that is difficult to exclude (Carneiro *et al.* 1981) is that the cell is oblique. While that does occur for the deuterium bilayer, Section 6.1.2, it seems to be ruled out for the ^{4}He bilayer by the absence of an intermediate peak in the neutron diffraction experiments (Lauter *et al.* 1987). Also, the layer coverages derived from specific heat measurements (Polanco and Bretz 1978) agree well with the densities inferred by assigning the diffraction peaks to two uncorrelated layers.

(Carneiro *et al.* 1981). For analyzing superfluidity in helium films, the solid layers are generally treated as inert and the question arises as to how many such layers there are (Brewer 1978; Zimmerli *et al.* 1992; Crowell and Reppy 1993, 1994). This depends on the temperature and is sensitive to the substrate. For a smooth and weak substrate, there may not be any solid layers before the onset of a thick fluid film, as discussed in Section 4.3.2.

The commensurate monolayer solid should have an energy gap in the spectrum of in-plane vibrations because the substrate corrugation breaks the translational symmetry of the monolayer center of mass. The effect is not very prominent in the specific heat, although Greywall's analysis (1993) shows there to be a gap of approximately 10 K. A clearer signature is observed by inelastic neutron scattering. Lauter *et al.* (1992) report the gap is 11 K; an earlier report (V. L. P. Frank *et al.* 1990) gave 13 K. There is a marginal level of agreement between these values and calculated values (Bruch 1994) of 13–16 K based on the corrugation derived from selective adsorption resonances (Boato *et al.* 1979; Derry *et al.* 1979, 1980).

Most modeling of the equation of state of the helium monolayer has been at zero temperature (Miller *et al.* 1972; Novaco 1973a,b,c; Novaco and Campbell 1975; K. S. Liu *et al.* 1976; Ni and Bruch 1986; Whitlock *et al.* 1988). There also have been a few path integral Monte Carlo calculations at finite temperatures (Abraham and Broughton 1987; Broughton and Abraham 1988; Abraham *et al.* 1990; Bernu *et al.* 1992).

6.1.2 *Hydrogen*

The isotopes of molecular hydrogen have been studied extensively in three dimensions because they provide an opportunity to follow quantum effects for systems with nearly the same intermolecular potentials, which are determined primarily by the electronic properties, and because H_2 and D_2 may be prepared both in spherical and rod-like forms. Additionally, for HD there is a distinction between the center of mass and the center of force. Here we will discuss phenomena for monolayers of the spherical variants, *para*-hydrogen (p-H_2) and *ortho*-deuterium (o-D_2). We omit discussion of the pinwheel ordering transition for the molecular axes of *ortho*-hydrogen in lattices on graphite, at temperatures below 1 K, which has analogies to orderings of the 3D solids (Kubik *et al.* 1985; Harris and Berlinsky 1979). Also, in spite of the historical importance of achieving equilibrium distributions of *ortho*- and *para*- hydrogen by catalysis with carbon black, we do not treat the studies of the slow conversion rates on the pure forms of graphite used for monolayer adsorption (Silvera 1980).

Analogy to 3D properties suggests that, at low temperatures, the isotopes of molecular hydrogen condense as monolayer solids. This is indeed the case for adsorption on graphite, where they form commensurate lattices (Nielsen *et al.* 1977; Freimuth *et al.* 1990). It is not a foregone conclusion for monolayer adsorption on other substrates and a search continues to identify a substrate on which molecular hydrogen can be brought to *ca.* 2 K as a liquid and thus to form a novel quantum liquid that might show superfluidity (Vilches 1992; F.-C. Liu *et al.* 1995; Y.-M. Liu *et al.* 1996).

We show in Figs. 6.5 and 6.6 the monolayer phase diagrams proposed by Wiechert (1991a) for p-H_2 and o-D_2 adsorbed on graphite. For reference, the triple point temperatures in 3D listed in Appendix A are 13.8 K for p-H_2 and 18.7 K for o-D_2. At low temperatures hydrogen condenses from a dilute 2D gas to a commensurate $\sqrt{3} \times \sqrt{3} R 30°$ lattice. With sufficient additional stress, the commensurate solid may be continuously compressed into a uniaxially incommensurate lattice (Cui *et al.* 1988) not yet observed for He/graphite. These *striped* lattices are studied by neutron diffraction (Freimuth *et al.*1990) and by LEED (Cui and Fain 1989) as discussed below. Upon further compression, there are first-order transitions to triangular incommensurate lattices that show non-trivial orientational alignment relative to the substrate.

As for helium, the monolayer solids are highly compressible: the nearest-neighbor spacing in the triangular commensurate lattice, 4.26 Å, is very dilated relative to that in the ground state 3D solid, 3.789 Å for H_2 and 3.605 Å for D_2. In the triangular lattices at monolayer completion L_{nn} for p-H_2 is 3.51 Å and for o-D_2 it is 3.40 Å (Nielsen *et al.* 1977). To reach $L_{nn} \sim 3.5$ Å in the low temperature 3D solid of H_2 requires a pressure of 1 kbar (Silvera 1980). Thus, the monolayer solids of the hydrogen isotopes span a wide range of separations on a scale set by lengths of the 3D solids. Another unusual feature is that the highly compressed monolayer of D_2 is a triangular lattice, but the bilayer forms as an oblique lattice (Wiechert 1991b; W. Liu and Fain 1993).

The structural properties of the monolayer solids of isotopic molecular hydrogen are known in much more detail than for adsorbed helium. The commensurate, striped, and orientationally aligned monolayer lattices on single-crystal graphite surfaces are tracked with LEED experiments, but great care is needed to avoid desorbing the adlayer by the electron beam.[*] The major puzzle in this body of data is the orientational epitaxy observed for the γ-phase of D_2, which does not follow elementary applications of the Novaco–McTague theory (Cui and Fain 1989; Grey and Bohr 1991; Novaco 1992,1994).

The striped lattice is observed both in the LEED experiments and in neutron diffraction. The diffraction patterns have peaks characteristic of the average compressed lattice and satellite peaks that arise from the spatial modulation and are displaced by misfit wavevectors, as discussed in Sections 3.4 and 5.2.2 (Nielsen *et al.* 1980b). The neutron diffraction intensities are more amenable to quantitative analysis because of the weak coupling of the neutron probe. The adlayer may be described as commensurate domains separated by parallel domain walls. The domain wall widths, defined as in Fig. 6.7, are adjusted to reproduce the observed intensity ratios (Freimuth *et al.* 1990). The satellite peaks are observed for densities of 1.06 to 1.15 times the commensurate lattice density $(0.0636/\text{Å}^2)$; this corresponds to superheavy walls separated by $16 - 42$ Å. The

---

[*]The kinetics of the desorption process have been analyzed in several papers (W. Liu and Fain 1992; Fain *et al.* 1994; Xia and Fain 1994).

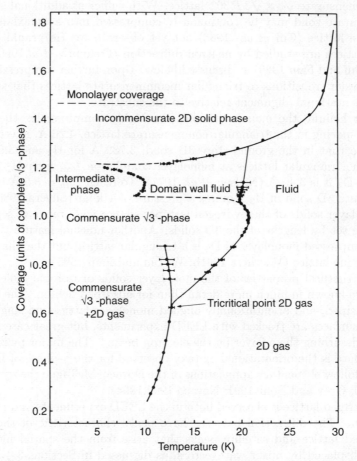

FIG. 6.5. Phase diagram of H_2/graphite, from Wiechert (1991a). Solid circles denote specific heat anomalies. Solid and dashed lines denote definite and tentative phase boundaries, respectively.

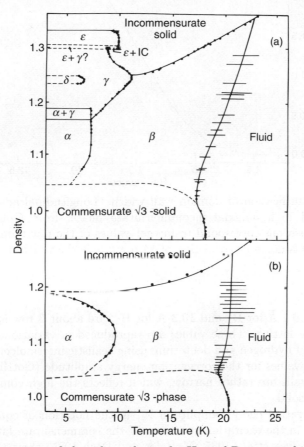

FIG. 6.6. Comparison of phase boundaries for H_2 and D_2 on graphite. Magnified view of commensurate region of phase diagram of (a) D_2 and (b) H_2 on graphite, from Wiechert (1991a). The meaning of dashed and solid lines is as in Fig. 6.5. Note the presence of additional phases for D_2 at densities intermediate between the commensurate and incommensurate solids.

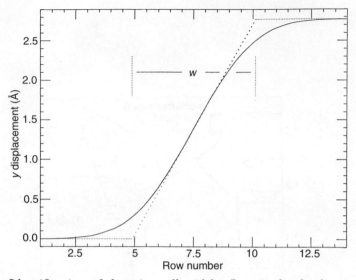

FIG. 6.7. Identification of domain wall width. Longitudinal adatom displacement field in a uniaxial incommensurate phase, from Gottlieb (1990b). Atomic positions correspond to integer values of the row number. The case depicted has domains consisting of 14 rows; w is the width of the wall.

fitted widths, 9.2 Å for D_2 and 10.3 Å for H_2, are about 3 row spacings of the commensurate lattice. These values are reproduced in variational calculations for 2D solids of hydrogen and deuterium using realistic intermolecular potentials and plausible values for the corrugation energy amplitude (Gottlieb and Bruch 1989). The walls are rather narrow, which reflects the high compressibility of the quantum solid.

The corrugation energy amplitude V_g, in the notation of eqn (3.9), enters most directly in the excitation spectrum of the commensurate lattice. Specific heat data (Chaves *et al.* 1985; Motteler and Dash 1985) show that there is a wide gap in the energy spectrum: the data for H_2 are fitted to a gap of 55 K. Inelastic neutron scattering for parallel momentum transfer corresponding to a reciprocal lattice vector of the adlayer gives a gap at the Brillouin zone center, the zone-center frequency gap, of 47 K for H_2 and 41 K for D_2 (V. L. P. Frank *et al.* 1988; Lauter 1990; Lauter *et al.* 1990). These values and the energy width of the density of states on the first Brillouin zone are reproduced in the self-consistent phonon calculations of Novaco (1992). The density of states for the hydrogen lowest phonon band is relatively narrow compared to the helium case.

The monolayer solids of hydrogen and deuterium fall well into the domain of quantum solid theory. The striped phase and the compressed monolayer near bilayer formation have been treated with Jastrow variational theory (Ni and Bruch 1986; Gottlieb and Bruch 1989). There is a good level of agreement with the experimental data. Novaco (1992) also found good agreement with experiment for calculations of the dynamical excitations.

The bilayer phase diagram is also quite rich (Wiechert 1991b). It has been studied as part of the search for a superfluid phase of adsorbed hydrogen (Vilches et al. 1994; F.-C. Liu et al. 1995; Y.-M. Liu et al. 1996).

6.1.3   Neon

Quantum effects are much smaller for neon than for helium and hydrogen but are still quantitatively significant. This is evident even for the 3D solid, where the nearest-neighbor distance L_{nn} in the ground state is 3.155 Å (Crawford 1977), while the spacing at the pair potential minimum is $R_m \sim 3.09$ Å. For a classical monatomic solid, L_{nn} is a little smaller than R_m, because it is compressed to enhance energy terms arising from farther neighbors, Section 5.2.1.1.

The lattice constant of the monolayer solid of Ne/graphite at condensation is found to be 3.25_2 Å at 1.7 K, from neutron diffraction (Tiby et al. 1982). This length is coincident with a higher-order superlattice (Huff and Dash 1976) on graphite, the $\sqrt{7} \times \sqrt{7} R 19.1°$ lattice with four neon atoms in the basis. The primitive vectors of the superlattice, in terms of Fig. 3.1, are $2\mathbf{a}_1 + \mathbf{a}_2$ and $3\mathbf{a}_2 - \mathbf{a}_1$; the first of these is at an angle of 19.1° relative to $\mathbf{a}_1$.[*] The dilation of L_{nn} from the 3D value of 3.16 Å to the average value of 3.25 Å for the superlattice mostly arises from increased zero-point energy terms and substrate-mediated dispersion forces in the monolayer; the latter contribute an increment of 0.04 Å (Bruch et al. 1984).

Besides the coincidence of lengths, additional evidence for the existence of the superlattice is that low-temperature specific heat data for the monolayer show an energy gap of 3.5 K (Huff and Dash 1976). The registry energy for this superlattice is small, estimated at −1 K (Bruch 1982), because it arises from the modulation of atomic positions in the 4-atom unit cell and is a second-order perturbation energy in terms of the leading corrugation amplitude V_g. A model calculation based on this mechanism obtained a gap of the observed size (Bruch 1988).

An orientational alignment angle is observed in LEED experiments for the incommensurate monolayer solid at about 16 K; the angle is 12°–13° relative to the primitive vector $\mathbf{a}_1$, for lattice constants of 3.09–3.23 Å (Calisti and Suzanne 1981; Calisti et al. 1982). This is in rough agreement with calculations for the perturbation theory (Novaco and McTague 1977a,b) of the rotation.[†]

---

[*]An alternative notation measures the angle from the direction of $\mathbf{a}_1 + \mathbf{a}_2$ and denotes this as $R10.9°$.

[†]The calculated values (Bruch 1982) are 1°–4° larger than the observed values, and there is a systematic trend to the discrepancy that indicates that the perturbation theory is inadequate. One resolution may be anharmonic effects which would be included in the more complete theory of Novaco (1979).

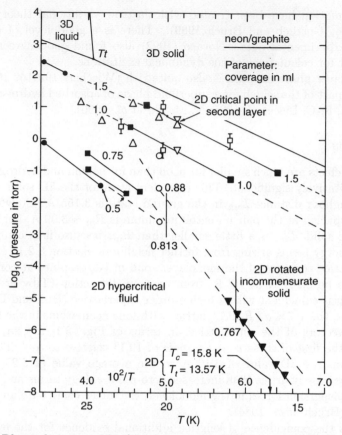

FIG. 6.8. Phase diagram of Ne/graphite, from Calisti *et al.* (1982). Not shown
are the $\sqrt{7}$ ordered phase at lower T and the 2D critical and triple points,
which occur at pressures that are too low to measure; the corresponding
temperatures are indicated, however. Data with filled symbols: triangles,
Calisti *et al.* (1982); circles, Antoniou (1976); squares, Lerner and Hanono
(1979). Data with unfilled symbols: squares, Lerner *et al.* (1972); circles,
Thomy *et al.* (1969); triangles, Huff and Dash (1976); inverted triangles,
Demetrio de Souza *et al.* (1984).

As a final comment on the monolayer solid, we note that the smallest spacing
observed for the incommensurate lattice is ~ 3.10 Å and that this is $\sim 2\%$
smaller than the spacing in the ground state of the 3D solid. The difference
is significant for discussions of limited layer growth because it means that the
compressed monolayer is a relatively poor template for epitaxial growth of bulk
neon. However, layering transitions and critical points are observed for three
layers at temperatures below the bulk triple point temperature (Lerner *et al.*

1985; Hanono *et al.* 1985).

A phase diagram for Ne/graphite in pressure–temperature variables is shown in Fig. 6.8. The distinct symbols marking the various phase boundaries are the results of different experiments and illustrate the point made at the beginning of this chapter that a phase diagram is the result of synthesizing many contributions. It is remarkable that, even so, the monolayer condensation to the Ne/graphite solid was not accessible in the equilibrium experiments: at these temperatures the corresponding 3D pressure is much lower than the experimental sensitivity. Also, the $\sqrt{7}$ superlattice is not indicated in the figure; it has been observed only below 5 K. The phase diagram indicates the locations of a gas–liquid critical point at $T_c = 15.8$ K and a gas–liquid–solid triple point at $T_t = 13.6$ K (Rapp *et al.* 1981). That is, at temperatures above 10 K, the phase diagram of the monolayer Ne/graphite begins to resemble that expected for 2D Lennard-Jonesium, Section 5.2.2.1. As for 3D, the corresponding states scalings for neon are influenced by quantum corrections (de Boer and Bird 1964; Nosanow 1977), see Fig.(5.8).

Modeling of the Ne/graphite system includes study of the monolayer equation of state (Bruch *et al.* 1984), of the initial layering transitions (Bruch and Ni 1985), and of the lattice dynamics of the monolayer solid (Hakim and Glyde 1988).

6.1.4 *Argon*

Dense phases of argon are among the most amenable cases for statistical mechanical modeling in terms of pairwise interactions (Barker 1989). The argon atom is massive enough that quantum effects are small and may be well approximated by quantization of the normal modes of the low-temperature solid. Also, the atom has a low polarizablity so that many-body interactions, e.g., the triple-dipole energy of Axilrod and Teller (1943) and Muto (1943), are a small part of the cohesive energy of the solid and liquid. In the quantum theory of corresponding states, for the gas–liquid critical point and the triple point, it is a nearly classical system. The P–T phase diagram for Ar/graphite constructed by Shaw and Fain (1979, 1980) is shown in Fig. 6.9.[*]

The monolayer solid of Ar/graphite was the first of the physical adsorption systems to have a wide range of structural and dynamic data. The isotope ^{36}Ar has a large neutron scattering cross-section that enabled structural and dynamical information to be obtained for temperatures up to monolayer melting (Taub *et al.* 1977). In spite of early concerns that an electron beam would desorb the adlayer, LEED experiments with single-crystal graphite were shown to be feasible and gave information for the orientation of the adlayer lattice relative to the substrate and data for the compressibility of the lattice (Shaw and Fain 1979, 1980). Calorimetric experiments are summarized by Migone *et al.* (1984). Also, the development of intense X-ray beams enabled high resolution studies of

---

[*]The diagram in coverage–temperature variables was given by Migone *et al.* (1984).

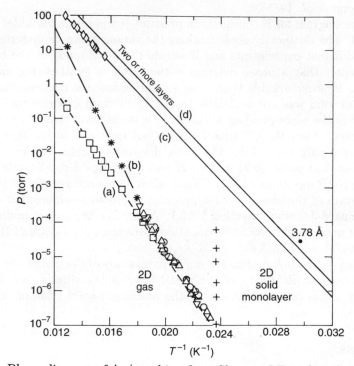

FIG. 6.9. Phase diagram of Ar/graphite, from Shaw and Fain (1979). The upper solid line denotes the bulk vapor pressure (Pollack 1964). The asterisks and long dashes denote the solid-to-fluid transition; the data are from Rouquerol *et al.* (1977), Larher (1978), and Prenzlow and Halsey (1957). The short dashes indicate Millot's fit (1979) of sublimation data. Other symbols denote data of Shaw and Fain.

the thermally expanded solid near melting (Nielsen *et al.* 1987; D'Amico *et al.* 1990).

The nearest-neighbor spacing in the ground state of the 3D solid of argon is 3.756 Å (Crawford 1977). The monolayer solid is a triangular lattice, with $L_{nn} = 3.86$ Å near $T = 0$ K (Taub *et al.* 1977); thermal expansion along the 2D sublimation curve leads to $L_{nn} = 3.97$ Å at $T = 49.7$ K (D'Amico *et al.* 1990). A model calculation of the monolayer-to-bilayer solid transition at low temperatures using quasi-harmonic lattice dynamics (Bruch and Wei 1980) showed the lattice constant there to be 3.77 Å. To achieve a change of 2% in the low temperature lattice constant of 3D solid argon requires a pressure of 1.6 kbar (Korpiun and Lüscher 1977).

The monolayer phase diagram for Ar/graphite derived from adsorption isotherms and confirmed by specific heat and structural measurements has gas, liquid, and incommensurate solid phases (only a triangular lattice is observed). There is a triple point at $T_t = 49.7$ K according to D'Amico *et al.* (1990) or at 47 K according to Migone *et al.* (1984); and a critical point at $T_c = 59$ K (Larher and Gilquin 1979) or 55 K (Migone *et al.* 1984). The early neutron diffraction experiments (Taub *et al.* 1977) already had indications that the solid-to-liquid transition is second order. This has been examined with much higher resolution in later thermodynamic experiments (Larher 1983; Migone *et al.* 1984; Larese and Zhang 1991) and by X-ray diffraction from a monolayer solid on a graphite single crystal. There is a transition from a solid with sharp Bragg spots to a liquid-like phase in which the Bragg spots broaden in both the radial and the azimuthal directions. The correlation lengths in the liquid-like phase reach 1500 Å close to the transition (D'Amico *et al.* 1990). A difference remains between the structural and thermodynamic experiments in the region of the melting transition: the thermodynamic experiments have signatures indicating two transitions closely spaced in temperature (Migone *et al.* 1984; Larese and Zhang 1991), while the structural experiments reveal only one transition (Nielsen *et al.* 1987; D'Amico *et al.* 1990).

The monolayer solid has an epitaxial alignment at $2°–3°$ from the $R\,30°$ axis of a $\sqrt{3}$ commensurate lattice (Shaw *et al.* 1978; D'Amico *et al.* 1990). The angle varies with lattice constant as shown in Fig. 6.10. The solid melts from such a configuration to a correspondingly 'rotated fluid phase'. The Ar/graphite solid was the first example in which the Novaco–McTague orientational alignment was observed and observations are in fair agreement with the perturbation theory (Shaw *et al.* 1978); better agreement is obtained for the larger lattice constant cases by going beyond the linear response approximation (Novaco 1979). In fact, there is reason to be surprised that the effect was seen in the early experiments. The theory indicates that the energy terms that drive the alignment are small enough that extrinsic effects such as steps on the substrate surface might determine the alignment.* The predicted alignment has not been seen in adsorption on Ag(111) (Unguris *et al.* 1979); it has been seen for adsorption on Pt(111), but there it was demonstrated that small degradations of the surface disrupt it (Kern *et al.* 1986c, 1988).

The first few layers of Ar/graphite apparently provide a good template for further layer growth (Seguin *et al.* 1983; Venables *et al.* 1984a,b). Although the multilayer phenomena (Ser *et al.* 1988; Youn and Hess 1990; Gay *et al.* 1991; Larese 1993; Youn *et al.* 1993) are beyond the scope of this monograph, we note that this system has been extensively studied for critical points in the further layering transitions (Ser *et al.* 1988) and for demonstration of the coupling of layer population changes at those transitions (Phillips *et al.* 1993).

There is modeling of the Ar/graphite monolayer by Hanson *et al.* (1977), Tsang (1979), Bruch and Wei (1980), and Nicholson and Parsonage (1986).

---

*It was recognized that terraces at least 1000 Å wide are needed for the energy gain of 0.2 K/atom, Fig. 3.11, on rotation from the aligned direction to outweigh estimated boundary energies at step edges (Shaw *et al.* 1978).

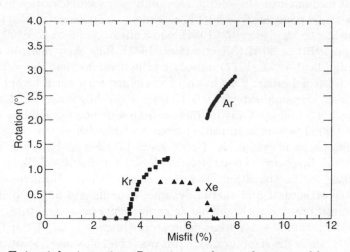

F<small>IG</small>. 6.10. Epitaxial orientation. Rotation angle as a function of fractional misfit for Ar, Kr, and Xe on graphite, from D'Amico *et al.* (1990). The angle shown is the increment from the 30° axis of the graphite. The misfits are negative for Ar and Kr and positive for Xe.

6.1.5 *Krypton*

Although the monolayer of Ar/graphite behaves as if it were a 2D system that is only slightly perturbed by the corrugation of the substrate, the 6% increase in characteristic length upon going to krypton leads to a system for which the monolayer phase diagram is dominated by effects of the substrate corrugation (Chinn and Fain 1977a,b; Price and Venables 1976; Butler *et al.* 1979, 1980; Fain *et al.* 1980; Suter *et al.* 1985; Specht *et al.* 1987). The ground state of the 3D solid has $L_{nn} = 3.993$ Å (Crawford 1977) and the monolayer condenses at low temperatures to the $\sqrt{3} \times \sqrt{3} R\,30°$ lattice. Two molecular examples with similar spacings of nearest centers in the 3D ground states are CH_4 with $L = 4.16$ Å (Baer *et al.* 1978) and N_2 with $L = 3.994$ Å (Scott 1976); both have quite robust $\sqrt{3}$ commensurate lattices. It is indicative of the importance of anharmonicity in the problem (Price and Venables 1976; Venables and Schabes-Retchkiman 1978) that the commensurate lattice is reached by a dilation from the characteristic spacing in the 3D solid: the 6.5% dilation of the krypton is readily achieved, but considerable additional stress is required to force the 2% compression needed to reach the corresponding lattice for Xe/graphite.

An overview of the monolayer phase diagram for Kr/graphite is shown in Fig. 6.11; an enlarged version, with significant differences in the region of the transitions from commensurate solid to fluid, is shown in Fig. 6.12, taken from the work of Specht *et al.* (1987). There is no gas–liquid critical point evident in

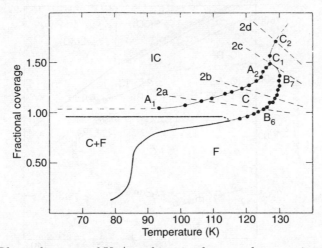

FIG. 6.11. Phase diagram of Kr/graphite, in the monolayer regime, in terms of fractional coverage and temperature, from Butler *et al.* (1980). C, F, and IC denote commensurate, fluid, and incommensurate phases, respectively. Dashes indicate paths of specific heat scans. There is a tricritical point at the extrapolated intersection of the IC to C + F and C to F boundaries and a multicritical point at C_1. See also Fig. 6.12.

these diagrams! In fact, this is proposed to be a system with an incipient triple point: the substrate corrugation has such a large effect on the monolayer fluid that the distinction between gas and liquid is lost (Butler *et al.* 1979; Migone *et al.* 1983a; Niskanen and Griffiths 1985). At low temperatures, the fluid that coexists with the solid has the very low density characteristic of gas at a sublimation curve, and at temperatures above 90 K it has a density characteristic of the coexistence at a melting curve.*

The density difference between the fluid and the commensurate solid narrows with increasing temperature and appears to vanish at a tricritical point at a temperature above 100 K (Larher and Terlain 1980; Suter *et al.* 1985). This is quite remarkable when we recall that the melting transition in 3D solids is first order, with no sign that the density difference between liquid and solid vanishes.[†]

---

*More precisely the changeover occurs at $T \approx 85$ K. The ratio of this temperature to the 3D critical temperature of krypton (209 K) is 0.41. For Lennard-Jonesium, the ratios of the 2D triple point and critical point temperatures to the 3D critical temperature are 0.31 and 0.40 respectively. This indicates that the absence of a triple point is more a matter of the corrugation effects raising the triple point than of lowering the critical point.

[†] For some materials, the melting curve in the $p - T$ diagram has a vertical tangent. Those are isolated conditions; here the monolayers of Kr/graphite and Xe/graphite appear to have a *range* of temperatures where the density difference of coexisting liquid and solid vanishes.

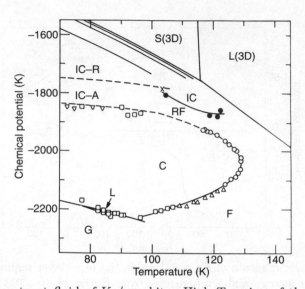

FIG. 6.12. Reentrant fluid of Kr/graphite. High T region of the Kr/graphite phase diagram, from Specht *et al.* (1987). The reentrant fluid (RF) phase is a domain wall fluid bounded by commensurate (C) and incommensurate (IC) regimes. S(3D) and L(3D) are the bulk solid and liquid phases. Solid lines denote first-order transitions and dashed lines denote continuous transitions.

The proposed phase diagram of Butler *et al.* (1980), shown in Fig. 6.11, has a tricritical point at ≈ 115 K on the fluid-commensurate solid phase boundary; at higher temperatures there is no longer a discontinuous jump in density upon solidification. As in the case of Ar/graphite, experimental techniques differ in the assignment of the phase boundary when the possible first order transition involves small discontinuities. Suter *et al.* (1985) find that the fluid to commensurate solid transition becomes continuous, to within their resolution, at 117 ± 2 K while the X-ray diffraction measurements show a small discontinuity up to 130 K (Specht *et al.* 1987).

The higher density region denoted as fluid in Fig. 6.12 is in fact a subtly disordered phase whose properties were first made apparent with high-resolution X-ray diffraction. Before it was recognized, results for the transition from commensurate to incommensurate solid were puzzling because it appeared to be a continuous transition to a hexagonal incommensurate lattice. Analysis in terms of solids with commensurate domains separated by arrays of domain walls had led to the expectation (Bak *et al.* 1979) that the alternatives were a continuous transition to a striped lattice, which was not observed, or a discontinuous transition to a hexagonal lattice, also not observed. The fluid phase between the two regions of solid in Fig. 6.12 is termed the domain wall fluid and it provides an escape from the contradiction. The well-separated walls in the uniaxial incommensurate lattice at small misfit are subject to large meanderings (Coppersmith et al. 1982) with a low level of thermal excitation and thus there is a disorder that is manifested in the details of the diffraction profile. The longitudinal (radial) width of the Bragg spot yields the correlation length in the reentrant fluid, while the transverse (azimuthal) width reflects the amplitude of orientational fluctuations.

Another effect of the proximity to a commensurate lattice is that the orientational epitaxy takes on a nonzero rotation only for a misfit larger than 3.5%, as shown in Fig. 6.10. The onset of rotation appears to be continuous: any discontinuity in the angle is less than 0.2° (D'Amico et al. 1984). Such an effect was predicted by Shiba (1979, 1980) in major papers that treated a hexagonal array of domain walls. The diffraction pattern of the aligned hexagonal incommensurate lattice has additional peaks at locations characteristic of a hexagonal array of well-separated domain walls (Stephens et al. 1984; Specht et al. 1987).

The incommensurate solid* can be compressed by increasing the 3D gas pressure at constant temperature or by decreasing the temperature at constant gas pressure. This process terminates with bilayer formation. At 47 K, Chinn and Fain (1977b) found the bilayer has $L_{nn} = 4.02 \pm 0.02$ Å, which coincides with the spacing in bulk Kr at this temperature. Figure 6.13 shows a portion of the Kr/graphite phase diagram where the second layer condensation line, second layer melting, and third layer condensation line are present (Gangwar and Suter 1990). There is an interplay between transitions which initially are assigned to distinct layers.

The low-temperature specific heat data for the commensurate solid are analyzed in terms of an energy gap in the spectrum of excitations. A gap of about 10 K is found (Shirakami et al. 1990), which corresponds to a corrugation amplitude large enough to stabilize the commensurate lattice as the ground state (Gooding et al. 1983). This corrugation amplitude has a magnitude that is consistent with a proposed enhancement by the dielectric anisotropy of the graphite (Carlos and Cole 1980a; Bruch 1991a).

Modeling of the Kr/graphite monolayer includes simulations of the coexistence of 2D gas and commensurate solid (Hanson and McTague 1980; Koch and Abraham 1986), quasi-harmonic theory of the solid (Gordon and Villain 1985; Shrimpton et al. 1988) and analysis of the domain wall network in the incommensurate solid at small misfit at zero temperature (Gordon and Lançon 1985; Gooding et al. 1983) and when thermally excited (Abraham et al. 1984; Koch et al. 1984). Hybridization of the normal modes of the commensurate monolayer solid with the substrate Rayleigh wave has been modeled by De Wette et al. (1983) and Shrimpton and Steele (1991) have evaluated the dynamic structure factor of the commensurate solid. Bhethanabotla and Steele (1988b) found that adjustments of the interaction model to include 4 substrate-mediated interac-

---

*A difference in the overlayer height of krypton in the commensurate and incommensurate solid phases has been sought but not detected (Guryan et al. 1988).

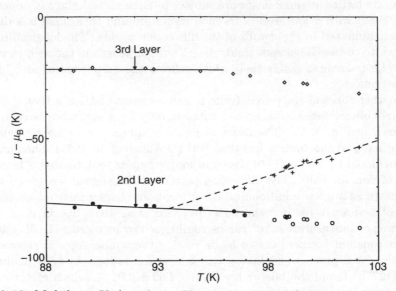

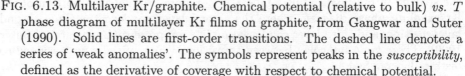

FIG. 6.13. Multilayer Kr/graphite. Chemical potential (relative to bulk) *vs. T* phase diagram of multilayer Kr films on graphite, from Gangwar and Suter (1990). Solid lines are first-order transitions. The dashed line denotes a series of 'weak anomalies'. The symbols represent peaks in the *susceptibility*, defined as the derivative of coverage with respect to chemical potential.

tions and substrate anisotropy in the corrugation led to good agreement with experimental data for the monolayer fluid near 100 K.

6.1.6 Xenon

The ground state of the 3D solid of xenon, with $L_{nn} = 4.336$ Å (Crawford 1977), has a length scale 8% larger than that of krypton. Thus the $\sqrt{3}$ lattice would correspond to $\sim 2\%$ compression, which requires 2 kbar for the 3D solid (Korpiun and Lüscher 1977). The monolayer Xe/graphite condenses as a hexagonal incommensurate lattice down to the lowest observed temperatures, with $L_{nn} = 4.32$ Å at 25 K (Hong *et al.* 1989). The $\sqrt{3}$ lattice is reached, before the bilayer forms, by compression at sufficiently low temperatures (Schabes-Retchkiman and Venables 1981; Hamichi *et al.* 1989; Nuttall *et al.* 1993). There does not seem to be a striped, uniaxially incommensurate solid phase between the triangular incommensurate and commensurate solids (Faisal *et al.* 1986). For most of the monolayer regime, the xenon shows the features expected for a 2D system that is only slightly perturbed by the substrate corrugation. The density–temperature phase diagram derived from adsorption isotherms, Fig. 4.4, has the appearance of a classical triple point system with $T_t = 99$ K and $T_c = 117$ K. However, there are surprises in following the melting curve to temperatures above 120 K.

The thermal expansion of the 2D solid along the monolayer sublimation curve has been followed by X-ray and electron diffraction. The nearest-neighbor spacing in the triangular lattice increases from 4.32 Å at 25 K to 4.34 Å at 60 K, 4.54 Å at 80 K, and 4.59 Å at 97 K, just below the triple point (Hong *et al.* 1989; Nuttall *et al.* 1993; E. M. Hammonds *et al.* 1980; Schabes-Retchkiman and Venables 1981). The lattice constant of the solid remains roughly constant at 4.55–4.59 Å along the melting curve up to 150 K (Heiney *et al.* 1983a; Dimon *et al.* 1985; Hong *et al.* 1989). For comparison, the values along the sublimation curve of the 3D solid are $L_{nn} = 4.42$ Å at 100 K and 4.49 Å at 160 K.

The melting transition at and near the triple point temperature (99 K) is found to be first order in both calorimetric (Litzinger and Stewart 1980) and X-ray diffraction experiments (E. M. Hammonds *et al.* 1980). The density difference between the coexisting solid and fluid at the melting line decreases as the temperature increases. A discontinuity is reported to persist to almost 150 K in calorimetric experiments (Tessier 1984; Collela and Suter 1986; Gangwar *et al.* 1989; Jin *et al.* 1989), but the X ray diffraction experiments find that the transition becomes continuous above about 125 K (Dimon *et al.* 1985; Nuttall *et al.* 1995). There is some evidence that there are two stages to the melting for temperatures in the range 110–120 K (Zerrouk *et al.* 1994). The dense fluid near the melting curve has strong orientational and positional correlations (Nagler *et al.* 1985; Zerrouk *et al.* 1994; Nuttall *et al.* 1994). While many of the features are similar to what would be expected in scenarios of continuous 2D melting, the roles of the substrate corrugation and of interchanges of atoms between the first and second layers of the xenon film add enough complications that it is not yet established that the melting proceeds by unbinding of dislocations.

The incommensurate solid has a complex evolution of the orientational alignment as a function of L_{nn}, as shown in Fig. 6.10. There is a small rotation angle at low temperatures, but it decreases to zero as the melting condition is approached. The xenon solid melts from the aligned (30°) phase. There is a pronounced hysteresis to the orientational alignment: nonzero values are obtained only on reducing the temperature for a monolayer solid, and not for the solid when it is formed by condensation at very low temperatures (Hong *et al.* 1987, 1989; Hamichi *et al.* 1989). Slow kinetics in the monolayer regime have been proposed as the resolution of the discrepancies between different experiments (Hamichi *et al.* 1989; Nuttall *et al.* 1993). There has been reason to anticipate that such problems would arise near the melting, because topological defects are proposed to have an important role. The suggestion is that they are more pervasive for the xenon layers.

Because the length scales and alignments of the xenon solid have such close correspondence to features of the graphite basal plane, there has been concern that the melting of the solid might be strongly influenced by the hexagonal external potential field of the substrate. Several related observations indicate that this effect is small. An X-ray diffraction measurement of the melting of Xe/Ag(111) showed (Greiser *et al.* 1987) quite similar development of azimuthal and radial broadening of the Bragg peak, although the orientational alignment was quite different for Xe/graphite. The triple point of Xe/Pt(111), which melts from a $\sqrt{3}$ commensurate lattice, is at about 98 K (Poelsema et al. 1985) and the triple point of Xe/MgO(001) is about 101 K (Coulomb et al. 1984). Tessier and Larher (1980) surveyed the triple point temperatures of xenon on several lamellar halides and found the values to be 100 ± 2 K.

As remarked at the beginning of this subsection, there is a domain of temperature and pressure where the commensurate solid is stable. It is observed below about 70 K for conditions just before bilayer condensation. The observations differ on whether the transition from the hexagonal incommensurate solid to the commensurate solid is first order (Nuttall et al. 1993) or continuous (Hamichi et al. 1989)* and on the value for the maximum temperature at which the commensurate solid occurs. More surprising is that there are quite different identifications for the structure that succeeds the commensurate solid upon further compression. Hamichi et al. find a mutually commensurate bilayer with $L_{nn} = 4.354 \pm 0.004$ Å, which is indistinguishable from the bulk value at the temperatures of observation (50–60 K). Hong and Birgeneau (1989) and Nuttall et al. (1993) find the bilayer to be mutually incommensurate below 60 K, with the first layer lattice locked into the $\sqrt{3}$ configuration.

Apart from harmonic zero point energy terms at very low temperatures, the dense phases of xenon are accurately treated by classical statistical mechanics. However, the xenon atoms are the most polarizable of the inert gas series and many-body interactions are the largest for xenon (Barker 1986, 1989, 1993). The triple-dipole interactions reduce the cohesive energy of 3D solid xenon by 10% and the McLachlan energy, which incorporates similar physics, reduces the 2D sublimation energy of monolayer xenon by 10 to 15% (Bruch and Ni 1985). Thus, the xenon dense phases have quantitatively significant contributions from many-body forces that remain the subject of discussion for the 3D inert gas solids. The principal many-body terms included in the modeling are those based on perturbation theory for dispersion forces (Barker 1989); while this clearly begs many questions, the procedure gives a quantitative account of an exceedingly large body of data.

Modeling of the Xe/graphite monolayer includes cell model approximations (Price and Venables 1976, Schabes-Retchkiman and Venables 1981); lattice dynamics for the commensurate solid (De Rouffignac et al. 1981); and computer simulations of the solid at relatively high temperatures (Abraham 1983a,b; Koch and Abraham 1983; Abraham 1984; Suh et al. 1989). Joos et al. (1983) examined the relative stability of commensurate and incommensurate configurations of monolayer xenon. Abraham and his coworkers found in their simulations of the melting at temperatures above 100 K that there is an appreciable density of second layer atoms, and that exchanges of first and second layer atoms are an important component of the thermally excited motions of the system. Suh et

---

*The Bak et al. (1979) analysis in terms of domain wall crossings leads to the conclusion it should be first order.

al. used a slightly different holding potential model that stabilizes the commensurate monolayer, and identified a transition to a rotated lattice (orientational epitaxy).

6.1.7 *Molecules*

Molecules have additional degrees of freedom that may enable certain experiments and that also lead to new classes of structures. We give a brief sketch here to indicate the possibilities for several molecules adsorbed on graphite. Patrykiejew and Sokołowski (1989), Knorr (1992), Steele (1993, 1996), and Marx and Wiechert (1996) have given extensive reviews for polar and nonpolar molecules adsorbed on graphite; Etters (1989) reviewed the 3D solids of diatomic molecules; much earlier, English and Venables (1974) systematized phenomena of the solids in terms of simple models that emphasized aspect ratios of the molecules. In general the molecular solids are not triangular lattices. The lattices have basis units of several molecules of distinct orientations at low temperatures. Then, at intermediate temperatures, there may be phase transitions to plastic solids in which the molecular orientations are disordered.

An underlying concern is that the adsorbate may become chemically active and hence that the holding potential would involve specific bonding mechanisms rather than the van der Waals interactions that dominate the inert gas adsorption. On graphite this does not occur for rather small molecules with strong internal bonds. Some information on the electronic states of adsorbed N_2 and O_2 is available from photoemission experiments (Tillborg *et al.* 1993).

In this connection, it is interesting to note the situation for the adsorption of water on graphite, which has been the subject of numerous studies because of its fundamental importance. It was realized long ago that untreated graphite adsorbs water strongly. Kiselev and Kovaleva (1956, 1959) showed that heat treatment reduces the saturation coverage to only a monolayer at ordinary temperatures. Studies by Miura and Morimoto (1991) indicate the importance of the edges of the basal planes with or without impurities. Under these circumstances, water is actually chemisorbed. The conclusion that 'clean' and flat graphite is not wet by water has its irony.

The organization of the remainder of this subsection is as follows: we first summarize the situation of methane, which is orientationally disordered at most temperatures, and of carbon dioxide, which is ordered over most of the range of the solid; then we review information for three diatomics with transitions to plastic solid phases; finally, we very briefly summarize studies on a series of adsorbed hydrocarbons.

6.1.7.1 *Methane* The nearly spherical methane molecule has rotational ordering in the 3D solid only at temperatures below 20 K for CH_4 (Press 1972) and 27 K for CD_4 (Press *et al.* 1970). In the monolayer, rotational diffusion has been demonstrated at temperatures near 50 K (Thorel *et al.* 1982; Quateman

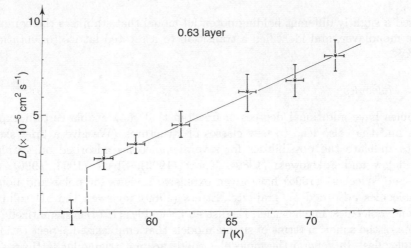

FIG. 6.14. Diffusion of 2D CH_4. Diffusion coefficient of 0.63 layer of methane on graphite, as measured by quasi-elastic neutron scattering, from Coulomb *et al.* (1981). The data are for the liquid–vapor coexistence regime, except for the data point for 2D solid methane at 56 K.

and Bretz 1984). There is a decrease in the rotational mobility in monolayer CH_4/graphite only below 17 K (Larese and Rollefson 1985). The solid is orientationally ordered at 4 K (Bomchil *et al.* 1980; Smalley *et al.* 1981). The monolayer solid is a triangular lattice that has L_{nn} both larger and smaller than the 4.26 Å of the $\sqrt{3}$ lattice (Vora *et al.* 1979; Beaume *et al.* 1984; H. K. Kim *et al.* 1986a; Gay *et al.* 1986). At low temperatures it condenses in the commensurate lattice and then thermally expands to an incommensurate lattice (Vora *et al.* 1979). It can also be compressed (Vora *et al.* 1979; Humes *et al.* 1988) to spacings less than 4.26 Å before the bilayer condenses. At intermediate temperatures it may form a *floating solid*, which has the 4.26 Å lattice constant but no energy gap in the excitation spectrum (Phillips 1984; Hakim *et al.* 1988). The evolution of the multilayer film with thickness has been studied for the stacking structure (Larese *et al.* 1988c), for the trends in the critical points of the layer condensations (Goodstein *et al.* 1984; Pettersen *et al.* 1986), and for surface melting (Bienfait *et al.* 1990).

The large incoherent neutron–proton scattering cross-section enables quasi-elastic neutron scattering measurements for adsorbed CH_4. Bienfait and coworkers have used this method to obtain diffusion constants of fluid hydrocarbon layers as a function of coverage and temperature (Coulomb *et al.* 1981; Coulomb and Bienfait 1986). An example is shown in Fig. 6.14.

Modeling of CH_4/graphite includes several approximations to the monolayer solid (Phillips 1984; Hunzicker and Phillips 1986; Bruch 1987; Hakim *et al.* 1988); study of layering transitions (Phillips 1986, 1989; Phillips and Story 1990);

and simulations of the monolayer melting (H.-Y. Kim and Steele 1992) and of
the commensurate–incommensurate transition (S. Jiang *et al.* 1993).

6.1.7.2 *Carbon dioxide* CO_2 is a linear molecule with a large quadrupole moment. Electrostatic energies are an important part of the cohesive energy of the
3D solid (Hirshfeld and Mirsky 1979) and may be responsible (Bruch 1983b) for
the fact that the monolayer solid does not wet the graphite basal plane below
100 K (Terlain and Larher 1983; Morishige 1993). This effect is expected to
depend sensitively on the single molecule interaction with graphite (Bottani *et
al.* 1994a). The 3D solid has ordering of the molecular axes until close to the
melting temperature. For modeling of CO_2/graphite, see K. D. Hammonds *et
al.* (1990).

6.1.7.3 *Diatomic molecules* The linear molecules N_2 and CO both have 3D
and 2D solids in which the molecular axes are ordered at low temperatures
and in which there are transitions to plastic solids with disordered molecular
axes at intermediate temperatures (Diehl *et al.* 1982; Morishige *et al.* 1985;
You and Fain 1985b; Feng and Chan 1993). The ordering is attributed to the
effect of the molecular electrostatic quadrupole moments, and the scaling of the
disordering temperatures in 3D (but not in 2D) is consistent with this. For O_2,
with a smaller quadrupole moment, magnetic orderings are proposed (Nielsen
and McTague 1979). The packing problem for the rod-like molecule has more
possibilities (Harris and Berlinsky 1979) than for the spherical inert gases. The
monolayers of N_2 and CO are incipient triple point systems. The portion of the
phase diagram for monolayer N_2/graphite which shows this behavior is presented
in Fig. 6.15. Marx and Wiechert (1996) have given a comprehensive review of
the ordering and phase transitions in adsorbed monolayers of N_2 and CO.
Nitrogen At low temperatures the N_2 monolayer condenses in a 2×1 commensurate herringbone lattice (Eckert *et al.* 1979; Diehl *et al.* 1982) and may be
compressed to a UIC lattice with herringbone order (Diehl and Fain 1983). Under
further compression it may form a nonplanar herringbone lattice (You and Fain
1985a) or a pinwheel lattice with similarities to the ordering in a close-packed
(111) plane of the ordered 3D lattice (Wang *et al.* 1989). Because of the very different ordering of the monolayer and the 3D solid (Scott 1976), arguments about
templates for epitaxial growth (Venables *et al.* 1984a,b) suggest that there may
be only limited layer growth for such a system. However Diehl and Fain (1982,
1983) report a second layer of solid forms, and model calculations (Kuchta and
Etters 1987a; Roosevelt and Bruch 1990) find that the low-temperature bilayer
is stable relative to the 3D solid.

The orientational disorder transition, at about 25 K, of the commensurate
N_2/graphite has received much attention because it is expected to be a first-
order transition according to the Domany *et al.* (1978) classification scheme.
The calorimetric (Migone *et al.* 1983b), LEED (Diehl and Fain 1983), and NMR
(Sullivan and Vassiere 1983) observations are consistent with it being a weakly
first-order transition. Computer simulations with quadrupole lattice models now

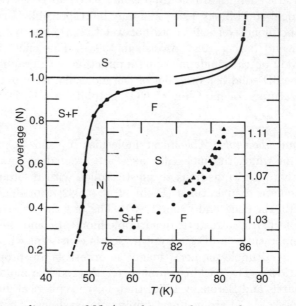

FIG. 6.15. Phase diagram of N_2/graphite, in fractional coverage and temperature, from Chan *et al.* (1984). There is only one fluid phase and this system is said to show incipient triple point behavior.

also reach this conclusion (Mouritsen and Berlinsky 1982; Opitz *et al.* 1993; Marx *et al.* 1994b).

The N_2/graphite monolayer phase diagram lacks a gas–liquid–solid triple point (Migone *et al.* 1983a; Chan *et al.* 1984), as does Kr/graphite (Butler *et al.* 1979). Thus, the substrate corrugation has a major effect on the phase diagram. The zone-center frequency gap of the commensurate solid has been measured with inelastic neutron scattering (Hansen *et al.* 1990; Lauter *et al.* 1990). Attempts to fit the data within the context of corrugation models used for the inert gases failed badly, underestimating the gap by 50% (Bruch 1991a). Models which include the interaction of the molecule's electrostatic quadrupole with estimated surface electric fields of the graphite (Vernov and Steele 1992) do better (Hansen *et al.* 1992; Hansen and Bruch 1995).

Because of the variety of monolayer solid structures and the strong dependence of the melting temperature on the coverage, the N_2/graphite system has been the subject of many model calculations. The structure of the solid has been investigated with potential energy constructions (Bruch 1983b; Kuchta and Etters 1987a,b), lattice dynamics theories (Cardini and O'Shea 1985; Roosevelt and Bruch 1990), and Monte Carlo simulations (Peters and Klein 1985; Bhethanabotla and Steele 1989). The melting of the submonolayer solid has been treated with molecular dynamics simulations (Joshi and Tildesley 1985; Hansen et al. 1995) and with Monte Carlo simulations (Etters et al. 1990, 1993). The spectrum of excitations for the monolayer solid both below and above the orientational disorder transition has been evaluated using molecular dynamics simulations (Lynden-Bell et al. 1985; Hansen and Bruch 1995). The spatial variation of orientational disorder in the uniaxially incommensurate UIC phase was demonstrated in Monte Carlo simulations (Peters and Klein 1983). Most of the work uses rather simple intermolecular force models, but the results are in good agreement with experiments both for the sequence of observed structures and for the disordering transitions in the thermally excited monolayer.

Carbon monoxide The phase diagram for CO/graphite is similar to that for N_2/graphite although the data are less extensive (Terlain and Larher 1980; Morishige et al. 1985; You and Fain 1985b; Bojan and Steele 1987b; Inaba et al. 1988, 1991; Wiechert and Arlt 1993; Feng and Chan 1993). A difference for CO/graphite is that there is end-to-end disorder of the molecule in the dense phases at most temperatures; a transition is observed below 5 K where this residual entropy is removed (Inaba et al. 1988; Wiechert and Arlt 1993; Pereyra et al. 1993). An unusual variant of the CO monolayer is the use (You et al. 1986; You and Fain 1986) of CO and Ar mixtures to create conditions favorable for a monolayer pinwheel structure in which the Ar would serve as the 'pin' atoms. There has been limited modeling of the CO/graphite system (Belak et al. 1985; Peters and Klein 1985; Marx et al. 1994a,c).

Oxygen O_2 has a magnetic dipole moment and only a small electrostatic quadrupole moment. This apparently is the basis for the large difference between adsorption second virial coefficients of O_2/graphite relative to those for N_2/graphite and CO/graphite (Bojan and Steele 1987b). Several monolayer lattices are observed (Nielsen and McTague 1979; Heiney et al. 1983b; Toney and Fain 1984, 1987). These include an example of orientational epitaxy for a nontriangular adlayer lattice (Toney et al. 1983). Magnetic ordering of the dipoles occurs (Nielsen and McTague 1979; Pan et al. 1982; Etters and Kobashi 1984; Etters and Duparc 1985; Duparc and Etters 1987). The melting of the monolayer solids occurs at temperatures that are rather low compared to those for N_2/graphite, especially when the strength of the intermolecular forces is considered; this has been studied with molecular dynamics simulations by Bhethanabotla and Steele (1987, 1988a, 1990).

6.1.7.4 *Hydrocarbon series* There also are calorimetric, structural and dynamical measurements for small hydrocarbon molecules (Taub 1980, 1988; Hansen and Taub 1987, 1992), following the series ethane (Coulomb et al. 1979; Taub et al. 1980; Suzanne et al. 1983; Coulomb et al. 1985b; Coulomb and Bienfait 1986; Osen and Fain 1987; Zhang and Migone 1989; Newton and Taub 1996), butane (Wang *et al.* 1985; Alkhafaji and Migone 1993; Herwig *et al.* 1994), and hexane (Krim *et al.* 1985). Changes in the intramolecular vibrations related to the adsorption have been observed by inelastic neutron scattering. The low-temperature solids for the hydrocarbon series include examples in which molecules are tilted out of the basal plane. The melting temperature depends strongly on the mechanism by which free volume is created and has been discussed in terms of the 'footprint' of the molecule (Hansen *et al.* 1993; Peters and Tildesley 1996).

Remarkably enough, observations for unsaturated hydrocarbons indicate that they behave as physically adsorbed layers on graphite. Adsorbed ethylene, C_2H_4/graphite, has been intensively studied because of its complex phase diagram in the multilayer regime. Calorimetry (Inaba and Morrison 1986; H. K. Kim *et al.* 1986b, 1988), X-ray diffraction (Mochrie *et al.* 1984), neutron scattering (Larese *et al.* 1988a,b), nuclear magnetic resonance (Larese and Rollefson 1983), and LEED (Eden and Fain 1991, 1992) have all been applied. There has also been modeling of the interactions in the dense monolayer (Moller and Klein 1988; Cheng and Klein 1992). There are a few reports of thermodynamic measurements for adsorbed acetylene, C_2H_2/graphite (Menaucourt *et al.* 1980; Peters *et al.* 1986; Alkhafaji and Migone 1992a; Trabelski and Larher 1992).

6.2   Physically adsorbed layers on metals

The phenomena in several examples of physically adsorbed layers of inert gases on metals parallel those in the graphite examples discussed in Section 6.1. There is much less direct thermodynamic information for the adsorption on metals, because the surface-to-volume ratios of samples of high homogeneity are usually small.* Scattering experiments with single crystal-substrates have been the main source of information. In this section, we survey adsorption phenomena on three face-centered cubic, fcc, metals, silver, platinum, and palladium, to show analogies to the graphite cases and to introduce some ideas that have been developed more for adsorption on metals than on graphite. We refer briefly to adsorption on copper and on tungsten. Xenon is a favored choice as the adsorbate in many experiments because it scatters X-rays and electrons strongly. It has a stronger binding to the substrate than the lighter inert gases and therefore has a lower vapor pressure at a given temperature. Effects such as electron-stimulated desorption are less troubling for xenon than for argon (Moog *et al.* 1983). There are indications that the holding potential for xenon on some metals has a substantial chemical component (Ishi and Viswanathan 1991; Müller 1990, 1993).

Table 6.2   *Parameters of some metal substrates.*[†]

Ni	Cu
$3d^84s^2$	$3d^{10}4s^1$
2.49	2.55
Pd	Ag
$4d^{10}5s^0$	$4d^{10}5s^1$
2.75	2.89
Pt	Au
$5d^{10}6s^0$	$5d^{10}6s^1$
2.77	2.88

---

*Techniques such as LEED isotherms (Unguris *et al.* 1979) and the quartz microbalance (Krim 1991) have been used to measure coverage on single-crystal surfaces as a function of 3D gas pressure. The largest gap is the absence of specific heat measurements; thus far *a.c.*-measurements have been applied only to helium and hydrogen adsorption (Kenny and Richards 1990 (see also Taborek 1990); Phelps *et al.* 1993).

[†]Atomic electron configuration and nearest-neighbor spacing, in Å, in the ground state fcc lattice of metals in Groups VIII and IB of the periodic table.

Table 6.3   *Survey of studies of inert gas physisorption on metal surfaces.*[†]

Adsorbate	Ag(111)	Pt(111)	Cu(110)	Pd(100)
Xe	LEED	HAS	LEED	LEED
	HAS		HAS	AES
Kr	LEED	HAS	LEED	LEED
	HAS		AES	
Ar	LEED	HAS	LEED	
	HAS		AES	

For reference, we show in Table 6.2 the two adjacent columns in the Periodic Table of the elements that contain most of the metals considered as substrates in this section. The most extensive studies of physical adsorption on metals have silver and platinum as the substrate. To date, the Ag(100) and Ag(111) surfaces are still regarded as nearly ideal, simple, terminations of the bulk stacking; this has recently been confirmed for Ag(111) by Statiris *et al.* (1994). For platinum and gold, reconstructions of both the (100) and (111) surfaces are observed (Abernathy *et al.* 1992, 1993; Barth *et al.* 1990; Sandy *et al.* 1991, 1992; Grübel *et al.* 1993). For gold, the (111) surface has a uniaxial reconstruction below 865 K (Harten *et al.* 1985; Sandy *et al.* 1991) that becomes an isotropic compression of the top layer above 880 K. For platinum the (111) surface is reconstructed at temperatures above 1330 K. The physisorption experiments on Pt(111) are all at temperatures that are below the threshold for the reconstruction.

Finally, we present in Table 6.3 a brief summary of the systems and methods discussed in more detail in the following subsections and in Table 6.4 a summary of the monolayer heats of adsorption for the systems listed in Table 6.3.

[†]The entries identify techniques used to study the monolayers. LEED indicates low energy electron diffraction measurements, including structure and isotherms; HAS denotes study by helium atomic acattering and includes diffraction and inelastic scattering; AES denotes study with either Auger or photoemission spectroscopies. Some 'leading references' for LEED are: Xe/Ag(111), Unguris *et al.* (1979); Kr/Ag(111) and Ar/Ag(111), Unguris *et al.* (1981); Xe/Cu(110), Glachant *et al.* (1981) and Berndt (1989); Kr/Cu(110), Glachant *et al.* (1981); Ar/Cu(110), Horn *et al.* (1982); Xe/Pd(100), Moog and Webb (1984) and Wandelt and Hulse (1984); Kr/Pd(100), Moog and Webb (1984). With HAS, for Xe, Kr, and Ar on Ag(111), see Gibson and Sibener (1988a); Xe/Pt(111), Kern *et al.* (1988); Kr/Pt(111), Kern *et al.* (1987a); Ar/Pt(111), Zeppenfeld *et al.* (1992a); Xe/Cu(110), Ramseyer *et al.* (1994b). Examples of AES are: Kr/Cu(110) and Ar/Cu(110), Horn *et al.* (1982); Xe/Pd(100), Wandelt and Hulse (1984) and Horn *et al.* (1978). More extensive citations and additional adsorbate/substrate combinations are given in the text.

Table 6.4   *Monolayers heats of adsorption*
q_1 *in kelvin for several inert gas/substrate*
combinations treated in this chapter.[†]

Substrate	Xe	Kr	Ar
Bulk	1907	1342	930
Ag(111)	2610±60	1750±60	1150±80
Pt(111)	~3700	~1800	~1100
Cu(110)	2620±70	1510±50	–
Graphite	2770±50	1990±25	1380±25

6.2.1   *Adsorption on the (111) face of silver*

The (111) face of the close-packed silver crystal is a triangular close-packed lattice, in contrast to the rather open honeycomb lattice of the basal plane surface of graphite. In addition, the delocalized conduction electrons of the silver help to make a smoother effective surface for an adsorbed inert gas atom than would arise for a dielectric with atoms of the same van der Waals radius as the silver atoms. Together, these arguments lead to the conclusion that the Ag(111) surface is likely to appear atomically smooth for an adsorbed inert gas atom. There is some experimental evidence that the Ag(111) surface is indeed quite smooth, from the analysis of diffracted intensities in helium scattering (Boato *et al.* 1978; Horne *et al.* 1980; Bruch and Phillips 1980) and from the analysis of friction for an adsorbed krypton layer measured with a quartz microbalance (Krim *et al.* 1991; Cieplak *et al.* 1994). However, such arguments must be regarded cautiously, as shown by the pronounced registry effects for inert gas adsorption on the (111) face of platinum, Section 6.2.2.

The inert gases Ar, Kr, and Xe form islands on Ag(111) at low temperatures. If a small dosage of gas amounting to a fraction, say $\frac{1}{2}$, of a nominal monolayer is put on the surface at vanishing 3D vapor pressure (low temperature), the adsorbate forms a close-packed lattice with nearest-neighbor spacing *ca.* 2% larger than that of its 3D solid at the same temperature (Chesters *et al.* 1973; Roberts and Pritchard 1976; Cohen *et al.* 1976). A low density 2D gas coexists with the solid islands. At temperatures as low as 15 K there is sufficient surface mobility for xenon atoms to order (Cohen *et al.* 1976). The energy and length scales of the lattice are close to those of the 3D solids, allowing for the change from 12 to 6 nearest neighbors in calculating the energy (Bruch *et al.* 1976a). The thermal expansion of the 2D solid along its sublimation curve is similar to that of the 3D solid and is much larger than that of the substrate. The 2D solids can be compressed in a controlled fashion by increasing the pressure of the 3D gas beyond that at which the monolayer condenses; simultaneous coverage

[†]Sources for the entries are: bulk values, Table A.3; Ag(111) and graphite cases, Unguris *et al.* (1981); Pt(111) cases, Kern and Comsa (1991); Xe/Cu(110), Ramseyer *et al.* (1994b); Kr/Cu(110), Glachant *et al.* (1981).

and lattice constant measurements show (Unguris *et al.* 1979) that the low temperature solid is essentially free of vacancies.

The compression of the 2D solid continues until a second layer condenses. Then the lateral nearest-neighbor spacing is identical to that in the 3D solid to within present experimental precision, which is ± 0.004 Å in the X-ray diffraction experiment of Dai *et al.* (1994). The fact that there is a negligible discontinuity in the lattice constant as the bilayer forms has the consequence (Bruch *et al.* 1985) that the chemical potential at monolayer completion is mainly determined by the polarization potential of a second layer atom interacting with the substrate. Thus, at a given temperature, the 3D gas pressure at which the bilayer forms is quite similar for Xe/Ag(111), Xe/Pt(111), Xe/graphite, and Xe/Pd(100).*

The smooth behavior is a great advantage in building a quantitative understanding of monolayers, since the lateral interactions are probed without complications from the periodic components of the adatom–substrate potential. It appears that the effect of the substrate may be accurately modelled by a planar semi-infinite uniform dielectric continuum with silver electrodynamics (Bruch and Phillips 1980; Phillips and Bruch 1985, 1988). The layer condensation transitions are first order, with discontinuities in the coverage. The 2D sublimation is unusual, because at temperatures above 70 K the density of what is believed to be an intrinsic 2D gas phase (Unguris *et al.* 1982) is much higher than analogies to the 3D solids would imply; densities up to 10% of the 2D solid density are observed. There is no direct observation of the 2D triple point, estimated (Phillips and Bruch 1985) to be near 95 K for Xe/Ag(111). X-ray diffraction experiments at the melting curve of monolayer Xe/Ag(111), at 125 K to 135 K, show (Greiser *et al.* 1987) an apparently continuous melting transition with quantitative similarities to the Xe/graphite case.[†]

The lattice constants of monolayer Ar, Kr, and Xe on Ag(111) are close to those of the same adsorbates on the basal plane surface of graphite at large misfits, where the corrugation of the graphite is expected to have small effect. The lateral interactions within the adsorbed layers appear to be very similar for inert gases on Ag(111) and graphite. In addition, the single adatom binding to the substrate is similar on the two substrates, as shown in Table 1.1; however, this is at least partially coincidental, because the holding potentials for He/graphite and He/Ag(110) differ by more than a factor of 2.

The data for Ar/Ag(111), Kr/Ag(111), and Xe/Ag(111) are fit with models

---

*A further consequence is that one anticipates the integral heat of adsorption, q_2, of the bilayer to be very similar for a given adsorbate on substrates where the lattice constant discontinuity at bilayer formation is small. An example that goes counter to this rule (Kern *et al.* 1986e) can be shown to be based on fitting the coexistence line over a narrow temperature range, with rather large attendant uncertainties in the slope.

[†]Phillips and Bruch (1988) reported that their model of the Xe/Ag(111) monolayer had a melting transition at a chemical potential that agreed with the value derived from the 3D gas pressure in the Greiser *et al.* experiment. However, the experiment is interpreted as showing a continuous melting transition (Greiser *et al.* 1987), while the calculation yields a weakly first-order transition.

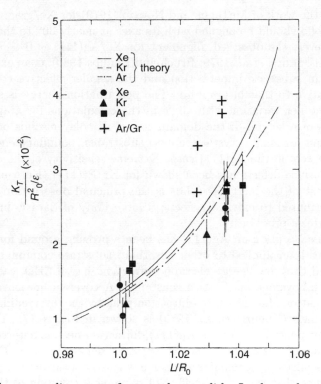

FIG. 6.16. Corresponding states for monolayer solids. Isothermal compressibility as a function of lattice constant, from Unguris *et al.* (1981). The data have been scaled by the appropriate interaction parameters in order to test for corresponding states behavior. Curves and filled symbols correspond to calculations and data for Xe, Kr, and Ar on Ag(111), while the + denotes data of Shaw and Fain (1980) for Ar on graphite.

based on the 3D inert gas pair potentials, supplemented by the McLachlan interaction (Bruch and Phillips 1980; Unguris *et al.* 1981). The absence of prominent substrate corrugation effects in the observations invites an attempt to systematize the observations for these three adsorbates in terms of corresponding states scalings, Section 5.2.2. Such a plot for the 2D compressibility is shown in Fig. 6.16, where the data for Ar/graphite are included, since registry effects are small in that system also.

Even for incommensurate monolayers, the effect of the geometry of the substrate can be manifested in the orientational epitaxy of Novaco and McTague, Section 3.6. The effect is observed for adsorption on graphite, Section 6.1, but it is not observed for the adsorption on Ag(111) (Unguris *et al.* 1979; Gibson and Sibener 1988a). Calculations for Xe/Ag(111) show (Bruch and Phillips

1980), as in the work of McTague and Novaco (1979) for Xe/graphite, that the monolayer solid should be aligned with its axes at nearly 30° to the axes of the silver. However, the observed alignment for Xe/Ag(111) is 0°. The result is rationalized (Unguris *et al.* 1979; Bruch and Phillips 1980) as an extrinsic effect resulting from substrate imperfection and the smaller effective corrugation of Ag(111) relative to that of graphite. The perturbation energy is second order in V_g and the net variation with angle in the calculations for graphite is only a few tenths of a kelvin. If the domain size of monolayer solids on Ag(111) is no larger than for single crystal graphite substrates, boundary energies would play a larger role in the Ag(111) case. Extreme sensitivity of the orientational epitaxy to surface defects has been shown for Kr/Pt(111) (Kern *et al.* 1986c). The alignment of the inert gas solids is the principal observation for Ag(111) which is attributed to extrinsic effects; a large body of data is understood in terms of intrinsic processes.

Inert gas adsorption on Ag(111) has been a proving ground for experimental techniques later applied to other adsorbate/substrate combinations. It was demonstrated that low energy electron diffraction, e.g., LEED, can be used to measure both coverage and lattice constants. A coverage measurement based on the attenuation of a substrate diffraction peak by the intervening incommensurate adsorbate (Unguris *et al.* 1979) is shown in Fig. 6.17. The electrons have sufficient penetration that a Ag(111) diffraction peak is detected even with a bilayer solid of adsorbate. The temperature dependence, or Debye–Waller factor, of the silver peak remains the same when the adsorbate is present, providing evidence that the average effect of dynamical coupling of the adsorbate and substrate is small.

High-resolution inelastic scattering experiments with thermal energy atomic helium beams show the evolution of the frequency spectrum of the adsorbed inert gas solids as a function of the number of layers (Gibson and Sibener 1988b; Gibson *et al.* 1985), as in Fig. 1.2. The spectra of the monolayer, bilayer, and trilayer are clearly distinct from each other and from that of the bulk adsorbate, so that few-layer films can be identified even when the lattice constants are too similar to be distinguished easily with diffraction experiments. Effects of the dynamical coupling of the xenon and krypton monolayers to the substrate are found for wavevectors near the Brillouin zone center (Gibson and Sibener 1985, 1988a). Calculations of the chemical potential at the third layer condensation of Xe/Ag(111) show that stability of the trilayer relative to the 3D solid depends on energy differences of *ca.* 10 K/atom, small compared to the total atom–atom potential energy but large compared to the difference in energy between the two closed packed lattices, fcc and hcp, of the 3D inert gas (Bruch and Ni 1985; Niebel and Venables 1976).

Another novel development is the measurement of the friction coefficient for sliding of xenon monolayers and bilayers on the Ag(111) surface. A preliminary report (Daly and Krim 1996) indicates that the dissipative processes in the substrate involve excitation of both electrons and phonons (Chen and Ying 1993; Persson 1993; Cieplak *et al.* 1994).

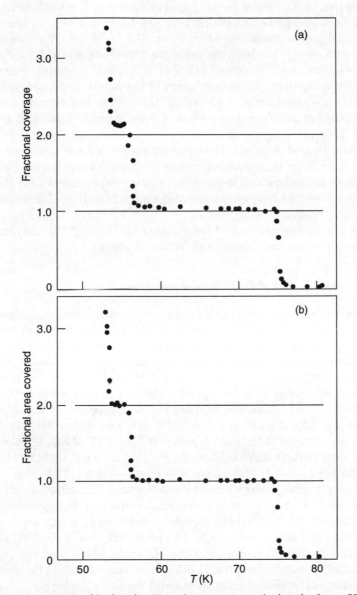

FIG. 6.17. 2D solid Xe/Ag(111). Xe adsorption on Ag(111), from Unguris *et al.* (1979). Fractional coverage in (a) is determined from the attenuation of the LEED substrate diffraction peak intensity. Fractional area in (b) is then deduced from measurements of the overlayer lattice constant. The incident gas flux is held constant in the experiment. Comparison of (a) and (b) demonstrates that the 2D solid is quite perfect, with few vacancies.

Adsorption of Xe and of Kr on Ag(111) decreases the work function of the metal. This is interpreted as arising from the formation of a dipole layer, with an effective dipole moment per atom of *ca.* 0.2 D for Xe. Associated shifts in the electronic energy levels of the inert gas atoms are studied for Xe/Ag(111) by photoelectron spectroscopies (Behm *et al.* 1986). Layering transitions are also evident in the data. Because so much of the initial stages of layer growth of Xe/Ag(111) is quantitatively understood, the system has become a subject for studies of the temperature dependence of layer growth (Qian and Bretz 1988; Dai *et al.* 1994).

The modeling of Ar/Ag(111) is less extensive and the database is smaller because of the large desorption effects in the LEED experiments (Moog *et al.* 1983). There are indications from photoemission experiments that the bonding of argon to silver may have qualitative differences from that of argon to graphite (Nilsson *et al.* 1993).

There is a limited amount of information on Xe/Ag(110), derived from inelastic helium scattering (Mason and Williams 1984).

6.2.2 *Adsorption on the (111) face of platinum*

After some early measurements by field emission spectroscopy (Nieuwenhuys *et al.* 1974), detailed studies of the adsorption of xenon, krypton, and argon on the (111) face of platinum, Pt(111), have been made with high resolution thermal energy helium atomic beams (Kern and Comsa 1989; Comsa *et al.* 1992).[*] Helium-scattering experiments are sensitive to small amounts of adsorbed gas because the scattering cross-section of a thermal helium atom by xenon is on the order of 100 Å^2. This has enabled the measurement of the adsorbed gas density at the 2D sublimation curve of Xe, Kr, and *Ar* on Pt(111) (Comsa *et al.* 1992); an example of the data is shown in Fig. 6.18. Although the one-atom binding energy to the surface is larger for Pt(111) than for Ag(111), the observed phenomena are similar to those on graphite and on Ag(111).

The larger binding energy carries the advantage that experiments can reach higher surface temperatures before appreciable thermal desorption occurs; thus the experiments for Xe/Pt(111) include a measure of a 2D triple point and critical point (Poelsema *et al.* 1985). The phase diagram of Xe/Pt(111) is much richer than a simple 2D analogue of the 3D xenon phase diagram (Kern *et al.* 1988). The observations for Pt(111) differ from those for Ag(111) principally in the occurrence of several commensurate and registry structures.

At monolayer completion, the lattice constants for Kr and Xe on Pt(111) are quite similar to those for bilayers of Xe and of Kr on Ag(111). There is also similarity in the few-layer frequency spectra (Hall *et al.* 1989). The spectrum of monolayer Kr/Pt(111) solid shows a dynamical coupling to the substrate

---

[*]The terrace widths on carefully prepared Pt(111) surfaces are larger than 2000 Å (Kern *et al.* 1986a); such widths are in the range of the largest terraces reported for cleaved single crystal graphite.

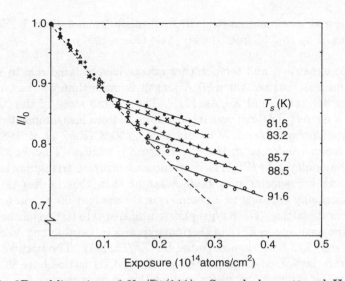

FIG. 6.18. 2D sublimation of Xe/Pt(111). Specularly scattered He atom intensity, relative to that from a bare Pt(111) surface, as a function of Xe coverage at specified temperatures, from Poelsema *et al.* (1983). The dashed line represents the scattering from a random array of isolated scatterers and is curved because it takes into account their statistical overlap. The data depart from this line when condensation occurs, because a submonolayer 2D solid scatters less than the same number of isolated atoms. Such plots have been analyzed to give the 2D latent heat of sublimation.

Rayleigh wave (Kern *et al.* 1987b) There is a great sensitivity of the adlayer orientation to minute concentrations of surface defects (Kern *et al.* 1986c). There are subtle higher order commensurate lattices of Kr/Pt(111) (Kern *et al.* 1987a) and of Ar/Pt(111) (Zeppenfeld *et al.* 1992a; Ramseyer *et al.* 1994a). The most extensive observations, and attempts at modeling, are for the Xe/Pt(111) case.

Four distinct monolayer solid phases of xenon are observed:*

1. A commensurate $\sqrt{3}\,R\,30°$ lattice, with nearest-neighbor spacing 4.80 Å (Poelsema *et al.* 1983, 1985).

2. A uniaxially incommensurate lattice of the type analyzed in Sec. 3.4, with misfits up to 6.6% (Kern 1987; Kern *et al.* 1987c).

3. A hexagonal incommensurate lattice aligned along the 30° axis (Kern *et al.* 1986a).

*The terminology *hexagonal* and *triangular* lattice is used interchangeably.

4. A hexagonal incommensurate lattice of misfit greater than 7.2% that has rotations of up to 3.3° from the 30° axis (Kern 1987).

Thermal expansion and temperature effects have a large role in stabilizing the commensurate lattice: the 4.80 Å length is more than 5% larger than the spacings for 3D xenon and Xe/Ag(111). The ground state of the Xe/Pt(111) lattice has a nearest-neighbor spacing *ca.* 4.48 Å, from measurements on unconstrained submonolayer islands of xenon down to 30 K (Kern *et al.* 1988). Above 60 K the xenon condenses in the commensurate lattice. That is, the ground state of submonolayer Xe/Pt(111) is an incommensurate triangular lattice with a nearest-neighbor spacing only 0.06 Å larger than that of Xe/Ag(111), but the layer thermally expands to a commensurate state at 60 K, via a uniaxially incommensurate lattice. The Kr/graphite commensurate lattice has been conjectured to exist as a consequence of thermal expansion (Gordon and Villain 1985), but the process has been demonstrated for Xe/Pt(111). The occurrence of the commensurate lattice, and the properties of the UIC lattice have drawn much attention in the attempts to create realistic interaction models for Xe/Pt(111) (Black and Janzen 1989a,b; Gottlieb 1990a,b; Gottlieb and Bruch 1991b; Barker and Rettner 1992; Rejto and Andersen 1993).

The latent heat of sublimation from the commensurate 2D solid to 2D gas is *ca.* 550 K/atom (Poelsema *et al.* 1983, 1985). Remarkably enough, it is only 10% larger than the value calculated with an interaction model that is based on the Xe/Ag(111) case and omits the substrate corrugation energy (Gottlieb and Bruch 1991b). The triple point coexistence between commensurate solid, 2D gas, and 2D liquid occurs at 98 K and the gas–liquid critical point at 120 K; these temperatures (Poelsema *et al.* 1985) are close to those for Xe/graphite.

The transition from the commensurate lattice to the UIC lattice is reported (Kern 1987) to be continuous, without a reentrant fluid phase. The transition from the uniaxial incommensurate lattice to the hexagonal incommensurate aligned lattice is assigned as a first-order transition, on the basis of observed two-phase coexistence (Kern *et al.* 1986a), while the transition from the hexagonal aligned to hexagonal rotated lattice is continuous (Kern 1987). The *UIC* and rotated hexagonal incommensurate lattices occur also for H_2/graphite and D_2/graphite (Cui and Fain 1989; Freimuth *et al.* 1990). It is noteworthy that, although the commensurate monolayer of Xe/Pt(111) is very dilated related to 3D xenon lattice constants, it is possible to grow a film many layers thick starting from the hexagonal incommensurate lattice at 25 K (Kern *et al.* 1986b).

There are direct signs of the modulation of the incommensurate monolayer by the substrate: a satellite to the average lattice peak of the UIC-Xe/Pt(111) at small misfit (Kern *et al.* 1987c) and a diffraction peak at the misfit wavevector of the hexagonal rotated Xe/Pt(111) (Kern *et al.* 1986d) and Kr/Pt(111) (Kern *et al.* 1987a). Peaks at satellite positions are observed for incommensurate layers of Kr/graphite and Xe/graphite. However, it is only in the atomic diffraction experiment that the peak at the misfit wavevector is observed.

The uniaxially incommensurate lattice of Xe/Pt(111) is observed over a wide range of misfits, in sharp contrast to the case of the compressed Kr/graphite monolayer, where the UIC lattice apparently is preempted by the domain wall fluid phase (Specht *et al.* 1987). As discussed in Section 5.2.2.2, the topography of the holding potential of an fcc(111) surface with degenerate potential minima at the fcc and hcp stacking sites differs from that of the basal plane graphite in the relative heights of the saddles and peaks above the minima of the potential surface. Gottlieb (1990a) noted that the existence of the observed modulation satellites in the UIC phase of Xe/Pt(111) excludes this as a model of the holding potential. He and Müller (1990) independently proposed that the energy minima for xenon atoms are atop surface platinum atoms; a model based on this idea is able to fit a great deal of data on the monolayer solid (Gottlieb and Bruch 1991b; Barker and Rettner 1992). Gottlieb's argument does not exclude the possibility that the modulation satellite arises by a splitting of the degeneracy of the holding potential at the fcc and hcp stacking sites (Zeppenfeld *et al.* 1992b).

Xenon atoms on the Pt(111) surface have been imaged in STM experiments, although the adsorption sites are not yet decided (Weiss and Eigler 1992; Zeppenfeld *et al.* 1994a). There is evidence for unusual interactions in the vicinity of steps on the platinum surface. Horch *et al.* (1995a,b) find, in STM observations, that submonolayer islands of Xe/Pt(111) display coarsening/annealing only at temperatures above 27 K. Diffusion in submonolayer Xe/Pt(111) has been measured by Meixner and George (1993a,b).

There also have been attempts to determine the adsorption site using photoelectron spin polarization experiments (Hilgers *et al.* 1991; Henk and Feder 1994; Potthoff *et al.* 1995). The analyses of the data now favor the three-fold site, rather than the atop site. However, the distance 4.2 Å from the commensurate xenon atoms to the platinum surface plane is surprisingly large compared to the value of 3.5 Å for Xe/Ag(111) (Stoner *et al.* 1978). This is especially remarkable in view of the much stronger binding of xenon to Pt(111) than to Ag (111), as given in Table 6.4. See also the discussion in Sections 2.3.3 and 2.4.1.

6.2.3   *Adsorption on the (100) face of palladium*

There are many similarities among the observations of inert gas adsorption on Ag(111), Pt(111), and the basal plane surface of graphite, but adsorption on Pd(100) presents great differences. We briefly summarize the observations and then discuss the modeling of the work function change for Xe/Pd(100). In spite of a very extensive set of experiments on the Xe/Pd(100) and Xe/Pd(111) monolayers (Palmberg 1971; Horn *et al.* 1978; Scheffler *et al.* 1979; Kaindl *et al.* 1980; Miranda *et al.* 1983a,b; Wandelt and Hulse 1984; Moog and Webb 1984; Hilgers *et al.* 1991, 1993, 1995), the situation remains confused. There is no understanding of the observed ordering (rather, the lack thereof in Xe/Pd(100)!).

Early work for Xe/Pd(100) (Palmberg 1971) at temperatures near 77 K indicated this might be a system for which the transition from 2D liquid (or dense disordered phase) to 2D solid could be followed on a single-crystal substrate by LEED to test theories of 2D melting. There are severe experimental problems associated with the effects of small concentrations of impurities in the palladium, and detailed agreement between observations in different laboratories is sometimes lacking (Moog and Webb 1984). However, after considerable effort, there is no observation of a first-order condensation to a monolayer solid of Xe/Pd(100) even at temperatures down to 15 K (Moog and Webb 1984; Miranda et al. 1983a). The adsorption isotherms and adsorption isobars are rounded even at the lowest experimental temperatures. Poorly defined diffraction peaks, that presumably reflect a 2D solid, are achieved by increasing the 3D gas pressure, i.e., by steric effects. There is evidence that crystalline order in the first layer is improved by condensation of a second layer. The Xe/Pd(100) solid may be one in which order is imposed from above by the second layer rather than from below as in commensurate monolayers. Even though the ordering in the first layer may be poor, multilayers of Xe/Pd(100) are reported (Miranda et al. 1983a).

Changes in the work function with adsorption are observed in both chemisorption and in physisorption on metals (Aruga and Murata 1989; Bonzel 1988). For chemisorbed species, it is usually attributed to charge transfer between the adsorbate and substrate, while for the physisorbed species the model is of a polarized adatom. The most detailed modeling in a physisorbed case is for Xe/Pd(100), which has a 2D fluid over a wide density range. The work function change is written (in esu)*

$$\Delta W = 4\pi n \mu(n). \tag{6.1}$$

For island-forming systems, such as Xe/Ag(111) and Xe/Pt(111), the work function varies linearly with the fractional coverage, and the effective dipole moment μ is taken at the density in the 2D solid islands, $\mu(n_o)$. For Xe/Pd(100), in contrast, the ratio $\Delta W/n$ is not constant as a function of coverage; the effective dipole moment depends on the density. Estimates of the electric fields from the neighboring adatoms show (Palmberg 1971) that these depolarizing fields are large enough to make observable changes in μ.† Observations of the density dependence can be fit by assigning a polarizability to the adatom that is of the order-of-magnitude of the 3D polarizability of the isolated atom.

Detailed modeling of the work function change includes effects of the screening response of the metal substrate to the electric fields from the adsorbate. The screening charge distribution effectively doubles the electrostatic interaction between the adsorption induced dipoles (Kohn and Lau 1976) and there is a corresponding contribution to the effective polarizability (Bruch and Ruijgrok 1979). The uncertainties caused by the use of continuum electrostatics methods to evaluate the screening for the small separation of the monolayer from the substrate are similar to those discussed in Section 2.3.2.3. The work function data

---

*The conversion from ΔW in volts and n in $1/\text{Å}^2$ to μ in debye units is $[\mu] = 0.0265[\Delta W]/[n]$.

†Bonissent and Bruch (1985) have calculated the depolarizing fields in a fluid; there are only small differences from the result of using lattice sums with a lattice constant fit to the average density, which is the so-called *Topping formula* (Palmberg 1971).

are not reproduced in detail between various laboratories (Palmberg 1971; Wandelt and Hulse 1984; Miranda *et al.* 1983a; Moog and Webb 1984). The effective adatom polarizability derived in recent experiments is closer to the isolated atom polarizability α_0 than to $2\alpha_0$.

Wandelt and Hulse (1984) made comparative studies of the adsorption of xenon on the (100), (110), and (111) faces of palladium. There are some commensurate adlayer solids, but there is no clarification of the lack of ordering on Pd(100).

Angle-resolved photoemission experiments show 2D band structures for the adatom electronic states in Xe/Pd(100) (Horn *et al.* 1978; Bradshaw and Scheffler 1979; Scheffler *et al.* 1979). The band structure is reproduced by electronic structure calculations based on the tight-binding picture of weakly overlapping electron wave functions centered on atomic sites of a regular lattice.

6.2.4   Other metals

We briefly note results for inert gas adsorption on copper and on tungsten.

The $\sqrt{3}$ commensurate lattice of Xe/Cu(111) would have a nearest-neighbor spacing 4.42 Å. This is nearly identical to the value for the intrinsic 2D system Xe/Ag(111) at low temperatures, and the commensurate lattice is indeed observed (Chesters *et al.* 1973).

The most extensive studies of inert gas adsorption on copper are for Xe/Cu(110), which has a series of commensurate structures in which the xenon atoms are arranged along the 'troughs' of the copper surface (Glachant *et al.* 1981, 1984; Berndt 1989; Ramseyer *et al.* 1994b; Zeppenfeld *et al.* 1994c). The more recent experiments indicate that there is a succession of superlattices. There was also an early helium scattering experiment (Mason and Williams 1983). Horn *et al.* (1982) studied the adsorption of Kr and of Ar on Cu(110) using LEED and angle-resolved photoemission experiments. Sticking coefficients have been measured by Zeppenfeld *et al.* (1994b). There are couplings between the Rayleigh mode of the copper and longitudinal modes of the adlayer that are understood in terms of the adlayer geometry (Zeppenfeld *et al.* 1994c).

The Xe/Cu(100) system appears to have an incommensurate monolayer (Glachant and Bardi 1979). It is noteworthy (Graham *et al.* (in press)) in that the longitudinal acoustic branch of the Xe monolayer solid has been detected with inelastic helium atom scattering. While the branch polarized perpendicular to the surface has a mixing with the Rayleigh mode of the copper, as seen for adsorption on Ag(111) and Pt(111), the wavevector dependence of the longitudinal acoustic branch is very much what would be anticipated for an ideal 2D solid.

Tungsten has a body-centered cubic lattice and the (110) face presents a relatively smooth surface. Even so, Engel *et al.* (1979) observed a set of commensurate lattices of Xe/W(110). Wang and Gomer (1980) measured the work function changes as a function of coverage and found that there are small adsorption-induced dipole moments on second layer xenon atoms.

6.3   Physically adsorbed layers on ionic crystals

The binding in ionic crystals has the property that cleavage leads to a sample exposing a single facet; powders of the ionic solid also expose primarily one facet. Atomic diffraction from the LiF(001) surface, Section 2.4.2, was historically important in confirming the wave character of matter and in the discovery of the phenomenon of selective adsorption. The alkali halides as well as MgO present square surface lattices, in contrast to the triangular and honeycomb adsorbate lattices encountered for adsorption on graphite. A major stimulus to recent studies of physical adsorption on ionic crystals has been the desire to examine critical phenomena corresponding to the new universality classes that become accessible for the square lattices. Recent advances in the use of helium atomic scattering (Hofmann and Toennies 1996) and infrared spectroscopies (Heidberg *et al.* 1995a) have led to a rapid expansion in the amount of information avaliable about adsorbates on insulating surfaces.

We emphasize adsorption on MgO(001) here and discuss the features that distinguish it from the preceding examples of this chapter. Homogeneous samples of MgO with large surface-to-volume ratios have been prepared and have been used as the substrate in physical adsorption experiments (Coulomb and Vilches 1984; Coulomb *et al.* 1984; Coulomb 1991). The MgO powders are homogeneous enough that sharp condensations, i.e., sharp risers in the adsorption isotherms, are observed for Ar, Kr, and Xe adsorption (Coulomb *et al.* 1984).

Near the surface of an ionic crystal there are strong electric fields that decrease rapidly with distance from the substrate, dropping by an order of magnitude in a distance equal to the van der Waals diameter of a krypton or xenon atom (Lennard-Jones and Dent 1928; Gready *et al.* 1978; Nijboer 1984). Thus the polarization of the adsorbate involves terms beyond the dipole polarizability, and a theory of linear nonuniform response applies (Guo and Bruch 1992). The electrostatic induction energy of an adatom is an additional term acting to stabilize registry and commensurate structures in the physical adsorption on ionic crystals; many such structures are observed (Meichel *et al.* 1988). Recently there has been a series of comparative studies of molecular adsorption of MgO(001) and NaCl(001) (Girardet *et al.* 1994) to help identify processes that depend on the surface electrostatic field.

There have been lengthy investigations as to how large a rumpling there is on the (001) surface of ionic crystals (De Wette *et al.* 1985; Martin and Bilz 1979). That is, there might be a differential relaxation of the positive and negative ions at the surface; the anions are expected to lie slightly outside the plane of the cations. However, as the theoretical modeling has become more sophisticated (Fowler and Tole 1988) it has been recognized that there are partially offsetting effects of the asymmetric electrostatic fields at the surface and of the overlap of charge clouds of the ions in the outermost layers. Zhou *et al.* (1994) find the rumpling to be negligible for MgO(001).

The most novel results from experiments with the MgO have been for methane adsorption. The monolayer solid is found to be a commensurate $c\,(2 \times 2)$ lattice over its entire domain (Coulomb *et al.* 1985a). Multilayers of CH_4/MgO are stable, counter to simple ideas of frustrated layer growth, apparently because the lattice parameter is close to that of the (100) plane of bulk methane. The fluid monolayer of CH_4 has a diffusivity* which is characterized as that of a lattice liquid and the fluid layers of multilayer films have been studied as an example of surface premelting (Bienfait *et al.* 1987, 1988; Bienfait 1987a; Bienfait and Gay 1991). The surface melting was later observed also for layers of $CH_4/graphite$ (Bienfait *et al.* 1990). There is a distinction in that the growth on MgO(001) presents a (100) face of methane, while on graphite it presents a (111) face. Simulations (Lynden-Bell 1990) lead to the expectation that surface melting occurs over a wider temperature range on the more open (100) face.

Another exploration has to do with the attempts to find a hydrogen layer in which the triple point is suppressed sufficiently that there might be a superfluid phase (Vilches 1992). Ma *et al.* (1988) determined the triple point of second layer $H_2/MgO(001)$.

A scan of the inert gas adsorption has also been performed, in analogy to the series described in Section 6.1 for the graphite substrate. Sullivan *et al.* (1985) examined $^4He/MgO(001)$. There is a question whether the monolayer ground state is a commensurate lattice or a 2D fluid; the evidence remains inconclusive (Kingsbury 1991). Coulomb *et al.* (1984) and Meichel *et al.* (1988) surveyed the adsorption of Ar, Kr, and Xe. There are commensurate lattices of Ar/MgO(001) and Kr/MgO(001), but not for Xe/MgO(001) which has a range of triangular incommensurate lattices (Degenhardt *et al.* 1989). The 2×3 commensurate lattice of Ar/MgO(001) has been the subject of modeling which includes hybridization of adlayer and substrate modes (Ramseyer *et al.* 1992; Girardet and Hoang 1993). There has also been a study of the perpendicular vibrations of the monolayer as a function of temperature and coverage and of the anisotropy of motions in the fluid phase of Ar/MgO(001) (Layet et al. 1993).

Structures and dynamics of a series of molecular adsorbates have been studied: CO (Panella *et al.* 1994; Minot *et al.* 1996), CO_2 (Meixner *et al.* 1992; Suzanne *et al.* 1993; Panella *et al.* 1994), acetylene (Coulomb *et al.* 1994; Ferry and Suzanne 1996), ethane (Trabelsi and Coulomb 1992; Hoang *et al.* 1993), water (Picaud and Girardet 1993), and NH_3 (Picaud *et al.* 1992, 1993b; Sidoumou *et al.* 1994). Girardet *et al.* (1994) made a comparative study of the adsorption of CO_2 on MgO(001) and on NaCl(001). There is a limited amount of information for the NaCl(001) systems: CO (Picaud *et al.* 1993a; Gerlach *et al.* 1995; Heidberg *et al.* 1995a), CO_2 (Lange *et al.* 1993, 1995; G.-Y. Liu *et al.* 1992; Picaud *et al.* 1995; Girardet *et al.* 1995), acetylene (Dunn and Ewing 1992; Quattrocci and Ewing 1992; Glebov *et al.* (in press), water (Bruch *et al.* 1995), CH_4 (Heidberg *et al.* 1995b), Kr (De Kieviet *et al.* 1996), and Xe (Schwennicke *et al.* 1993; Ramseyer and Girardet 1995). A combination of helium atomic scat-

---

*There is also a rotational diffusion. Rotational tunneling has been observed by Larese *et al.* (1991). Gay *et al.* (1992) report a rotational diffusion corresponding to a dipod-down orientation of the CH_4.

tering and polarization infrared spectroscopy enables the orientational structure to be determined through the analysis of the external and internal vibrational molecular modes (Weiss 1995). Hofmann and Toennies (1996) have reviewed recent measurements of the low-frequency modes of such adsorbates.

There is some uncertainty (Birkenheuer *et al.* 1994; Lakhlifi and Girardet 1991; Sidoumou *et al.* 1994) about the effective charge state and polarizability to assign to the oxygen ion at the MgO(001) surface; this is an issue that does not arise in the preceding examples in this chapter and is a complicating factor for modeling the holding potential. However, the existence of large electrostatic surface fields leads to interesting possibilities for exploring the orientational geometry of admolecules that have permanent dipole moments.

Although the MgO(001) surface is complex, modeling analogous to the atom–atom sum calculations for inert gases on graphite (Steele 1974) described in Chapter 2 leads to holding potentials for Ar/MgO and CH_4/MgO that are in fair agreement with experimental data (Girard and Girardet 1987). Such models are applied to a series of adsorbates, including Ar, Xe, N_2, NH_3, and CH_4 (Lakhlifi and Girardet 1991; Picaud *et al.* 1993b; Ramseyer *et al.* 1992), with a good level of agreement with experimental data.

6.4 Other systems

Finally we mention four classes of adsorption not covered under the previous headings: adsorbed mixtures, adsorption on quantum solid surfaces, adsorption on boron nitride, and adsorption on alkali metals.

6.4.1 *Adsorbed mixtures*

There are rather few examples of mixtures of inert gases in dense 3D phases. This is because the van der Waals radii of the inert gases are different enough that packing of the spheres involves a large level of *frustration*. The problem of achieving mixing is anticipated to be even more severe for the monolayer regime because of the large differences in holding potentials of the inert gases on a given substrate. Nevertheless, there are ranges of temperature and concentration where the mixing occurs.

For adsorption on graphite, Bohr *et al.* (1983) observed mixtures of argon and xenon and Ceva *et al.* (1986) observed mixtures of krypton and xenon.

For Pt(111), Zeppenfeld *et al.* (1993a,b) observed mixing and phase separation of argon and krypton; Yanuka *et al.* (1993) observed binary mixtures of krypton and xenon.

For modeling of the phase separation in 2D mixtures of Lennard-Jones 'atoms', see Leptoukh *et al.* (1995).

6.4.2 *Quantum solid substrate*

At times, it is desired to passivate a surface by plating it with a material that presents a very weak holding potential to another 3D gas. Such is the case in experiments that attempt to achieve a substantial density of nuclear-spin-polarized 3D ^{3}He gas. The occurrence of an adsorbed phase of the ^{3}He would be a major channel for depolarizing processes. Thus Lefèvre et al. (1985) used a system in which surfaces were plated with molecular hydrogen. The system ^{4}He/H$_2$ has been studied by Mugele et al. (1994) and by Mochel and Chen (1994). These both have a very weak holding potential for the helium atoms (Pierre et al. 1985; Wagner and Ceperley 1992, 1994).

Perhaps the most extensively studied example is the preplating of surfaces with ^{4}He to create a very weakly adsorbing substrate in experiments that attempt to stabilize a moderately dense gas of electron-spin-polarized atomic hydrogen. In the absence of preplating, the dominant channel for recombination and depolarization of the hydrogen atoms is the reaction of adsorbed atoms. With preplating, the rate can be brought down to the scale set by three-body collisions in the 3D gas phase. While this apparently is no longer the most promising route to a metastable spin-polarized atomic hydrogen gas of sufficient density to display Bose condensation (Walraven and Hijmans 1994; Silvera 1995), there was a 10 year effort that clarified many features of the extremely weakly bound H–He system. These have been reviewed by Walraven (1992).

6.4.3 Boron nitride

Migone and his coworkers have prepared large surface area samples of boron nitride (Shrestha et al. 1994) and have made comparative studies of adsorption on boron nitride and on graphite. The expectation is that there should be many similarities, because the surface lattices of the substrates are quite similar. The argon adsorptions are similar (Alkhafaji and Migone 1991, 1992b; Migone et al. 1993), as are the nitrogen adsorptions (Alkhafaji et al. 1994), but the situation for krypton is unclear (Dupont-Pavlovsky et al. 1985). Recent work (Li et al. 1996) shows that krypton condenses as an incommensurate solid on BN, in contrast to the commensurate lattice on graphite and that this is correlated to the presence of local quadrupole moments at the BN surface. The case of CO is similar to N$_2$ and different from Kr (Meldrim and Migone 1995). This is in contrast to the similarities of the monolayer phase diagrams of Kr/graphite, CO/graphite, and N$_2$/graphite. The potential wells for CO and N$_2$ on BN appear to be shallower and less corrugated than on graphite; recent evidence is reviewed by Marx and Wiechert (1996).

6.4.4 Adsorption on alkali metals

A most remarkable adsorption behavior is found for the case of alkali metal substrates; see Cole (1995) and Hallock (1995) for reviews. While known to be chemically reactive, these surfaces are the most inert substrates for inert gases. The reason for this distinct behavior is that the loosely bound conduction electrons of the metal extend far out in space, preventing the inert gas atoms from coming close enough to benefit from the van der Waals attraction. Typical

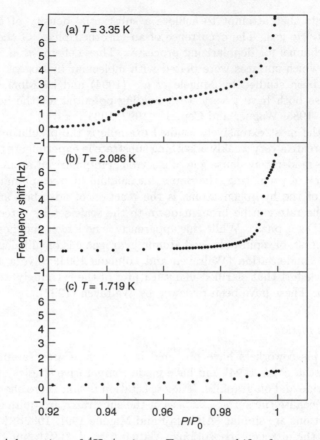

FIG. 6.19. Adsorption of ^{4}He/cesium. Frequency shift of quartz microbalance
    due to the adsorption of helium on cesium. The shift is proportional to the
    film coverage, which is shown over a very small range of pressure relative to
    saturation. From (a) to (b) to (c), one observes wetting, prewetting, and
    nonwetting behavior. From Rutledge and Taborek (1992).

equilibrium distances are 5 to 7 Å above the top layer of alkali nuclei, depending
on the system.

An example of the dramatic consequences of these weak interactions is the
anomalous wetting behavior of ^{4}He on Cs, depicted in Fig. 6.19. At low T,
virtually no gas adsorbs on the surface over the entire range of pressure up to
saturation. For $T > 2.5$ K, the adsorption is qualitatively identical to that
of a continuously wetting fluid. In a narrow regime $2 < T < 2.5$ K, instead,
there occurs, very close to saturation, a virtual discontinuity in coverage. This
so-called *prewetting* transition is the 2D analog of the liquid–vapor transition
familiar in 3D (Cahn 1977; Ebner and Saam 1977). Its critical behavior is

believed to be Ising-like, with a prewetting critical point at the end of a first-order line. Similar transition behavior has been seen for hydrogen (Mistura *et al.* 1994) and neon (Hess *et al.* (in press)) adsorption on alkali metals. The specific behavior is consistent with expectations (E. Cheng *et al.* 1993a,b) based on the potential strength: Li is the most attractive and Cs is the least attractive surface. While the details of this phenomenon are complicated, the origin is at least qualitatively straightforward. At low T, the weak adsorption potential provides little incentive for a film to adsorb below saturation pressure. At sufficiently high T, however, wetting should occur, because the surface tension cost becomes sufficiently low that the presence of even weak attraction suffices to attract a film. The transition between these regimes of wetting and nonwetting can take the form of this intermediate prewetting behavior in certain circumstances, which are not yet known in general. The inert gas/alkali metal system is thus far the unique example in adsorption science of this wetting behavior.*

---

*Recently prewetting has been observed in other systems: surface melting of an organic crystal (Chandavarkar *et al.* 1992) and a binary fluid at the consolute point (Kellay *et al.* 1993).

APPENDIX A

PARAMETERS OF THE 3D PHASES

Table A.1 Parameters of the critical point of 3D fluids.[†]

Gas[‡]	T_c (K)	V_c (cm^3)	P_c (MPa)	Z_c
^{3}He	3.38	73.6	0.123	0.321
^{4}He	5.21	57.7	0.229	0.304
Ne	44.4	41.7	2.71	0.306
Ar	150.7	74.9	4.88	0.292
Kr	209.4	92.2	5.50	0.291
Xe	289.7	113.7	5.87	0.277
H_2	32.99	65.45	12.8	0.305
D_2	38.26	60.3	16.4	0.311
N_2	126.3	89.9	3.41	0.292
O_2	154.6	74.6	5.05	0.293
CO	133.4	93.1	3.50	0.294
CH_4	190.7	98.9	4.63	0.289
CF_4	227.5	140.5	3.74	0.278
CO_2	304.2	94.8	7.39	0.277
C_2H_2	309.2	112.9	6.25	0.274
C_2H_4	282.8	127.2	5.11	0.276
C_2H_6	305.4	145.7	4.88	0.280

[†] T_c, V_c, and P_c are the bulk critical temperature, molar volume, and pressure. The compressibility factor PV/RT at the critical point is denoted Z_c; it would be 0.375 according to the van der Waals equation.

[‡] Entries for H_2 and D_2 are from Hoge and Lassiter (1951). All other data are from Simmrock et al. (1986).

Table A.2   Pair potentials. Parameters of the potential minimum for homonuclear inert gas pairs.[†]

Gas	r_m	ϵ	α	C_6	$\langle \mathbf{p}^2 \rangle$[‡]
He	2.967	11.01	1.383	1.461	2.21
Ne	3.091	42.25	2.66	6.87	5.46
Ar	3.76	143.2	11.1	67.2	13.1
Kr	4.011	201.3	16.7	142	17.5
Xe	4.366	282.8	27.3	340	24.8

[†] ϵ is the depth in kelvin and r_m is the separation in Å, from Aziz (1984). Also listed are the polarizability α and the coefficient of the dispersion energy C_6 in atomic units, from Standard and Certain (1985) and Tang and Toennies (1986).

[‡] The mean-square dipole moment, in atomic units, is calculated from the sum-rule data for S(-1) of Berkowitz (1979) using $\langle \mathbf{p}^2 \rangle = 3[S(-1)/(e^2/2a_0)]$.

Table A.3 Parameters of the three-dimensional solids of several atoms and molecules whose physically adsorbed monolayers are treated in Chapter 6.[†]

System[‡]	T_t (K)	E_o (K)	L_o (Å)	T_t/T_c
^{3}He	none	2.52	3.75	...
^{4}He	none	7.14	3.65	...
H_2	13.81	89.8	3.789	0.419
D_2	18.69	132.8	3.605	0.488
CH_4	90.694	1126	4.16	0.475
CD_4	89.78	1156	4.14	0.475
Ne	24.55	232	3.144	0.552
Ar	83.81	930	3.756	0.556
Kr	115.76	1342	3.992	0.553
Xe	161.39	1907	4.336	0.557
N_2	63.148	831	3.994	0.500
O_2	54.34	1043	3.20	0.351
CO	68.13	1000	3.99	0.511
CO_2	216.57	3160	3.925	0.712
CF_4	89.5	2115	4.23	0.394
C_2H_2	192	2200	4.34	0.622
C_2H_4	104	2300	4.10	0.368
C_2H_6	90.35	2460	4.46	0.296

[†] T_t is the triple-point temperature; E_0 is the cohesive energy (ground state energy) per molecule or atom and L_0 is the zero temperature lattice constant. For each helium isotope, the entry is the nearest-neighbor spacing in the solid at the melting curve. For the molecules, the entry is the spacing of nearest molecular centers and the crystal structure is in general complex. The ratio of the triple point temperature to the critical temperature (taken from Table A.1) is 0.54 for 3D Lennard-Jonesium.

[‡] Sources of the values for T_t, E_0, and L_0 are: He, Wilks (1967); H_2, D_2, Woolley et al. (1948) and Silvera (1980); CH_4, Colwell et al. (1963) and Greer et al. (1969); CD_4, Colwell et al. (1963), Grigor and Steele (1968) [with $T_c(CD_4)$ = 189.2 K], and Baer et al. (1978); Ne, Crawford (1977), McConville (1974) and Korpiun and Lüscher (1977); Ar, Kr, Crawford (1977) and Korpiun and Lüscher (1977); Xe, Crawford (1977), Tessier et al. (1982), and Korpiun and Lüscher (1977); N_2, Scott (1976); O_2, Knapp et al. (1987), English and Venables (1974), and Stephens et al. (1983); CO, Knapp et al. (1987) and Janssen et al. (1991); CO_2, Knapp et al. (1987) and Bruch (1983b); CF_4, Nosé and Klein (1983); C_2H_2, C_2H_6, Knapp et al. (1987) and Landolt–Börnstein III/5a (1971); C_2H_4, Knapp et al. (1987) and Mochrie et al. (1984). For C_2H_2, C_2H_4, and C_2H_6, the listed value of E_0 is actually the enthalpy of sublimation at T_t.

APPENDIX B

THERMODYNAMIC ANALYSIS

We have emphasized an atomic scale theoretical description and the goal of a statistical mechanical account of properties of extended monolayers. However, there is a useful thermodynamic analysis of the adsorbed systems in terms of macroscopic functions such as energy, entropy, and density (Larher 1971; Steele 1974; Price and Venables 1976; Bruch *et al.* 1979, 1985; Webb and Bruch 1982). The purposes of this appendix are to (1) outline the thermodynamic analysis, (2) provide definitions of heats of adsorption and compressibilities which may be derived from calorimetric experiments, and (3) discuss the commensurate–incommensurate transition using such concepts.

The combination of an adsorbed film coupled to the low-density 3D gas and to the substrate is a highly nonuniform thermodynamic system. However, the situation for thin physically adsorbed layers may usefully be viewed as the coexistence of two distinct thermodynamic phases, bulk gas and a 2D system, each of which may be approximated relatively simply. First, there are some preliminary observations, which is where 'the cat goes into the bag', to create a situation where the power of thermodynamic reasoning takes effect. The significant idealizations incorporated in the statistical mechanics treatment of Chapters 4 and 5 are reviewed briefly. We emphasize that there is an important domain of experimental physics where the idealizations are valid, namely for low temperatures and the first stages of film formation. Clearly, it also is easy to envision situations where the idealizations are quite inappropriate, such as a temperature high enough that the Boltzmann factor for one atom in the potential field of the substrate, $\exp(-\beta V(r, z))$ where $\beta = 1/(k_B T)$, does not have a sharp maximum which can be used to define the region of the monolayer.

In physical adsorption, the substrate is usually described as if it were inert.[*] The contributions of known processes by which the adlayer is coupled to the electrodynamics and the lattice dynamics of the substrate are assigned to the adlayer. The McLachlan interaction for physically adsorbed atoms is a substrate-mediated dispersion energy that involves electronic charge fluctuations of the substrate; it is assigned to the energy function of the adsorbed layer. The presence of the adsorbate must influence the positions and dynamics of substrate surface atoms, but the fact that the presence of a physisorbed monolayer has no observed effect on the Debye–Waller factor of the substrate surface is an indication that the

---

[*]There are relaxations of the outermost layers of the substrate, even without the adsorbate (Allen and de Wette 1969; Allen *et al.* 1969; Kobashi and Etters 1985a,b).

overall influence is limited.[†] The hybridization of monolayer solid normal modes with those of the substrate surface occurs in a qualitatively understood fashion and the associated shifts might lead to small corrections to the dynamical free energy of the adlayer. Elastic distortions of the substrate by surface structures are significant for semiconductor surfaces, but the corresponding energies for physically adsorbed layers are smaller than the McLachlan energies.

The adsorbed layer is treated as a distinct thermodynamic phase that coexists with a 3D gas phase and in which there may be transitions between various 2D phases. As in Section 1.1, consider the situation of a substrate surface exposed to the addition of atoms or molecules in a high vacuum environment. Formally, there is a thermodynamic limit process for defining surface-excess thermodynamic functions. The adsorbate–substrate combination is expressed as a sum of functions of the bulk substrate and of the 3D gaseous phase of the adsorbate and of a component proportional to the interface area which combines contributions of the perturbed substrate surface and of the adlayer. Such a procedure is essential for complex cases of liquid–solid interfaces (Griffiths 1980) and has been useful for clarifying the mathematical structure of the theory of multilayer adsorption (Pandit *et al.* 1982). However, we simply assert that we may identify functions associated with the thin layer in the case of physical adsorption. For the monolayer under highly anisotropic stress, this includes the large stress in the plane and essentially discards the much smaller stresses perpendicular to the plane arising from nearby 3D gas.

Phase equilibrium in 3D has three conditions: mass transfer equilibrium (equal chemical potentials), thermal equilibrium (equal temperatures), and mechanical equilibrium (equal pressures). Phase equilibrium between the 3D gas and the adsorbed layer has only the first two of these three conditions. Phase equilibrium within the layer has three conditions, as for the 3D phase: equal chemical potentials, temperatures, and spreading pressures.[*] The counting of the number of degrees of freedom F at phase coexistence, i.e., the Gibbs phase rule, goes as follows for a one component system. For P coexisting 3D and adsorbed phases, there are $P - 1$ conditions of chemical potential equality and three thermodynamic coordinates (say, pressure, temperature, and spreading pressure). Then the number of degrees of freedom is

$$F = 3 - (P - 1) = 4 - P . \tag{B.1}$$

Thus, the coexistence of 3D gas and a single surface phase $(P = 2)$ has two degrees of freedom and corresponds to an area in a pressure–temperature $(P\text{-}T)$ diagram. Similarly, a condensation transition in the adsorbed layer has three

[†]A noteworthy exception is the system He/H_2 treated by Wagner and Ceperley (1994).

[*]There are subtleties associated with the use of a spreading pressure. One must distinguish surface tension and surface stress (Wolf and Griffiths 1985). For thick films, it is appropriate to introduce a tensor that reflects the anisotropy of stress parallel and perpendicular to the substrate surface (Steele 1974; Rowlinson and Widom 1989).

coexisting phases and one degree of freedom, a line in the P–T diagram. A triple point of the adsorbate corresponds to a point in the P–T diagram.

The notations for the principal thermodynamic functions of a monolayer phase 'x' are: internal energy, u_x; enthalpy, h_x; entropy, s_x; number density, n_x; chemical potential, μ_x; spreading pressure, ϕ_x; and the temperature (T). The first four of these are 'densities' constructed from extensive thermodynamic variables, while the last three are intensive variables. We use the spreading pressure only for the formulation of the equation of state of a monolayer on a structureless surface, when $h_x = u_x + (\phi/n_x)$. For the treatment of the commensurate lattice, the grand potential (Ω) is introduced later in this section. There has been no assumption that the number density n_x reflects a single plane of adsorbate at the substrate surface. There may be a second layer of gas and there may be vacancies in a first-layer solid. The density n_x is the projection of the adsorbate quantity onto unit area of the surface; the notation θ is frequently used for the coverage or density when relative coverages are known but an absolute normalization is not known.

The functions which enter for the 3D gas are the chemical potential, μ_g, the enthalpy h_g, the number density n_g and the pressure p. The 3D gas which coexists with a physically adsorbed monolayer is at such a low density that it is essentially an ideal gas and Boyle's law holds:

$$p = n_g k_B T ,\tag{B.2}$$

where k_B is Boltzmann's constant. If the 3D gas is monatomic, the enthalpy is

$$h_g(\text{monatomic}) = \frac{5}{2} k_B T .\tag{B.3}$$

A closed form expression for the chemical potential of the 3D monatomic ideal gas of spin zero atoms is, with the thermal deBroglie wavelength defined as in eqn (4.2):

$$\mu_g = k_B T \ln(n_g \lambda^3) ,\tag{B.4}$$

The differential of the chemical potential is

$$d\mu = -s_g dT + (1/n_g) dp ,$$

and a form convenient for following phase coexistences is

$$d(\mu_g/T) = h_g d(1/T) + k_B d\ln p ,\tag{B.5}$$

where the ideal gas law eqn (B.2) has been used.

The corresponding thermodynamic differentials for the monolayer are

$$d\mu_x = -s_x\, dT + (1/n_x)\, d\phi ,$$

and

$$d(\mu_x/T) = h_x\, d(1/T) + (1/n_x T)\, d\phi .\tag{B.6}$$

The spreading pressure is defined to be the negative of the derivative of the free energy with respect to area per adatom; it has been evaluated that way in calculations for thin layers on a structureless surface that included the effects of vacancies and second-layer gas (Wei and Bruch 1981).

The spreading pressure differential is also available empirically in the so-called Gibbs adsorption isotherm, using the equality of $d(\mu_g/T)$ and $d(\mu_x/T)$ for equilibrium under changes at constant temperature:

$$d\phi = (k_B T n_x) d\ln p \ . \tag{B.7}$$

The isothermal compressibility of an adsorbed phase 'x'

$$K_T = [n_x \partial\phi/\partial n_x|_T]^{-1} \tag{B.8}$$

can be derived by measurements of the variation of the 3D gas pressure

$$\partial \ln p/\partial n_x|_T = 1/[n_x^2 k_B T K_T] \ . \tag{B.9}$$

The monolayer equation of state thus is accessible via isotherm measurements (Elgin and Goodstein 1974; Taborek and Goodstein 1979; Shaw and Fain 1980; Unguris et $al.$ 1979,1981; Glachant et $al.$ 1984; Gangwar et $al.$ 1989). An example of the comparison of calculations and experiments for low-temperature ^{4}He/graphite is shown in Fig. B.1.

Adsorption phase diagrams are usually plotted with $\ln p$ and $1/T$ as the axes, because layer condensation lines and phase boundaries then tend to be straight lines. The reason is the near constancy of heats of condensation, sometimes also called integral heats of adsorption, defined by

$$q_i = -k_B \frac{\partial \ln p}{\partial(1/T)}|_{i-layer} \ . \tag{B.10}$$

The notation for the heat of first-layer condensation is q_1 and for the heat of second-layer condensation is q_2. Similarly, we may define heats corresponding to the line for monolayer melting or other phase boundaries noted in the $\ln p$–$(1/T)$ plane. Another defined heat is the isosteric heat, following a line of constant coverage in a single phase region:

$$q_{st} = -k_B \frac{\partial \ln p}{\partial(1/T)}|_n \ . \tag{B.11}$$

A first example of the relations imposed by thermodynamic calculus is to use the relation

$$\partial z/\partial x|_y \partial x/\partial y|_z = -\partial z/\partial y|_x$$

to transform eqn (B.9) to

$$K_T = (T/q_{st}) \frac{\partial}{\partial T} (1/n_x)|_p \ . \tag{B.12}$$

This transformation is useful because several diffraction experiments on monolayer solids follow an isobar (constant p). Then the right-hand side of eqn (B.12) is proportional to the temperature derivative of the lattice constant $\partial a/\partial T|_p$.

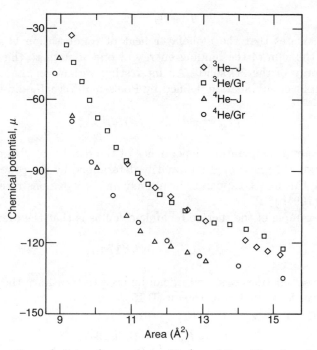

FIG. B.1. Equation of state of monolayer He/graphite. The chemical potential μ (in K) of monolayer solids of helium on graphite as a function of area per atom in Å^2, from Ni and Bruch (1986) The data are extrapolations to zero temperature and are taken from Taborek and Goodstein (1979) for ^{4}He (circles) and from Greif (1982) for ^{3}He (squares). The calculations, denoted J, use Jastrow trial functions. Notice that the commensurate phase occurs at larger area than shown here and that no consideration has been given to a possible striped solid.

Another example is to use the definition of q_1 and the coexistence condition for the phases at the 2D sublimation curve ($x = 2s$ and $x = 2g$)

$$[h_{2s} - h_{2g}]\, d(1/T) = [(1/Tn_{2g}) - (1/Tn_{2s})]\, d\phi \qquad (\text{B.13})$$

to get

$$q_1 = h_g - (n_{2g}h_{2g} - n_{2s}h_{2s})/(n_{2g} - n_{2s}) \qquad (\text{B.14})$$

$$= h_g - (n_{2g}u_{2g} - n_{2s}u_{2s})/(n_{2g} - n_{2s})\,. \qquad (\text{B.15})$$

Over much of the range of the sublimation curve the 2D gas density is far less, by at least a factor of 100, than the density of the 2D solid. Then eqn (B.15) simplifies to

$$q_1 \simeq h_g - u_{2s} \ . \tag{B.16}$$

This demonstrates that the monolayer heat of condensation is a quite direct measure of the sum of the binding energy of one adatom to the solid and the cohesive energy of the 2D solid. An interesting variant on the analysis is the heat of monolayer sublimation defined by Poelsema, Verheij, and Comsa (1983)

$$q_{sub} = -d\ln(n_{2g}/\beta)/d\beta|_{coex} \ , \tag{B.17}$$

using a measured temperature dependence of the density of 2D gas at the sublimation curve. The heat q_{sub} is primarily determined by the cohesive energy of the 2D solid; this may be apparent intuitively and was demonstrated by Gottlieb and Bruch (1991b).

A final example of the thermodynamic calculus is that the Maxwell relation

$$\partial\phi/\partial T|_n = \partial s/\partial(1/n)|_T$$

for the adlayer functions $\phi, s,$ and n, can be used to transform the isosteric heat to a derivative form analogous to eqn (B.15):

$$q_{st} = h_g - \frac{\partial}{\partial n_x}(n_x u_x)|_T \ . \tag{B.18}$$

Comparing eqn (B.16) and eqn (B.18) shows that the difference between q_1 and q_{st} at the point where an isostere intersects the monolayer condensation line in the $\ln p$–$1/T$ diagram gives a direct measure of the lateral stress in the monolayer.

The thermodynamic analysis of a commensurate solid is singular because at low temperatures, where there are very few vacancies in the solid and very little second-layer gas, the solid is essentially incompressible. An approach which bypasses this pathology and which leads to an identification of the principal terms at the commensurate–incommensurate transition can be based on the grand potential of the adlayer,

$$\Omega = F_a(\overline{N}_a) - \mu\overline{N}_a = \overline{N}_a(f - \mu) \ . \tag{B.19}$$

The chemical potential of the coexisting gas is μ, and F_a is the helmholtz free energy of the $\overline{N}_a$ atoms adsorbed on a constant area A. The latter is related to the number of atoms N_c in the perfect commensurate lattice and the mean misfit $\overline{m}$ by

$$\overline{N}_a = N_c/(1 - \overline{m}) \ . \tag{B.20}$$

To illustrate the use of these quantities, consider the case of a system such as H_2/graphite which condenses to a commensurate lattice at low temperatures. Then if f_c denotes the helmholtz free energy per atom in the commensurate solid, the surface phase is a 2D gas until the chemical potential is increased to $\mu_c = f_c$. The surface phase remains the commensurate lattice with further increase of chemical potential until a value is reached where the following equation can be satisfied:

$$\mu = f(\overline{m}) + (1 - \overline{m})\partial f/\partial\overline{m}|_T .$$ (B.21)

The derivative entering in eqn (B.21) has been evaluated in model calculations for a UIC lattice at small misfit, where it is proportional to the energy per unit length of a domain wall (Gottlieb and Bruch 1989, 1991a).

APPENDIX C

SURFACE RESPONSE FUNCTION

To complete the derivation of the Lifshitz formula, eqn (2.83), we required the zero wavevector limit of the surface response function $g(q, iu)$, eqn (2.76). Here we first give an expression for g in terms of the surface charge density driven by an external charge sheet and then treat the $q \to 0$ limit.

First, consider the substrate response substrate to an external charge sheet

$$\rho_{ext}(\mathbf{r}, t) = \delta(z - Z) \exp[\imath \mathbf{q} \cdot \boldsymbol{\rho} - \imath \omega t] \tag{C.1}$$

in the form of a density wave confined to the $z = Z$ plane. Then, using eqn (2.62), the resulting electrostatic potential is

$$\phi_{ext}(\mathbf{r}, t) = \frac{2\pi}{q} \exp[-q|z - Z| + \imath \mathbf{q} \cdot \boldsymbol{\rho} - \imath \omega t]. \tag{C.2}$$

For $z < Z$, which includes the $z < 0$ half-space occupied by the substrate, the amplitude is

$$\phi_{ext}(z, q, \omega) = \frac{2\pi}{q} e^{-qZ} e^{qz}. \tag{C.3}$$

Thus, the external potential decays exponentially into the substrate with an exponent given by the 2D wavevector q. The induced substrate electron density has

$$\delta n_s(\mathbf{r}, \omega) = \int d\mathbf{r}' \chi_s(\mathbf{r}, \mathbf{r}', \omega) \frac{2\pi}{q} e^{-qZ} e^{qz'} \exp[\imath \mathbf{q} \cdot \boldsymbol{\rho}']. \tag{C.4}$$

For explicitness, we use the jellium response function χ_s with translational symmetry given in eqn (2.73) and Fourier transform defined in eqn (2.74). Then the induced charge has

$$\delta n_s(\mathbf{r}, \omega) = \delta n_s(z, q, \omega) \exp[\imath \mathbf{q} \cdot \boldsymbol{\rho}], \tag{C.5}$$

with

$$\delta n_s(z, q, \omega) = \frac{2\pi}{q} e^{-qZ} \int dz' \, e^{qz'} \chi_s(z, z', q, \omega). \tag{C.6}$$

The electrostatic potential from this induced charge is

$$\phi_{ind}(z, q, \omega) = -\frac{2\pi}{q} \int dz' \, e^{-q|z-z'|} \delta n_s(z', q, \omega), \tag{C.7}$$

and at $z > 0$ it must correspond to

$$\phi_{ind}(z, q, \omega) = -\frac{2\pi}{q} e^{-qZ} e^{-qz} g(q, \omega), \tag{C.8}$$

where $g(q, \omega)$ is given by eqn (2.76).

Therefore, the total electrostatic potential for $0 < z < Z$ is

$$\phi_{tot}(z, q, \omega) = \frac{2\pi}{q} e^{-qZ} \left[e^{qz} - g(q, \omega) e^{-qz} \right] . \tag{C.9}$$

The first term may be interpreted as an 'incident' evanescent wave and the second as a 'reflected' evanescent wave with amplitude specified by the surface response function $g(q, \omega)$. Further, we have obtained an alternative expression for $g(q, \omega)$ in terms of the induced surface charge density δn_s:

$$g(q, \omega) = e^{qZ} \int dz \, e^{qz} \, \delta n_s(z, q, \omega) . \tag{C.10}$$

We use eqn (C.10) to obtain the $q = 0$ limit of $g(q, \omega)$:

$$g_0(\omega) = \int_{-\infty}^{\infty} dz \, \delta n_s(z, 0, \omega) . \tag{C.11}$$

$\delta n_s(z, 0, \omega)$ is the charge density induced by a uniform external electric field of amplitude -2π that is oriented perpendicular to the surface. Thus, $g_0(\omega)$ is the total surface charge density $Q(\omega)$. The continuity equation,

$$-i\omega\delta n_s(z, 0, \omega) + \frac{d}{dz} j_z(z, 0, \omega) = 0 , \tag{C.12}$$

relates δn_s to $j_z(z, 0, \omega)$, the normal component of the current density in the substrate. Using this in eqn (C.11) gives

$$g_0(\omega) = \frac{i}{\omega} j_z(-\infty, 0, \omega) , \tag{C.13}$$

That is, the accumulated surface charge is fed by the currents deep within the substrate.

This connection enables us to relate $g_0(\omega)$ to the bulk conductivity and hence to the dielectric function of the substrate. By Gauss' law, the electric field inside the substrate is

$$E_{in} = -2\pi + 2\pi Q(\omega) = -2\pi \left[1 - g_0(\omega) \right] . \tag{C.14}$$

It generates a bulk current density given by

$$j_z(z, 0, \omega) = -\sigma(\omega) E_{in} = 2\pi\sigma(\omega) \left[1 - g_0(\omega) \right] , \tag{C.15}$$

where $\sigma(\omega)$ is the bulk conductivity. Comparison with eqn (C.13) yields, finally,

$$\begin{aligned} g_0(\omega) &= \frac{2\pi i\sigma(\omega)/\omega}{1 + 2\pi i\sigma(\omega)/\omega} \\ &= \frac{\epsilon(\omega) - 1}{\epsilon(\omega) + 1} , \end{aligned} \tag{C.16}$$

where we have introduced the usual definition of the frequency dependent dielectric function

$$\epsilon(\omega) = 1 + \frac{4\pi i}{\omega} \sigma(\omega) . \tag{C.17}$$

APPENDIX D

DERIVATION OF THE FRAGMENT FORMULA

We present here a derivation of eqn (2.98) in the Hartree–Fock approximation. This follows the derivation of Zaremba and Kohn (1977) and is based on eqns (2.91)–(2.97) of Section 2.2.2.

We define a set of orthonormal orbitals $|\tilde{\imath}\rangle$ that will be used to obtain a variational estimate of the HF energy. The new orbitals satisfy the equation

$$(\hat{T} + \hat{V}_{ext} + \hat{V}_{sup}) \, |\tilde{\imath}\rangle = \tilde{\epsilon}_i \, |\tilde{\imath}\rangle \; , \tag{D.1}$$

where $\hat{V}_{ext}$ and $\hat{V}_{sup}$ are defined at eqns (2.91) and (2.97). These operators are given in terms of the separated-fragment HF orbitals, so that the solution of eqn (D.1) does not involve a self-consistency procedure. Regardless of how the new orbitals are defined, they can be used in a Slater determinant $|\tilde{\Psi}\rangle$ to provide a variational estimate of the many-electron energy,

$$\tilde{E}_{HF} = 2 \sum_i \langle \tilde{\imath} | \hat{h} | \tilde{\imath} \rangle + \sum_i \langle \tilde{\imath} | \tilde{V}_{HF} | \tilde{\imath} \rangle + E_{NUC} \; . \tag{D.2}$$

Note that $\tilde{V}_{HF}$ is the nonlocal HF potential operator constructed from the states $|\tilde{\imath}\rangle$. By the variational theorem, eqn (D.2) is an upper bound to the true HF energy and is in error only at order $(\delta\Psi)^2$, where $\delta\Psi = \Psi - \tilde{\Psi}$. Since this is the error in the *total* HF energy, it is difficult to predict the error in the quantity of real interest, the repulsive interaction. However, we argued in Section 2.2.2 that $\hat{V}_{sup}$ is close to $\hat{V}_{HF}$ for physisorption, so that we expect the error to be small. This is supported by experience.

The unperturbed atomic and metallic potentials each consist of external and HF potentials and can be written as $\hat{V}_a = \hat{V}_{ext}^a + \hat{V}_{HF}^a$ and $\hat{V}_s = \hat{V}_{ext}^s + \hat{V}_{HF}^s$. Using these conventions, the first term on the right-hand side of eqn (D.2) can be rearranged:

$$\begin{aligned} \langle \tilde{\imath} | \hat{h} | \tilde{\imath} \rangle &= \langle \tilde{\imath} | \left[\hat{T} + \hat{V}_{ext} + \hat{V}_{sup} - \hat{V}_{HF}^a - \hat{V}_{HF}^s \right] | \tilde{\imath} \rangle \\ &= \tilde{\epsilon}_i - \langle \tilde{\imath} | \hat{V}_{HF}^a + \hat{V}_{HF}^s | \tilde{\imath} \rangle \; . \end{aligned} \tag{D.3}$$

Substituting eqn (D.3) into eqn (D.2) gives an important intermediate result:

$$\tilde{E}_{HF} = 2 \sum_i \tilde{\epsilon}_i - 2 \sum_i \langle \tilde{\imath} | \hat{V}_{HF}^a + \hat{V}_{HF}^s | \tilde{\imath} \rangle$$

$$+ \sum_i \langle \tilde{\imath} | \tilde{V}_{HF} | \tilde{\imath} \rangle + E_{NUC} . \tag{D.4}$$

Subtracting the HF ground state energies of the isolated adsorbate and substrate, eqns (2.95) and (2.96), from eqn (D.4) yields the expression

$$V_{rep}(\mathbf{R}) \simeq 2 \sum_i \delta \tilde{\epsilon}_i - 2 \sum_i \langle \tilde{\imath} | \hat{V}_{HF}^a + \hat{V}_{HF}^s | \tilde{\imath} \rangle + \sum_i \langle \tilde{\imath} | \tilde{V}_{HF} | \tilde{\imath} \rangle$$

$$+ \sum_a \langle a | \hat{V}_{HF}^a | a \rangle + \sum_s \langle s | \hat{V}_{HF}^s | s \rangle + \Delta E_{NUC} , \tag{D.5}$$

where $\delta \tilde{\epsilon}_i = \tilde{\epsilon}_i - \epsilon_i$ is the difference between the energy eigenvalues in eqn (D.1) and the unperturbed eigenvalues ϵ_i. The index i denotes substrate and adsorbate levels. We have implicitly assumed that there is a one-to-one correspondence between the states $| \tilde{\imath} \rangle$ and those of the separated fragments.

The approximate nature of the estimate in eqn (D.5) arises from using the states $| \tilde{\imath} \rangle$ to calculate the HF energy of the coupled system. However, by definition, eqn (D.5) is an upper bound to the HF repulsion and its utility is revealed by making one additional simplification. Each matrix element in eqn (D.5) has the form $\sum_A \langle A | V_{HF}^B | A \rangle$ and can be expressed as a sum of direct and exchange terms,

$$\sum_A \langle A | V_{HF}^B | A \rangle = \sum_A \left[\langle A | V_C^B | A \rangle + \langle A | V_X^B | A \rangle \right]$$

$$= \frac{1}{2} \int n_A \phi_B - \int \rho_A \rho_B v . \tag{D.6}$$

We are using a short-hand notation for the direct and exchange terms

$$\int n_A \phi_B \equiv \int d\mathbf{r} \, n_A(\mathbf{r}) \left[\int d\mathbf{r}' \frac{n_B(\mathbf{r}')}{|\mathbf{r} - \mathbf{r}'|} \right] , \tag{D.7}$$

$$\int \rho_A \rho_B v \equiv \int d\mathbf{r} \int d\mathbf{r}' \frac{\rho_A^*(\mathbf{r}, \mathbf{r}') \rho_B(\mathbf{r}, \mathbf{r}')}{|\mathbf{r} - \mathbf{r}'|} , \tag{D.8}$$

with the single-particle density matrix (per spin)

$$\rho(\mathbf{r}, \mathbf{r}') = \sum_i \phi_i(\mathbf{r}) \phi_i^*(\mathbf{r}') \tag{D.9}$$

and electron number density

$$n(\mathbf{r}) = 2 \rho(\mathbf{r}, \mathbf{r}) = 2 \sum_i |\phi_i(\mathbf{r})|^2 . \tag{D.10}$$

To simplify eqn (D.5), we define deviations from superposition

$$\tilde{\rho} = \rho_a + \rho_s + \delta\rho$$

$$\tilde{n} \;=\; n_a + n_s + \delta n \,. \tag{D.11}$$

Collecting all electrostatic matrix elements in eqn (D.5), the net change in electrostatic energy is

$$
\begin{aligned}
\Delta E_{es} \;&=\; -\int \tilde{n}(\phi_a + \phi_s) + \frac{1}{2}\int \tilde{n}\tilde{\phi} \\
&\quad + \frac{1}{2}\int n_a\phi_a + \frac{1}{2}\int n_s\phi_s \\
&=\; -\int n_a\phi_s + \mathcal{O}(\delta n^2) \,.
\end{aligned}
\tag{D.12}
$$

Similarly, the exchange terms combine to give

$$
\begin{aligned}
\Delta E_x \;&=\; 2\iint \tilde{\rho}(\rho_a + \rho_s)v - \iint \tilde{\rho}\tilde{\rho}v \\
&\quad - \iint \rho_a\rho_a v - \iint \rho_s\rho_s v \\
&=\; -2\sum_a \langle a|\hat{V}^s_X|a\rangle + \mathcal{O}(\delta\rho^2),
\end{aligned}
\tag{D.13}
$$

where $\hat{V}^s_X$ is the exchange part of $\hat{V}^s_{HF}$. Equations (D.12) and (D.13) are correct to second order in the deviations δn and $\delta\rho$. The HF repulsive potential, eqn (D.5), is therefore given by

$$V_{rep}(R) \;=\; 2\sum_i \delta\tilde{\epsilon}_i \;-\; 2\sum_a \langle a|\hat{V}^s_{HF}|a\rangle \;+\; \Delta E_{NUC} \;+\; \mathcal{O}(\delta\rho^2) \,. \tag{D.14}$$

This expression remains symmetric with respect to the adsorbate and substrate, because there is an identity $\sum_a \langle a|\hat{V}^s_{HF}|a\rangle = \sum_s \langle s|\hat{V}^a_{HF}|s\rangle$.

The approximate HF eigenvalues $\tilde{\epsilon}_i$ can be separated into two groups: those corresponding to substrate orbitals $\{\tilde{\epsilon}_s\}$ and a smaller group corresponding to the adsorbate orbitals $\{\tilde{\epsilon}_a\}$. Equation (D.14) becomes

$$V_{rep}(R) \;=\; \Delta E_s + \Delta E_a + \Delta E_+ + \mathcal{O}(\delta\rho^2), \tag{D.15}$$

where

$$\Delta E_s \;=\; 2\sum_s \delta\tilde{\epsilon}_s \,, \tag{D.16}$$

$$\Delta E_a \;=\; 2\sum_a \delta\tilde{\epsilon}_a \;-\; 2\sum_a \langle a|\hat{V}_s|a\rangle \,, \tag{D.17}$$

and

$$\Delta E_+ \;=\; 2\sum_a \langle a|\hat{V}^s_{ext}|a\rangle \;+\; \Delta E_{NUC} \;=\; \int d\mathbf{r}\,\Phi_a(\mathbf{r})n_+(\mathbf{r}) \,. \tag{D.18}$$

ΔE_+ is the electrostatic interaction between the total atomic charge density and the *positive* ionic background of the substrate and vanishes if the atomic electron density does not overlap with the ionic background.

We now examine the eigenvalue shifts in more detail, exploiting the fact that the atomic orbitals are only weakly perturbed. The isolated HF atomic orbitals $|a\rangle$ satisfy the equation

$$(\hat{T} + \hat{V}_a) |a\rangle = \epsilon_a |a\rangle , \tag{D.19}$$

while the orbitals $|\tilde{\imath}\rangle$ for the potential superposition satisfy eqn (D.1) which contains the additional term $\hat{V}_m$. Estimating the shifts in the atomic eigenvalues by perturbation theory, we have

$$\delta\tilde{\epsilon}_a = \tilde{\epsilon}_a - \epsilon_a = \delta\tilde{\epsilon}_a^{(1)} + \delta\tilde{\epsilon}_a^{(2)} + \cdots , \tag{D.20}$$

where

$$\delta\tilde{\epsilon}_a^{(1)} = \langle a|\hat{V}_s|a\rangle \tag{D.21}$$

and

$$\delta\tilde{\epsilon}_a^{(2)} = \sum_{a'} \frac{|\langle a|\hat{V}_s|a'\rangle|^2}{\epsilon_a - \epsilon_{a'}}. \tag{D.22}$$

If the second-order shift is neglected, ΔE_a vanishes and eqn (D.15) reduces to

$$V_{rep}(\mathbf{R}) \simeq 2 \sum_s \delta\tilde{\epsilon}_s + \Delta E_+ + \mathcal{O}(\delta\rho^2) . \tag{D.23}$$

If in addition the electrostatic energy ΔF_+ is small, the final expression for the repulsive interaction takes the remarkably simple form

$$V_{rep}(\mathbf{R}) \simeq 2 \sum_s \delta\tilde{\epsilon}_s . \tag{D.24}$$

APPENDIX E

ADATOM–SUBSTRATE DISPERSION ENERGIES

It was shown in Chapter 2 that both the van der Waals energy of one adatom with a substrate and the adsorption-induced modification of the van der Waals interaction between the adatoms may be expressed in terms of the dynamic polarizability of the adatom $\alpha(\imath\omega)$ and the frequency-dependent dielectric response of the substrate $g(\imath\omega)$ through the use of the Casimir–Polder transformation, as in eqns (2.59)–(2.61). Besides providing a decomposition into factors that depend on the subsystems, the fact that these functions are evaluated at imaginary frequency has the advantage that approximations are made on much smoother functions than the possibly-resonant responses at real frequency. Approximations for $\alpha(\imath\omega)$ were reviewed by Tang *et al.* (1976). The dielectric response is given in eqn (2.18), which is based on Kramers–Kronig relations. There are extensive tabulations of the input data, e.g., the compilations of Standard and Certain (1985) and Weaver *et al.* (1981). However, it is often useful to be able to make quick estimates of the values of the coefficients C_3, C_{s1}, and C_{s2}. There are simple approximations to the frequency dependences that enable estimates to within 20% or so. In this appendix, we extend the tabulations provided by Rauber *et al.* (1982).

The approximation to the dynamic polarizability is based on the Drude model of the atom, in which there is a single harmonic oscillator coordinate:

$$\alpha(\imath\omega) = \alpha_0/[1 + (\omega/E_a)^2] . \tag{E.1}$$

α_0 is the static polarizability and E_a is the excitation frequency. The value of E_a is derived from the static polarizability and the atom–atom van der Waals coefficient C_6 by using eqn (E.1) in eqn (2.8)

$$C_6 = \frac{3}{4}\alpha_0^2 E_a . \tag{E.2}$$

Values for many species treated in physical adsorption are listed in Table E.1.

The dielectric response function of the substrate is modeled by

$$
\begin{aligned}
g(\imath\omega) &= \frac{[\epsilon(\imath\omega) - 1]}{[\epsilon(\imath\omega) + 1]} \\
&= g_0/[1 + (\omega/E_s)^2] .
\end{aligned}
\tag{E.3}
$$

This is the form if the response is dominated by the plasma resonance of a metal. However, it also provides a smooth fit to the response function when

Table E.1 Parameters of the van
der Waals dipole–dipole interaction
for atoms and molecules in physical
adsorption.[†]

Gas[‡]	α_0	C_6	E_a
H	4.50	6.50	0.4280
H_2	5.439	11.49	0.5179
He	1.384	1.46	1.016
Ne	2.67	6.43	1.203
Ar	11.08	64.2	0.6972
Kr	16.78	127.9	0.6056
Xe	27.11	290.5	0.5270
N_2	11.74	73.39	0.710
CO	13.1	81.4	0.6324
O_2	10.59	62.01	0.7372
CO_2	17.5	158.7	0.6909
CH_4	17.27	129.6	0.5794
CF_4	25.9	276	0.5486
C_2H_2	22.96	204.1	0.5162
C_2H_4	27.7	400.6	0.5224
C_2H_6	29.6	422.6	0.6431
C_6H_6	67.79	1723	0.500
H_2O	9.64	45.4	0.6514
NH_3	14.56	89.1	0.5604

[†] The static polarizability α_0, pair potential coefficient C_6, and characteristic excitation energy E_a are given in atomic units where 1 au$(\alpha) = 0.1482$ Å^3/atom, 1 au$(C_6) = 0.9573 \times 10^{-12}$ erg Å$^6 = 0.5976$ eV Å^6, 1 au$(E_a) = 4.360 \times 10^{-11}$ erg $= 27.21$ eV.

[‡] For molecules the entries are spherical averages. Sources of the data are: CO and CO_2, Jhanwar and Meath (1982); CF_4, Bruch (1987); C_2H_2 and C_6H_6, Kumar and Meath (1992); C_2H_4, Jhanwar et al. (1983) and Meath (unpublished); C_2H_6, Jhanwar and Meath (1980); H_2O and NH_3, Zeiss and Meath (1977); all others, from a compilation by Rauber et al. (1982).

more complex responses such as interband transitions are involved. Interband transitions usually contribute significantly, because the characteristic excitation energies of inert gas atoms are in the range of 15 eV; thus, the substrate is driven at frequencies much higher than its plasma frequency. Parameters from fits of eqn (E.3) to the response of several metals and dielectrics are listed in Table E.2.

Table E.2 Parameters of the substrate dielectric response.[†]

Substrate[‡]	g_0	E_a
Al	0.976	0.473
Ni	0.74	0.70
Cu	0.857	0.652
Pd	0.785	0.629
Ag	0.812	0.800
Pt	0.80	0.83
Au	0.840	0.888
Graphite	0.619	0.667
Si	0.84	0.43
Ge	0.88	0.44
LiF	0.31	0.74
NaCl	0.42	0.56
MgO	0.47	0.66
BN	0.38	0.71

[†] Parameters for eqn (E.3), with energy in au, fitted to calculated coefficients C_3 and C_{s1}.
[‡] The entries are taken from: Ni and Pt, this work; Si and Ge, Vidali and Cole (1981); LiF and BN, Karimi and Vidali (1986); NaCl and MgO, Chung and Cole (1984); all others, Rauber *et al.* (1982).

The coefficients C_3, C_{s1}, and C_{s2} from eqns (2.17) and (2.132) are then given by (Vidali and Cole 1981; Rauber *et al.* 1982)

$$C_3 = (g_0 \alpha_0 E_s / 8)/(1 + x) \tag{E.4}$$

$$C_{s1} = g_0 C_6 x(2 + x)/(1 + x)^2 \tag{E.5}$$

$$C_{s2} = g_0^2 C_6 x(x^2 + 3x + 1)/(1 + x)^3 , \tag{E.6}$$

where $x = E_s/E_a$.

APPENDIX F

2D FOURIER TRANSFORMS

In eqns (2.120) and (2.121), we presented the Fourier components of the adatom–substrate potential formed by summation over Lennard-Jones (12,6) pair potentials. In this appendix, we collect results for the 2D Fourier transforms in eqn (2.119) for other functional forms which have been used in place of eqn (2.116).

First, for more general choices of inverse power law potentials, the 2D Fourier transform is

$$F_n(g,z) = \int d^2r \exp(i\mathbf{g}\cdot\mathbf{r})/(r^2+z^2)^{n+1}$$

$$= 2\pi(g/2z)^n \frac{1}{n!}K_n(gz),\tag{F.1}$$

where K_n is a modified Bessel function.

For an exponential repulsion, the 2D Fourier transform is (Belak et $al.$ 1985)

$$F_0(g,\alpha,z) = \int d^2r \exp(-\alpha\sqrt{r^2+z^2}+i\mathbf{g}\cdot\mathbf{r})$$

$$= \frac{2\pi\alpha}{g^2+z^2}[z+\frac{1}{\sqrt{g^2+\alpha^2}}]$$

$$\times \exp(-z\sqrt{g^2+\alpha^2}).\tag{F.2}$$

A generalization of eqn (F.2) is needed when the Tang–Toennies damping functions are used. Mann et $al.$ (1987) have shown that the identity

$$\frac{\exp(-\alpha R)}{R^n} = \int_\alpha^\infty dx \frac{(x-\alpha)^{n-1}}{(n-1)!}\exp(-Rx)\tag{F.3}$$

can be used to simplify the 2D Fourier transform as

$$F_n(g,\alpha,z) = \int d^2r \exp(-\alpha\sqrt{r^2+z^2}+i\mathbf{g}\cdot\mathbf{r})/(r^2+z^2)^{n/2}$$

$$= -2\pi\int_\alpha^\infty dx \frac{(x-\alpha)^{n-1}}{(n-1)!}$$

$$\times \frac{\partial}{\partial x}\frac{\exp(-z\sqrt{x^2+g^2})}{\sqrt{x^2+g^2}}.\tag{F.4}$$

For $n=1$, the integration is elementary. For $n>1$, the integral is no longer elementary, but it is of a form quite easily evaluated by numerical means. For $g\neq 0$, eqns (F.2) and (F.4) show that the corrugation amplitudes V_g vary as $\exp[-z\sqrt{g^2+\alpha^2}]$ unless z is very small.

APPENDIX G

LATTICE DYNAMICS EXAMPLES

We present here some explicit solutions for the normal mode spectra of triangular and square lattices, to illustrate the content of the formalism developed in Sections 4.2.3 and 5.2.1. The results are for the 2D theory of Bravais lattices with one atom (of mass m) per unit cell and include the modifications which arise when the lattice is commensurate with the substrate. We assume that the atoms interact by spherically symmetric pair potentials so that the force-constant matrix in the normal mode theory can be expressed in terms of the tensor second gradient as

$$\begin{align}
\nabla\nabla\phi &= (\hat{\mathbf{r}}\,\hat{\mathbf{r}})\,p_2(r) + \mathbf{I}\,p_1(r), \\
p_1(r) &= \frac{1}{r}\frac{d}{dr}\phi, \\
p_2(r) &= \frac{d^2}{dr^2}\phi - \frac{1}{r}\frac{d}{dr}\phi.
\end{align} \tag{G.1}$$

In eqn (G.1), $\hat{\mathbf{r}}$ denotes a unit vector and $\mathbf{I}$ denotes the unit matrix.

G.1 Triangular lattice

Let the primitive vectors of this lattice be

$$\begin{align}
\mathbf{a}_1 &= L(\frac{\sqrt{3}}{2}\hat{\mathbf{x}} + \frac{1}{2}\hat{\mathbf{y}}) \\
\mathbf{a}_2 &= L(\frac{\sqrt{3}}{2}\hat{\mathbf{x}} - \frac{1}{2}\hat{\mathbf{y}}).
\end{align} \tag{G.2}$$

There are simple explicit solutions for the normal mode frequencies for wavevectors $\mathbf{q}$ along the x and y axes. In the conventional terminology used for the Brillouin zone, these are the ΓM and ΓK directions, respectively. Denote by ω_0 the zone-center frequency gap in the case of a commensurate lattice:

$$\omega_0 = \sqrt{-3g^2 V_g/m}, \tag{G.3}$$

where it has been assumed that the commensurate lattice is similar to the krypton/graphite case with atoms located at honeycomb centers on the graphite. There is a similar result for a commensurate lattice of sites at the 3-fold sites of an fcc(111) surface (Bruch 1988). For the 'floating' incommensurate triangular lattice $\omega_0 = 0$.

The normal mode frequencies along the ΓM direction are then given in terms of sums over all neighbors on the lattice:

$$m\left[\omega(q)^2 - \omega_0^2\right] = \sum_r [1 - \cos(qr_x)](\hat{r}_x^2 p_2 + p_1) \ \ (LA)$$

$$= \sum_r [1 - \cos(qr_x)](\hat{r}_y^2 p_2 + p_1) \ \ (TA) \ . \qquad (G.4)$$

The notation LA and TA denotes longitudinal and transverse acoustic modes, respectively, as these become acoustic branches (frequency varying linearly with q at small q) for the incommensurate lattice. Even for the commensurate lattice, the branches have 'clean' polarization for this direction of the wavevector. The wavevector at the 'M' point is $q = 2\pi/L\sqrt{3}$.

The expressions for the normal mode frequencies along the ΓK direction are obtained from eqn (G.4) by replacing the factor $[1-\cos(qr_x)]$ by $[1-\cos(qr_y)]$ and interchanging the identification of the LA and TA branches. The polarizations are again 'clean' along the ΓK axis. The wavevector at the 'K' point is $q = 2\pi/L$ and the frequencies of the LA and TA branches are equal there (i.e., degenerate).

Another feature of the triangular lattice is that the dispersion relation is isotropic (frequency independent of the direction of q) at small wavenumber:

$$\omega_\alpha(q)^2 - \omega_0^2 \simeq c_\alpha q^2 \ , \quad q \to \infty \ . \qquad (G.5)$$

The speeds of longitudinal and transverse sound are then given by the lattice sums:

$$c_\ell^2 = \frac{1}{16m} \sum_r r^2 [3p_2 + 4p_1]$$

$$c_t^2 = \frac{1}{16m} \sum_r r^2 [p_2 + 4p_1] \ . \qquad (G.6)$$

If it were not for the presence of the p_1 term in the summand, the sound speeds would satisfy the relation for a Cauchy solid, as discussed in Section 5.2.2. This term is proportional to the spreading pressure of the solid and the analysis for the triangular lattice at small wavenumber amounts to an explicit treatment of the pressure renormalization of the Lamé constants (Barron and Klein 1965; Stewart 1974).

G.2   Square lattice

Consider a commensurate square lattice where the zone-center frequency gap is given by a relation analogous to eqn (G.3):

$$\omega_0 = \sqrt{-2g^2 V_g/m} \ \ (\text{Square}) \ . \qquad (G.7)$$

Let primitive vectors be $\mathbf{a}_1 = L\hat{x}$ and $\mathbf{a}_2 = L\hat{y}$. Now assume that the two-body interactions are restricted to nearest-neighbors.

Then the spectrum for wavevectors along the ΓM axis (i.e., the x-axis) is given by

$$
\begin{aligned}
\omega(q)^2 &= \omega_0^2 + (2/m)\left[1 - \cos(qL)\right] p_1(L) \qquad \text{(TA)} \\
&= \omega_0^2 + (2/m)\left[1 - \cos(qL)\right] (p_2 + p_1) \quad \text{(LA)} .
\end{aligned} \qquad \text{(G.8)}
$$

That is, it has a very weakly dispersing branch. On the other hand, for wavevectors along the ΓK direction (i.e., the $45°$ axis $\hat{\mathbf{x}} + \hat{\mathbf{y}}$), the spectrum degenerates to a single branch:

$$
\omega(q)^2 = \omega_0^2 + (2/m)\left[1 - \cos(qL)\right] (p_2 + 2p_1) , \quad \Gamma K . \qquad \text{(G.9)}
$$

APPENDIX H

SYSTEMS OF UNITS

The fact that physical adsorption has contributions from many fields is reflected in the variety of units that are used for the same physical quantities. We have been guilty of this in our own presentation, because although we asserted in Chapter 1 that we would follow cgs/esu, we have adopted other 'convenient' sets of units later. Much of Chapter 2 is presented in atomic units, in which $e = m = \hbar = 1$.* In this Appendix, we state several conversion factors for length, energy, dipole moment, etc.

Length　The characteristic length scale for dense monolayers is the angstrom. The unit of length in atomic units is the Bohr radius a_0.

$$
\begin{aligned}
1\,\text{Å} &= 10^{-8}\,\text{cm} = 0.1\,\text{nm} \\
1\,a_0 &= 0.5292\,\text{Å}
\end{aligned}
\qquad\qquad (\text{H.1})
$$

Energy　The cohesive energies of physically adsorbed layers are conveniently stated in millielectron volts, meV, per atom or molecule. Other units are kelvins per atom, kilojoules per mole, kilocalories per mole, and atomic units (hartree).

$$
\begin{aligned}
1\,\text{meV} &= 11.6\,\text{K} \\
1\,\text{kJoule} &= 10.4\,\text{meV} \\
1\,\text{kcal} &= 43.4\,\text{meV} \\
1\,\text{THz} &= 4.13\,\text{meV} \\
1\,\text{au} &= 27.21\,\text{eV}
\end{aligned}
\qquad\qquad (\text{H.2})
$$

Dipole moment　The characteristic scale of induced and fluctuating dipole moments is the debye, an esu term.

$$
\begin{aligned}
1\,\text{debye} &= 1 \times 10^{-18}\,\text{esu cm} = 3.33 \times 10^{-30}\,\text{coul m} \\
1\,e\,a_0 &= 2.54\,\text{debye}
\end{aligned}
\qquad\qquad (\text{H.3})
$$

Quadrupole moment　Molecular quadrupole moments are frequently quoted in buckingham.

$$
1\,\text{buckingham} = 1 \times 10^{-26}\,\text{esu cm}^2 = 3.33 \times 10^{-40}\,\text{coul m}^2 \qquad (\text{H.4})
$$

---

*Electron charge, electron mass, and the reduced Planck constant.

Pressure The SI unit of pressure, 1 newton/meter^2, is termed the pascal Pa. However, other traditional units such as the atmosphere (atm) and the torr still are widely used.

$$
\begin{aligned}
1\,\text{Pa} &= 10\,\text{dyne/cm}^2 \\
1\,\text{atm} &= 1.01 \times 10^5\,\text{Pa} \\
1\,\text{torr} &= 133\,\text{Pa}
\end{aligned}
\tag{H.5}
$$

A related quantity is the unit of exposure: 1 langmuir $= 10^{-6}$ torr sec.

Dispersion energy coefficients The coefficients C_3, C_6, C_{s1}, and C_{s2} are expressed in atomic units in the theoretical development in the text. Some useful conversions for these and for the dipole polarizability are

$$
\begin{aligned}
1\,\text{au}\,(C_3) &= 4.032\,\text{eV\AA}^3 \\
1\,\text{au}\,(C_6) &= 1\,\text{au}\,(C_{s1}) = 1\,\text{au}\,(C_{s2}) = 0.5976\,\text{eV\AA}^6 \\
1\,\text{au}\,(\alpha) &= 0.1482\,\text{\AA}^3/\text{atom}
\end{aligned}
\tag{H.6}
$$

REFERENCES

The numbers in square brackets refer to the text sections where the reference is cited.

Abernathy, D. L., Mochrie, S. G. J., Zehner, D. M., Grübel, G., and Gibbs, D. (1992). *Physical Review B*, **45**, 9272–91. [3.6, 5.2.1.2, 6.2.0].

Abernathy, D. L., Gibbs, D., Grübel, G., Huang, K. G., Mochrie, S. G. J., and Sandy, A. R. (1993). *Surface Science*, **283**, 260–76. [6.2.0].

Abraham, F. F. (1980a). *Physical Review Letters*, **44**, 463–6. [4.3, 5.2.1.6].

Abraham, F. F. (1980b). *Journal of Chemical Physics*, **72**, 1412–3. [5.2.2.1].

Abraham, F. F. (1981a). *Physics Reports*, **80**, 339–74. [4.2.3, 4.3, 5.2.1.6].

Abraham, F. F. (1981b). *Physical Review B*, **23**, 6145–8. [5.2.2.1].

Abraham, F. F. (1982). *Reports on Progress in Physics*, **45**, 1113–61. [4.3, 5.2.1.6].

Abraham, F. F. (1983a). *Physical Review Letters*, **50**, 978–81. [6.1.6].

Abraham, F. F. (1983b). *Physical Review B*, **28**, 7338–41. [6.0, 6.1.6].

Abraham, F. F. (1984). *Physical Review B*, **29**, 2606–10. [6.1.6].

Abraham, F. F. (1986). *Advances in Physics*, **35**, 1–111. [1.7.4.4, 4.3, 5.2.1.6, 5.2.2.2].

Abraham, F. F. and Broughton, J. Q. (1987). *Physical Review Letters*, **59**, 64–7. [5.2.1, 6.1.1].

Abraham, F. F., Rudge, W. E., Auerbach, D. J., and Koch, S. W. (1984). *Physical Review Letters*, **52**, 445–8. [5.2.1.6, 5.2.2.2, 6.1.5].

Abraham, F. F., Broughton, J. Q., Leung, P. W., and Elser, V. (1990). *Europhysics Letters*, **12**, 107–12. [5.2.1, 6.1.1].

Ahlrichs, R., Böhm, H. J., Brode, S., Tang, K. T., and Toennies, J. P. (1988). *Journal of Chemical Physics*, **88**, 6290–302; **98**, 3579(E) (1993). [2.1.1.2, 2.3.1.3].

Alder, B. J. and Wainwright, T. E. (1967). *Physical Review Letters*, **18**, 988–90. [4.4].

Alder, B. J. and Wainwright, T. E. (1970). *Physical Review A*, **1**, 18–21. [4.4].

Alerhand, O. L., Vanderbilt, D., Meade, R. D., and Joannopolous, J. D. (1988). *Physical Review Letters*, **61**, 1973–6. [2.3.2.4, 5.2.2.2].

Alexander, S. (1975). *Physics Letters A*, **54**, 353–4. [5.1.1.1, 5.1.3.1, 6.1.1].

Alkhafaji, M. T. and Migone, A. D. (1991). *Physical Review B*, **43**, 8741–3. [6.4.3].

Alkhafaji, M. T. and Migone, A. D. (1992a). *Physical Review B*, **45**, 5729–32. [6.1.7.4].

Alkhafaji, M. T. and Migone, A. D. (1992b). *Physical Review B*, **45**, 8767–9. [6.4.3].

Alkhafaji, M. T. and Migone, A. D. (1993). *Physical Review B*, **48**, 1761–4. [6.1.7.4].

Alkhafaji, M. T., Shrestha, P., and Migone, A. D. (1994). *Physical Review B*, **50**, 11088–92. [6.4.3].

Allen, M. P. and Tildesley, D. J. (1987). *Computer simulation of liquids*. Clarendon Press, Oxford. [1.7.4.4, 5.2.1.6].

Allen, R. E. and de Wette, F. W. (1969). *Physical Review*, **179**, 873–85. [B].

Allen, R. E., de Wette, F. W., and Rahman, A. (1969). *Physical Review*, **179**, 887–92. [B].

Anderson, J. B., Traynor, C. A., and Boghosian, B. M. (1993). *Journal of Chemical Physics*, **99**, 345–51. [2.1.1.1, 2.1.1.3].

Annett, J. F. and Haydock, R. (1984a). *Physical Review Letters*, **53**, 838–41; **57**, 1382(E) (1986). [2.3.3.1].

Annett, J. F. and Haydock, R. (1984b). *Physical Review B*, **29**, 3773–6. [2.3.3.1].

Annett, J. F. and Haydock, R. (1986). *Physical Review B*, **34**, 6860–68. [2.3.3.1].

Antoniewicz, P. R. (1974). *Physical Review Letters*, **32**, 1424–5. [2.3.2.3].

Antoniou, A. A. (1976). *Journal of Chemical Physics*, **64**, 4901–11. [6.1.3].

Aruga, T. and Murata, Y. (1989). *Progress in Surface Science*, **31**, 61–130. [1.6.4, 2.3.2.3, 6.2.3].

Asada, H. (1990). *Surface Science*, **230**, 323–8. [5.1.1.3].

Aubry, S. and LeDaeron, P. Y. (1983). *Physica D*, **8**, 381–422. [5.2.2.2].

Avgul, N. N. and Kiselev, A. V. (1970). In *Chemistry and Physics of Carbon*, Vol.6, (ed. P. Walker), pp.1–124. Dekker, New York. [1.7.3, 6.1].

Axilrod, B. M. and Teller, E. (1943). *Journal of Chemical Physics*, **11**, 299–300. [6.1.4].

Aziz, R. A. (1984). In *Inert gases*, (ed. M. L. Klein), pp.5–86. Springer, Berlin. [2.1.1.2, 2.4.1, 2.4.4, A.2].

Aziz, R. A., Buck, U., Jónsson, H., Ruiz-Suárez, J.-C., Schmidt, B., Scoles, G., Slaman, M. J., and Xu, J. (1989). *Journal of Chemical Physics*, **91**, 6477–92; **93**, 4492(E) (1990). [2.3.1].

Aziz, R. A., Janzen, A. R., and Moldover, M. R. (1995). *Physical Review Letters*, **74**, 1586–9. [2.1.1.3].

Baer, D. R., Fraass, B. A., Riehl, D. H., and Simmons, R. O. (1978). *Journal of Chemical Physics*, **68**, 1411–7. [6.1.5, A.3].

Bagchi, K., Andersen, H. C., and Swope, W. (1996). *Physical Review Letters*, **76**, 255–9. [4.3, 5.2.1.6].

Baierlein, R. (1968). *American Journal of Physics*, **36**, 625–9. [5.2.2.1].

Bak, P. (1982). *Reports on Progress in Physics*, **45**, 587–629. [5.2.2.2].

Bak, P. and Bohr, T. (1983). *Physical Review B*, **27**, 591–3. [5.1.2.2].

Bak, P. and Bruinsma, R. (1982). *Physical Review Letters*, **49**, 249–51. [5.2.2.2].

Bak, P., Mukamel, D., Villain, J., and Wentowska, K. (1979). *Physical Review B*, **19**, 1610–3. [3.4, 5.2.2.2, 6.1.5, 6.1.6].

Bakker, A. F., Bruin, C., and Hilhorst, H. J. (1984). *Physical Review Letters*, **52**, 449–52. [4.3.1, 5.2.2.1].

Baldereschi, A. (1973). *Physical Review B*, **7**, 5212–4. [5.2.1.2].

Bardeen, J. (1940). *Physical Review*, **58**, 727–36. [2.1.2.1].

Barker, J. A. (1963). *Lattice theories of the liquid state*. Pergamon, New York. [5.2.1.3].

Barker, J. A. (1981). *Proceedings of the Royal Society (London) A*, **377**, 425–9. [5.2.1.4, 5.2.2.1].

Barker, J. A. (1986). *Physical Review Letters*, **57**, 230–3. [2.3.1, 6.1.6].

Barker, J. A. (1989). In *Simple molecular systems at very high density*, (ed. A. Polian, P. Loubeyre, and N. Boccara), pp.331–51. Plenum, New York. [2.3.1, 6.1.4, 6.1.6].

Barker, J. A. (1993). *Molecular Physics*, **80**, 815–20. [2.3.1, 6.1.6].

Barker, J. A. and Auerbach, D. J. (1985). *Surface Science Reports*, **4**, 1–99. [1.6.2.1, 1.6.3, 1.7.4.2, 4.1.2].

Barker, J. A. and Henderson, D. (1976). *Reviews of Modern Physics*, **48**, 587–671. [5.2.1.4, 5.2.2.1].

Barker, J. A. and Rettner, C. T. (1992). *Journal of Chemical Physics*, **97**, 5844–50. [2.3.3.1, 2.3.3.2, 2.4.4, 5.2.2.2, 6.2.2].

Barker, J. A., Henderson, D., and Abraham, F. F. (1981). *Physica A*, **106**, 226–38. [4.3.1, 5.2.1.3, 5.2.2.1].

Barnes, R. F. (1967). Unpublished Ph.D. thesis, University of Wisconsin-Madison. [3.5].

Barron, T. H. K. and Klein, M. L. (1965). *Proceedings of the Physical Society (London)*, **85**, 523–32. [4.3.1, 5.2.2.2, G]

Barth, J. V., Brune, H., Ertl, G., and Behm, R. J. (1990). *Physical Review B*, **42**, 9307–18. [5.2.2.2, 6.2.0].

Burton, G. (1979). *Reports on Progress in Physics*, **42**, 963–1016. [1.6.4].

Baskin, Y. and Meyer, L. (1955). *Physical Review*, **100**, 544. [6.1.0].

Batra, I. P. (1984). *Surface Science*, **148**, 1–20. [2.3.3.1].

Batra, I. P., Bagus, P. S., and Barker, J. A. (1985). *Physical Review B*, **31**, 1737–43. [2.1.2.2, 2.3.3.1].

Bauer, E. (1982). *Applications of Surface Science*, **11/12**, 479–94. [3.6].

Baxter, R. J. (1982a). *Journal of Physics A*, **15**, 3329–40. [5.1.1.1, 5.1.1.2].

Baxter, R. J. (1982b). *Exactly solved models in statistical mechanics*. Academic, New York. [5.1.1, 5.1.1.1, 5.1.1.2].

Baxter, R. J., Temperley, H. N. V., and Ashley, S. E. (1979). *Proceedings of the Royal Society (London) A*, **358**, 535–59. [5.1.1.1].

Beaume, R., Suzanne, J., Coulomb, J. P., and Glachant, A. (1984). *Surface Science*, **137**, L117–22. [6.1.7.1].

Behm, R. J., Brundle, C. R., and Wandelt, K. (1986). *Journal of Chemical Physics*, **85**, 1061–73. [1.6.3, 1.6.4, 6.2.1].

Belak, J., Kobashi, K., and Etters, R. D. (1985). *Surface Science*, **161**, 390–410. [2.3.1, 5.2.1.1, 6.1.7.3, F].

Benedek, G. (ed.). (1992). *Surface properties of layered materials*. Kluwer, Dordrecht. [1.7.3].

Benedek, G. and Onida, G. (1993). *Physical Review B*, **47**, 16471–6. [6.1.0].

Benedek, G. and Valbusa, U. (ed.) (1982). *Dynamics of gas–surface interaction*. Springer series in Chemical Physics, Vol.**21**. [1.7.2].

Benedek, G., Celli, V., Cole, M. W., Toigo, F., and Weare, J. (ed.) (1984). *Gas–surface interactions and physical adsorption*. Surface Science Vol.**148**. [1.7.2].

Berker, A. N., Ostlund, S., and Putnam, F. A. (1978). *Physical Review B*, **17**, 3650–65. [5.1.1, 5.1.1.1, 5.1.1.2, 5.1.3.2].

Berkhout, J. J., Wolters, E. J., van Roijan, R., and Walraven, J. T. M. (1986). *Physical Review Letters*, **57**, 2387–90. [4.1.2].

Berkowitz, J. (1979). *Photoabsorption, photoionization, and photoelectron spectroscopy*. Academic, New York. [A.2].

Berndt, W. (1989). *Surface Science*, **219**, 161–76. [6.2.0, 6.2.4].

Bernu, B., Ceperley, D., Lhuillier, C., and Pierre, L. (1992). In *Recent progress in many-body theories*, **3** (ed. T. L. Ainsworth, C. E. Campbell, B. E. Clements, and E. Krotscheck), pp.459–67. Plenum, New York. [6.1.1].

Bhethanabotla, V. R. and Steele, W. A. (1987). *Langmuir*, **3**, 581–7. [2.3.1, 2.4.4, 6.1.7.3].

Bhethanabotla, V. R. and Steele, W. A. (1988a). *Canadian Journal of Chemistry*, **66**, 866–74. [2.4.4, 6.1.7.3].

Bhethanabotla, V. R. and Steele, W. A. (1988b). *Journal of Physical Chemistry*, **92**, 3285–91. [6.1.5].

Bhethanabotla, V. R. and Steele, W. A. (1989). *Journal of Chemical Physics*, **91**, 4346–52. [6.1.7.3].

Bhethanabotla, V. R. and Steele, W. A. (1990). *Physical Review B*, **41**, 9480–7. [6.1.7.3].

Bienfait, M. (1987a). *Europhysics Letters*, **4**, 79–84. [6.3].

Bienfait, M. (1987b). In *Dynamics of molecular crystals*, (ed. J. Lascombe), pp.353–63. Elsevier, Amsterdam. [6.1.0].

Bienfait, M. and Gay, J.-M. (1991). In *Phase transitions in surface films 2*, (ed. H. Taub, G. Torzo, H. J. Lauter, and S. C. Fain, Jr.), pp.307–25. Plenum, New York. [1.6.3, 6.3].

Bienfait, M., Coulomb, J. P., and Palmari, J. P. (1987). *Surface Science*, **182**, 557–66. [1.6.3, 5.2.1.4, 6.3].

Bienfait, M., Gay, J. M., and Blank, H. (1988). *Surface Science*, **204**, 331–44. [6.3].

Bienfait, M., Zeppenfeld, P., Gay, J. M., and Palmari, J. P. (1990). *Surface Science*, **226**, 327–38. [6.1.7.1, 6.3].

Binder, K. (1992a). *Annual Review of Physical Chemistry*, **43**, 33–59. [1.7.4.3, 5.1, 5.1.2].

Binder, K. (ed.) (1992b). *The Monte Carlo method in condensed matter physics*, Springer Topics in Applied Physics Vol.**71**, Springer, Berlin. [1.7.4.4, 5.2.1.6].

Binder, K. and Heermann, D. W. (1992). *Monte Carlo simulation in statistical physics. An introduction*, (2nd edn). Springer, Berlin. [1.7.4.4, 5.2.1.6].

Binder, K. and Landau, D. P. (1989). *Advances in Chemical Physics*, **76**, 91–152. [1.7.4.3, 5.1, 5.2.1.6].

Binder, K., Ferrenberg, A. M., and Landau, D. P. (1994). *Berichte der Bunsen-Gesellschaft für Physikalische Chemie*, **98**, 340–5. [5.0].

Birkenheuer, U., Boettger, J. C., and Rösch, N. (1994). *Journal of Chemical Physics*, **100**, 6826–36. [6.3].

Black, J. E. and Janzen, A. (1989a). *Physical Review B*, **39**, 6238–40. [2.3.3.2, 6.2.2].

Black, J. E. and Janzen, A. (1989b). *Surface Science*, **217**, 199–232. [6.2.2].

Black, J. E. and Mills, D. L. (1990). *Physical Review B*, **42**, 5610–20. [5.2.2.2].

Black, J. E. and Tian, Z.-J. (1993). *Physical Review Letters*, **71**, 2445–8. [5.2.1.4].

Blumstein, C. and Wheeler, J. C. (1973). *Physical Review B*, **8**, 1764–76. [5.2.1.2].

Boato, G. and Cantini, P. (1983). *Advances in Electronics and Electron Physics*, **60**, (ed. P. W. Hawkes), 95–160. [1.7.4.2].

Boato, G., Cantini, P., and Tatarek, R. (1978). *Physical Review Letters*, **40**, 887–9. [1.6.2.1, 2.3.3.1, 6.2.1].

Boato, G., Cantini, P., Guidi, C., Tatarek, R., and Felcher, G. P. (1979). *Physical Review B*, **20**, 3957–69. [2.3.3.2, 2.4.2, 6.1.1].

Bohr, J., Nielsen, M., Als-Nielsen, J., and Kjaer, K. (1983). *Surface Science*, **125**, 181–7. [6.4.1].

Bojan, M. J. and Steele, W. A. (1987a). *Langmuir*, **3**, 116–20. [2.4.4, 5.2.1.4].

Bojan, M. J. and Steele, W. A. (1987b). *Langmuir*, **3**, 1123–7. [2.4.4, 5.2.1.4, 6.1.7.3].

Bomchil, G., Hüller, A., Rayment, T., Roser, S. J., Smalley, M. V., Thomas, R. K., and White, J. W. (1980). *Philosophical Transactions of the Royal Society B*, **290**, 537–52. [6.1.7.1].

Bonino, G., Pisani, C., Ricca, F., and Roetti, C. (1975). *Surface Science*, **50**, 379–87. [2.3.3.2].

Bonissent, A. and Bruch, L. W. (1985). *Journal of Chemical Physics*, **83**, 2373–5. [5.2.1.4, 6.2.3].

Bonzel, H. P. (1988). *Surface Science Reports*, **8**, 43–125. [6.2.3].

Bortolani, V. and Levi, A. C. (1986). *La Rivista del Nuovo Cimento*, **9**, #11, 1–77. [1.7.4.2]

Bortolani, V., March, N. H., and Tosi, M. P. (ed.) (1990). *Interactions of atoms and molecules with solid surfaces*. Plenum, New York. [1.7.3].

Bottani, E. J. and Bakaev, V. A. (1994). *Langmuir*, **10**, 1550–5. [5.2.1.6].

Bottani, E. J., Ismail, I. M. K., Bojan, M. J., and Steele, W. A. (1994a). *Langmuir*, **10**, 3805–8. [6.1.7.2].

Bottani, E. J., Bakaev, V., and Steele, W. (1994b). *Chemical Engineering Science*, **49**, 2931–9. [2.3.1].

Bradshaw, A. M. and Scheffler, M. (1979). *Journal of Vacuum Science and Technology*, **16**, 447–54. [6.2.3].

Brady, G. W., Fein, D. B., and Steele, W. A. (1977). *Physical Review B*, **15**, 1120–35. [1.6.2.2, 6.0].

Brami, B., Joly, F., and Lhuillier, C. (1994). *Journal of Low Temperature Physics*, **94**, 63–76. [6.1.1].

Bretz, M. (1977). *Physical Review Letters*, **38**, 501–5. [5.1.2.2, 5.1.3.1, 6.1.1].

Bretz, M. and Chung, T. T. (1974). *Journal of Low Temperature Physics*, **17**, 479–88. [6.1.1].

Bretz, M., Dash, J. G., Hickernell, D. C., McLean, E. O., and Vilches, O. E. (1973). *Physical Review A*, **8**, 1589–615. [1.6.1.2, 4.2.2.2, 5.1.3.1, 6.1.0, 6.1.1].

Brewer, D. F. (1978). In *The physics of liquid and solid helium*, Part II (ed. K. H. Bennemann and J. B. Ketterson), pp.573–673. Wiley, New York. [6.1.1].

Brooks, F. C. (1952). *Physical Review*, **86**, 92–7. [2.1.1.1].

Broughton, J. Q. and Abraham, F. F. (1988). *Journal of Physical Chemistry*, **92**, 3274–7. [5.2.1, 6.1.1].

Bruch, L. W. (1978). *Physica*, **93A**, 95–113. [5.2.2.1].

Bruch, L. W. (1982). *Surface Science*, **115**, L67–70. [6.1.3].

Bruch, L. W. (1983a). *Surface Science*, **125**, 194–217. [2.0, 2.3.2.2, 2.3.2.4].

Bruch, L. W. (1983b). *Journal of Chemical Physics*, **79**, 3148–56. [2.2.1, 6.1.7.2, 6.1.7.3. A.3].

Bruch, L. W. (1985). *Surface Science*, **150**, 503–37. [5.2.1.2, 5.2.2.2].

Bruch, L. W. (1987). *Journal of Chemical Physics*, **87**, 5518–27. [4.2.3, 5.2.1.2, 5.2.1.3, 6.1.7.1, E.1].

Bruch, L. W. (1988). *Physical Review B*, **37**, 6658–62. [3.3, 5.2.1.2, 6.1.3, G].

Bruch, L. W. (1991a). In *Phase transitions in surface films 2*, (ed. H. Taub, G. Torzo, H. J. Lauter, and S. C. Fain, Jr.), pp.67–82. Plenum, New York. [2.3.3.2, 6.1.5, 6.1.7.3].

Bruch, L. W. (1991b). *Surface Science*, **250**, 267–78. [5.2.2.2].

Bruch, L. W. (1992). *Physical Review B*, **45**, 7161–4. [4.3.1].

Bruch, L. W. (1993). *Physical Review B*, **48**, 1765–78. [3.4, 5.2.1.2, 5.2.2.2].

Bruch, L. W. (1994). *Physical Review B*, **49**, 7654–9. [6.1.1].

Bruch, L. W. and Gottlieb, J. M. (1988). *Physical Review B*, **37**, 4920–9. [5.2.2.1].

Bruch, L. W. and Ni, X.-Z. (1985). *Faraday Discussions of the Chemical Society (London)*, **80**, 217–26. [5.2.1.2, 6.1.3, 6.1.6, 6.2.1].

Bruch, L. W. and Osawa, T. (1980). *Molecular Physics*, **40**, 491–507. [2.3.2.3].

Bruch, L. W. and Phillips, J. M. (1980). *Surface Science*, **91**, 1–23. [6.2.1].

Bruch, L. W. and Ruijgrok, Th. W. (1979). *Surface Science*, **79**, 509–48. [2.1.2, 6.2.3].

Bruch, L. W. and Venables, J. A. (1984). *Surface Science*, **148**, 167–86. [2.3.1, 3.3, 3.6, 5.2.1.2].

Bruch, L. W. and Watanabe, H. (1977). *Surface Science*, **65**, 619–32. [2.3.2.1, 2.4.2].

Bruch, L. W. and Wei, M. S. (1980). *Surface Science*, **100**, 481–506. [4.2.3, 5.2.2.1, 6.1.4].

Bruch, L. W., Cohen, P. I., and Webb, M. B. (1976a). *Surface Science*, **59**, 1–16. [2.4.4, 5.2.1.1, 5.2.2.1, 6.2.1].

Bruch, L. W., Schick, M., and Siddon, R. L. (1976b). *American Journal of Physics*, **44**, 1007–8. [5.2.2.1].

Bruch, L. W., Unguris, J., and Webb, M. B. (1979). *Surface Science*, **87**, 437–56. [5.2.1.1, B].

Bruch, L. W., Phillips, J. M., and Ni, X.-Z. (1984). *Surface Science*, **136**, 361–80. [2.3.1, 6.1.3].

Bruch, L. W., Gay, J. M., and Krim, J. (1985). *Journal de Physique (Paris)*, **46**, 425–33. [5.2.1.1, 6.2.1, B].

Bruch, L. W., Glebov, A., Toennies, J. P., and Weiss, H. (1995). *Journal of Chemical Physics*, **103**, 5109–20. [2.3.1.3, 6.3].

Brueckner, K. A. (1966). In *Many-body theory: 1965 Tokyo summer lectures in theoretical physics*, (ed. R. Kubo), pp.152–60. Benjamin, New York. [6.0].

Brunauer, S., Emmett, P. H., and Teller, E. (1938). *Journal of the American Chemical Society*, **60**, 309–19. [1.3, 4.1.1, 5.1].

Bruschi, L. and Torzo, G. (1991). In *Phase transitions in surface films 2*, (ed. H. Taub, G. Torzo, H.J. Lauter, and S.C. Fain, Jr.), pp.425–35. Plenum, New York. [1.6.1.1].

Buckingham, A. D., Fowler, P. W., and Hutson, J. M. (1988). *Chemical Reviews*, **88**, 963–88. [2.1.1.2].

Butler, D. M., Litzinger, J. M., Stewart, G. A., and Griffiths, R. B. (1979). *Physical Review Letters*, **42**, 1289–92. [6.1.0, 6.1.5, 6.1.7.3].

Butler, D. M., Litzinger, J. M., and Stewart, G. A. (1980). *Physical Review Letters*, **44**, 466–8. [6.1.5].

Caflisch, R. G., Berker, A. N., and Kardar, M. (1985). *Physical Review B*, **31**, 4527–37. [5.1.3.2].

Cahn, J. W. (1977). *Journal of Chemical Physics*, **66**, 3667–72. [6.4.4].

Calisti, S. and Suzanne, J. (1981). *Surface Science*, **105**, L255–9. [6.1.3].

Calisti, S., Suzanne, J., and Venables, J. A. (1982). *Surface Science*, **115**, 455–68. [3.3, 6.1.3].

Campbell, C. E. and Schick, M. (1972). *Physical Review A*, **5**, 1919–25. [5.1.1.3].

Campbell, C. E., Krotscheck, E., and Pang, T. (1992). *Physics Reports*, **223**, 1–42. [2.1.2.2].

Campbell, J. H. and Bretz, M. (1985). *Physical Review B*, **32**, 2861–9. [1.6.1.2].

Campbell, J. H., Bretz, M., and Chan, M. H. W. (1980). In *Ordering in two dimensions*, (ed. S. K. Sinha), pp.295–6. North Holland, New York. [1.6.1.2].

Cardillo, M. J. (1981). *Annual Review of Physical Chemistry*, **32**, 331–57. [1.7.4.2].

Cardillo, M. J. and Tully, J. C. (1984). In *Dynamics on surfaces*, (ed. B. Pullman, J. Jortner, A. Nitzan, and B. Gerber), pp.169–80. Reidel, Dordrecht. [4.1.2].

Cardini, G. and O'Shea, S. F. (1985). *Surface Science*, **154**, 231–53. [5.2.1.2, 6.1.7.3].

Cardini, G. G., O'Shea, S. F., Marchese, M., and Klein, M. L. (1985). *Physical Review B*, **32**, 4261–3. [5.2.1.2].

Carlos, W. E. and Cole, M. W. (1980a). *Surface Science*, **91**, 339–57. [2.3.1, 2.3.3, 2.3.3.2, 6.1.5].

Carlos, W. E. and Cole, M. W. (1980b). *Physical Review B*, **21**, 3713–20. [2.3.3.2, 4.2.2.2].

Carneiro, K., Passell, L., Thomlinson, W., and Taub, H. (1981). *Physical Review B*, **24**, 1170–6. [2.4.4, 3.5, 6.1.1].

Carraro, C. and Cole, M. W. (1992). *Physical Review B*, **45**, 12930–5. [4.1.2].

Carraro, C. and Cole, M. W. (1995). *Zeitschrift für Physik B*, **98**, 319–21. [2.1.2.1, 4.1.2].

Casimir, H. B. G. and Polder, D. (1946). *Nature*, **158**, 787–8. [2.1.2.1, 2.2.1].

Casimir, H. B. G. and Polder, D. (1948). *Physical Review*, **73**, 360–72. [2.1.2.1, 2.2.1].

Celli, V. (1992). In *Helium atom scattering from surfaces*, (ed. E. Hulpke), pp.25–40. Springer, Berlin. [2.1.2.2, 2.3.3.1].

Celli, V., Eichenauer, D., Kaufhold, A., and Toennies, J. P. (1985). *Journal of Chemical Physics*, **83**, 2504–21. [2.3.1.3].

Cencek, W., Komasa, J., and Rychlewski, J. (1995). *Chemical Physics Letters*, **246**, 417–20. [2.1.1.1].

Ceva, M., Goldmann, M., and Marti, C. (1986). *Journal de Physique (Paris)*, **47**, 1527–32. [6.4.1].

Chabal, Y. J. (1988). *Surface Science Reports*, **8**, 211–357. [1.6.3].

Chadi, D. J. and Cohen, M. L. (1973). *Physical Review B*, **8**, 5747–53. [5.2.1.2].

Chae, D. G., Ree, F. H., and Ree, T. (1969). *Journal of Chemical Physics*, **50**, 1581–9. [5.2.1.4].

Chan, M. H. W. (1991). In *Phase transitions in surface films 2*, (ed. H. Taub, G. Torzo, H. J. Lauter, and S. C. Fain, Jr.), pp.1–10. Plenum, New York. [6.1.0].

Chan, M. H. W., Migone, A. D., Miner, K. D., and Li, Z. R. (1984). *Physical Review B*, **30**, 2681–94. [6.1.7.3].

Chandavarkar, S. and Diehl, R. D. (1989). *Physical Review B*, **40**, 4651–5. [4.3.1].

Chandavarkar, S., Geertman, R. M., and De Jeu, W. H. (1992). *Physical Review Letters*, **69**, 2384–7. [6.4.4].

Chaves, F. A. B., Cortez, M. E. B. P., Rapp, R. E., and Lerner, E., (1985). *Surface Science*, **150**, 80–8. [6.1.1, 6.1.2].

Chen, K., Kaplan, T., and Mostoller, M. (1995). *Physical Review Letters*, **74**, 4019–22. [4.3.1, 5.2.2.1].

Chen, L. J. C., Knackestedt, M., Robert, M., and Shukla, K. P. (1991). *Physica A*, **172**, 53–76. [5.2.2.1].

Chen, L. Y. and Ying, S. C. (1993). *Physical Review Letters*, **71**, 4361–4. [5.2.1.4, 6.2.1].

Chen, R., Trucano, P., and Stewart, R. F. (1977). *Acta Crystallographica*, **A33**, 823–8. [2.3.3.2, 6.1].

Cheng, A. and Klein, M. L. (1992). *Langmuir*, **8**, 2798–803. [2.3.1, 6.1.7.4].

Cheng, A. and Steele, W. A. (1989). *Langmuir*, **5**, 600–7. [2.3.1].

Cheng, A. and Steele, W. A. (1990). *Journal of Chemical Physics*, **92**, 3858–66. [2.3.1].

Cheng, E., Ihm, G., and Cole, M. W. (1989). *Journal of Low Temperature Physics*, **74**, 519–28. [4.3.2].

Cheng, E., Chizmeshya, A., Cole, M. W., Klein, J. R., Ma, J., Saam, W. F., and Treiner, J. (1991a). *Physica A*, **177**, 466–73. [2.1.2.3, 2.4.3, 5.2.2.1].

Cheng, E., Cole, M. W., Saam, W. F., and Treiner, J. (1991b). *Physical Review Letters*, **67**, 1007–10. [2.1.2.3, 2.4.3].

Cheng, E., Cole, M. W., Saam, W. F., Treiner, J. (1992). *Physical Review B*, **46**, 13967–82. [1.6.3].

Cheng, E., Cole, M. W., Dupont-Roc, J., Saam, W. F., and Treiner, J. (1993a). *Reviews of Modern Physics*, **65**, 557–67. [5.1.2.2, 6.4.4].

Cheng, E., Mistura, G., Lee, H. C., Chan, M. H. W., Cole, M. W., Carraro, C., Saam, W. F., and Toigo, F. (1993b) *Physical Review Letters*, **70**, 1854–7. [1.1, 2.1.2.3, 6.4.4].

Cheng, E., Cole, M. W., Saam, W. F., and Treiner, J. (1994). *Physical Review B*, **48**, 18214–21. [1.1, 2.1.2.3].

Chester, G. V., Fisher, M E., and Mermin, N. D. (1969). *Physical Review*, **185**, 760–2. [4.3.2].

Chesters, M. A., Hussain, M., and Pritchard, J. (1973). *Surface Science*, **35**, 161–71. [6.2.1, 6.2.4].

Ching, W. Y., Huber, D. L., Lagally, M. G., and Wang, G.-C. (1978). *Surface Science*, **77**, 550–60. [5.1.1.1].

Chinn, M. D. and Fain, S. C. Jr. (1977a). *Physical Review Letters*, **39**, 146–9. [6.1.5].

Chinn, M. D. and Fain, S. C. Jr. (1977b). *Journal de Physique (Paris) Colloq*, **38**, C4-99 to C4-104. [6.1.5].

Chizmeshya, A. and Zaremba, E. (1989). *Surface Science*, **220**, 443–70. [2.2.2.2, 2.2.2.3, 2.4.3].

Chizmeshya, A. and Zaremba, E. (1992). *Surface Science*, **268**, 432–56. [2.1.1.2, 2.1.2, 2.2.0, 2.2.1.3, 2.2.2.2, 2.2.2.3].

Christmann, K. (1991). *Introduction to surface physical chemistry*. Steinkopff Verlag, Darmstadt. [1.7.3].

Chudley, C. T. and Elliott, R. J. (1961). *Proceedings of the Physical Society (London)*, **77**, 353–61. [5.2.1.4].

Chung, S. and Cole, M. W. (1984). *Surface Science*, **145**, 269–80. [E.2].

Chung, S., Holter, N., and Cole, M. W. (1986a). *Surface Science*, **165**, 466–76. [2.3.1].

Chung, S., Kara, A., and Frankl, D. R. (1986b). *Surface Science*, **171**, 45–54. [2.4.2].

Cieplak, M., Smith, E., and Robbins, M. O. (1994). *Science*, **265**, 1209–12. [6.2.1].

Clements, B. E., Forbert, H., Krotscheck, E., and Saarela, M. (1994). *Journal of Low Temperature Physics*, **95**, 849–81. [5.2.1.5].

Clougherty, D. P. and Kohn, W. (1992). *Physical Review B*, **46**, 4921–37. [4.1.2].

Cohen, P. I., Unguris, J., and Webb, M. B. (1976). *Surface Science*, **58**, 429–56. [2.4.4, 6.2.1].

Cole, M. W. (1995). *Journal of Low Temperature Physics*, **101**, 25–30. [2.4.3, 6.4.4].

Cole, M. W. and Toigo, F. (1981). *Physical Review B*, **23**, 3914–9. [2.3.3.2].

Cole, M. W. and Toigo, F. (1982). In *Interfacial aspects of phase transformations*, (ed. B. Mutaftschiev), pp.223–60. Reidel, Dordrecht. [4.1.2].

Cole, M. W. and Toigo, F. (1985). *Physical Review B*, **31**, 727–9. [2.1.2.2].

Cole, M. W., Frankl, D. R., and Goodstein, D. L. (1981). *Reviews of Modern Physics*, **53**, 199–210. [1.7.4.2, 2.3.3.2, 2.4.4, 6.1.1].

Cole, M. W., Toigo, F., and Tosatti, E. (ed.) (1983). *Statistical mechanics of adsorption. Surface Science*, Vol.**125**. [1.7.2].

Cole, M. W., Cheng, E., Carraro, C., Saam, W. F., Swift, M. R., and Treiner, J. (1994). *Physica B*, **197**, 254–9. [2.1.1.1, 2.1.2.3, 5.2.1.5].

Colella, N. J. and Suter, R. M. (1986). *Physical Review B*, **34**, 2052–5. [6.1.6].

Colwell, J. L., Gill, E. K., and Morrison, J. A. (1963). *Journal of Chemical Physics*, **39**, 635–53. [A.3].

Comsa, G., Kern, K., and Poelsema, B. (1992). In *Helium atom scattering from surfaces*, (ed. E. Hulpke), pp.243–63. Springer, Berlin. [6.2.2].

Conner, M. W. and Ebner, C. (1987). *Physical Review B*, **36**, 3683–92. [5.1.1.1].

Conrad, E. R. and Webb, M. B. (1983). *Surface Science*, **129**, 37–58. [4.2.1].

Constabaris, G. and Halsey, G. D. Jr. (1957). *Journal of Chemical Physics*, **27**, 1433–4. [1.4].

Constabaris, G., Singleton, J. H., and Halsey, G. D. Jr. (1959). *Journal of Physical Chemistry*, **63**, 1350–5. [2.3.1].

Constabaris, G., Sams, J. R. Jr., and Halsey, G. D. Jr. (1961). *Journal of Physical Chemistry*, **65**, 367–69. [2.3.1].

Coppersmith, S. N., Fisher, D. S., Halperin, B. I., Lee, P. A., and Brinkman, W. F. (1982). *Physical Review B*, **25**, 349–63. [5.2.2.2, 6.1.5].

Cortona, P., Dondi, M. G., and Tommasini, F. (1992a). *Surface Science*, **261**, L35–8. [2.1.2.2, 2.3.3.1].

Cortona, P., Dondi, M. G., Lausi, A., and Tommasini, F. (1992b). *Surface Science*, **276**, 333–40. [2.3.3.1].

Cotterill, R. M. J. and Pederson, L. B. (1972). *Solid State Communications*, **10**, 439–41. [5.2.2.1].

Coulomb, J. P. (1991). In *Phase transitions in surface films 2*, (ed. H. Taub, G. Torzo, H. J. Lauter, and S. C. Fain, Jr.), pp.113–34. Plenum, New York. [6.3].

Coulomb, J. P. and Bienfait, M. (1986). *Journal de Physique (Paris)*, **47**, 89–96. [6.1.7.1, 6.1.7.4].

Coulomb, J. P. and Vilches, O. E. (1984). *Journal de Physique (Paris)*, **45**, 1381–9. [1.6.1.1, 6.3].

Coulomb, J. P., Biberian, J. P., Suzanne, J., Thomy, A., Trott, G. J., Taub, H., Danner, H. R., and Hansen, F. Y. (1979). *Physical Review Letters*, **43**, 1978–81. [6.1.7.4].

Coulomb, J. P., Bienfait, M., and Thorel, P. (1981). *Journal de Physique (Paris)*, **42**, 293–306. [1.6.3, 5.2.1.4, 6.1.7.1].

Coulomb, J. P., Sullivan, T. S., and Vilches, O. E. (1984). *Physical Review B*, **30**, 4753–60. [6.1.6, 6.3].

Coulomb, J. P., Madih, K., Croset, B., and Lauter, H. J. (1985a). *Physical Review Letters*, **54**, 1536–8. [6.3].

Coulomb, J. P., Bienfait, M., and Thorel, P. (1985b). *Faraday Discussions of the Chemical Society (London)*, **80**, 79–90. [6.1.7.4].

Coulomb, J. P., Larher, Y., Trabelsi, M., and Mirebeau, I. (1994). *Molecular Physics*, **81**, 1259–64. [6.3].

Cowley, E. R. and Barker, J. A. (1980). *Journal of Chemical Physics*, **73**, 3452–5. [5.2.1.3].

Crawford, R. K. (1977). Chap. 11, In *Rare gas solids*, Vol. 2, (ed. M. L. Klein and J. A. Venables), pp.663–728. Academic, London. [6.1.3, 6.1.4, 6.1.5, 6.1.6, A.3].

Crowell, P. A. and Reppy, J. D. (1993). *Physical Review Letters*, **70**, 3291–4. [1.6.3, 6.1.1].

Crowell, P. A. and Reppy, J. D. (1994). *Physica B*, **197**, 269–77. [6.1.1].

Cui, J. and Fain, Jr. S. C. (1989). *Physical Review B*, **39**, 8628–42. [1.6.2.1, 3.6, 5.2.2.2, 6.1.2, 6.2.2].

Cui, J., Fain, S. C. Jr., Freimuth, H., Wiechert, H., Schildberg, H. P., and Lauter, H. J. (1988). *Physical Review Letters*, **60**, 1848–51. [3.3, 5.2.2.2, 6.1.2].

Cui, J., Jung, D. R., and Diehl, R. D. (1992). *Physical Review B*, **45**, 9375–81. [5.2.1.2].

Cunningham, S. L. (1974). *Physical Review B*, **10**, 4988–94. [5.2.1.2].

Currat, R. and Janssen, T. (1988). *Solid State Physics*, **41**, 201–302. [5.2.2.2].

Dai, P., Angot, T., Ehrlich, S. N., Wang, S.-K., and Taub, H. (1994). *Physical Review Letters*, **72**, 685–8. [3.5, 6.2.1].

Dalgarno, A. and Lewis, J. T. (1956). *Proceedings of the Physical Society (London)*, **A69**, 57–64. [2.1.1.1].

Daly, C. and Krim, J. (1996). *Physical Review Letters*, **76**, 803–6. [6.2.1].

D'Amico, K. L., Moncton, D. E., Specht, E. D., Birgeneau, R. J., Nagler, S. E., and Horn, P. M. (1984). *Physical Review Letters*, **53**, 2250–3. [5.2.1.2, 6.1.5].

D'Amico, K. L., Bohr, J., Moncton, D. E., and Gibbs, D. (1990). *Physical Review B*, **41**, 4368–76. [4.3.1, 6.1.4].

Dash, J. G. (1975). *Films on solid surfaces*. Academic, New York. [1.6.1.2, 1.7.1].

Dash, J. G. (1978). *Physics Reports*, **38C**, 177–226. [1.7.4.3, 6.1.1].

Dash, J. G. and Ruvalds, J. (ed.) (1980). *Phase transitions in surface films*. Plenum, New York. [1.7.2].

Dash, J. G. and Schick, M. (1978). In *The physics of liquid and solid helium*, Part II, (ed. K. H. Bennemann and J. B. Ketterson), pp.497–571. Wiley, New York. [1.7.4.3, 6.1.1].

Dash, J. G., Schick, M., and Vilches, O. E. (1994). *Surface Science*, **299/300**, 405–24. [4.2.2.2].

Davis, H. T. (1996). *Statistical mechanics of phases, interfaces, and thin films.* VCH Publishers, New York. [1.7.3].

Daunt, J. G. and Lerner, E. (ed.) (1973). *Monolayer and submonolayer helium films.* Plenum, New York. [1.7.2].

Daw, M. S. and Baskes, M. I. (1984). *Physical Review B*, **29**, 6443–53. [2.1.2.2].

De Boer, J. (1948). *Physica*, **14**, 139–48. [5.2.2.1].

De Boer, J. and Bird, R. B. (1964). Chap. 6, In *The molecular theory of gases and liquids*, 2nd printing. (Hirschfelder et al.) [5.2.2.1, 6.1.3].

De Boer, J. H. (1968). *The dynamical character of adsorption*, (2nd edn). Clarendon Press, Oxford. [1.7.1].

Degenhardt, D., Lauter, H. J., and Frahm, R. (1989). *Surface Science*, **215**, 535–54; **221**, 619 (E) (1989). [6.3].

De Jong, F. and Jansen, A. P. J. (1994). *Surface Science*, **317**, 1–7. [5.2.1.4].

De Kieviet, M. F. M., Bahatt, D., Scoles, G., Vidali, G., and Karimi, M. (1996). *Surface Science*, **365**, 789-800. [6.3]

Demetrio de Souza, J. L. M., Rapp, R. E., de Souza, E. P., and Lerner, E. (1984). *Journal of Low Temperature Physics*, **55**, 273–82. [6.1.3].

Den Nijs, M. (1988). In *Phase transitions and critical phenomena*, (ed. C. Domb and J. L. Lebowitz), Vol. **12**, pp.220–333. Academic, New York. [5.2.2.2].

De Oliveira, M. J. and Griffiths, R. B. (1978). *Surface Science*, **71**, 687–94. [4.1.1, 5.0, 5.1.1.1, 5.1.1.2].

De Rouffignac, E., Alldredge, G. P., and De Wette, F. W. (1981). *Physical Review B*, **24**, 6050–9. [5.2.1.2, 6.1.6].

Derry, G., Wesner, D., Carlos, W., and Frankl, D. R. (1979). *Surface Science*, **87**, 629–42. [6.1.1].

Derry, G., Wesner, D., Vidali, G., Thwaites, T., and Frankl, D. R. (1980). *Surface Science*, **94**, 221–31. [2.3.3.2, 2.4.2, 6.1.1].

Deville, G., Valdes, A., Andrei, E. Y., and Williams, F. I. B. (1984). *Physical Review Letters*, **53**, 588–91. [4.3.1].

De Wette, F. W. and Nijboer, B. R. A. (1965). *Physics Letters*, **18**, 19–20. [5.2.2.1].

De Wette, F. W., Firey, B., De Rouffignac, E. P., and Alldredge, G. P. (1983). *Physical Review B*, **28**, 4744–52. [6.1.5].

De Wette, F W., Kress, W., and Schröder, U. (1985). *Physical Review B*, **32**, 4143–57. [6.3].

Diehl, R. D. (1991). In *Phase transitions in surface films 2*, (ed. H. Taub, G. Torzo, H. J. Lauter, and S. C. Fain, Jr.), pp.97–112. Plenum, New York. [3.6].

Diehl, R. D. and Fain, S. C. Jr. (1982). *Journal of Chemical Physics*, **77**, 5065–72. [6.1.7.3].

Diehl, R. D. and Fain, S. C. Jr. (1983). *Surface Science*, **125**, 116–53. [3.1, 6.1.7.3].

Diehl, R. D., Toney, M. F., and Fain, S. C. Jr. (1982). *Physical Review Letters*, **48**, 177–89. [6.1.7.3].

Dimon, P., Horn, P. M., Sutton, M., Birgeneau, R. J., and Moncton, D. E. (1985). *Physical Review B*, **31**, 437–47. [6.1.6].

Doering, D. L. and Semancik, S. (1986). *Surface Science*, **175**, L730–6. [3.6].

Domany, E., Schick, M., Walker, J. S., and Griffiths, R. B. (1978). *Physical Review B*, **18**, 2209–17. [5.1, 5.1.2, 5.1.2.2, 6.1.7.3].

Domb, C. and Green, M. S. (ed.) (1972). *Phase transitions and critical phenomena*, Vol. 2. Academic, New York. [5.0].

Domb, C. and Green, M. S. (ed.) (1976). *Phase transitions and critical phenomena*, Vol. 6. Academic, New York. [5.0, 5.1].

Duke, C. B. (ed.) (1994). *Surface Science; the first thirty years, Surface Science* **299/300**. [1.7.3].

Dunn, S. K. and Ewing, G. E. (1992). *Journal of Physical Chemistry*, **96**, 5284–90. [6.3].

Duparc, O. M. B. and Etters, R. D. (1987). *Journal of Chemical Physics*, **86**, 1020–5. [6.1.7.3].

Dupont-Pavlovsky, N., Bockel, C., and Thomy, A. (1985). *Surface Science*, **160**, 12–22. [6.4.3].

Dupont-Roc, J., Himbert, M., Pavloff, N., and Treiner, J. (1990). *Journal of Low Temperature Physics*, **81**, 31–44. [5.2.1.5].

Dykstra, C. E. (1993). *Chemical Reviews*, **93**, 2339–53. [2.3.1.2].

Dzyaloshinskii, I. E., Lifshitz, E. M., and Pitaevskii, L. P. (1961). *Advances in Physics*, **10**, 165–209. [1.2, 2.1.2.1].

Ebner, C. (1981). *Physical Review A*, **23**, 1925–9. [5.1.1.2].

Ebner, C. (1983). *Physical Review B*, **28**, 2890–2. [5.1.1.1].

Ebner, C. and Saam, W. F. (1977). *Physical Review Letters*, **38**, 1486–9. [5.2.1.5, 6.4.4].

Ebner, C., Saam, W. F., and Stroud, D. (1976). *Physical Review A*, **14**, 2264–73. [5.2.1.5].

Ecke, R. E. and Dash, J. G. (1983). *Physical Review B*, **28**, 3738–52. [5.1.3.1, 6.1.1].

Ecke, R. E., Shu, Q.-S., Sullivan, T. S., and Vilches, O. E. (1985). *Physical Review B*, **31**, 448–55. [5.1.3.1, 6.1.1].

Eckert, J., Ellenson, W. D., Hastings, J. B., and Passell, L. (1979). *Physical Review Letters*, **43**, 1329–32. [3.1, 6.1.7.3].

Edberg, R., Evans, D. J., and Morriss, G. P. (1986). *Journal of Chemical Physics*, **84**, 6933–9. [5.2.1.6].

Eden, V. L. and Fain, S. C. Jr. (1991). *Physical Review B*, **43**, 10697–705. [6.1.7.4].

Eden, V. L. and Fain, S. C. Jr. (1992). *Journal of Vacuum Science and Technology A*, **10**, 2227–30. [6.1.7.4].

Ehrlich, G. (1994). *Surface Science*, **299/300**, 628–42. [5.2.1.4].

Eigler, D. M. and Schweizer, E. K. (1990). *Nature (London)*, **344**, 524–6. [1.6.5].

Eigler, D. M., Weiss, P. S., Schweizer, E. K., and Lang, N. D. (1991). *Physical Review Letters*, **66**, 1189–92. [1.6.5].

Einstein, T. L. (1997). In *Physical Structure of Solid Surfaces*, (ed. W. N. Unertl). Elsevier, Amsterdam. [2.3.2.2].

Elgin, R. L. and Goodstein, D. L. (1974). *Physical Review A*, **9**, 2657–75. [6.1.1, B].

Elliott, G. S., Wei, D. H., Wu, K. J., and Kevan, S. D. (1993). *Journal of Chemical Physics*, **99**, 4152–9. [1.6.3, 2.4.4, 5.2.1.4].

Ellis, D. E. (1995). *Density functional theory of molecules, clusters, and solids.* Kluwer Academic, Dordrecht. [2.1.2.2].

Ellis, J. P. and Toennies, J. P. (1993). *Physical Review Letters*, **70**, 2118–21. [5.2.1.4].

Elrud, M. J. and Saykally, R. J. (1994). *Chemical Reviews*, **94**, 1975–97. [2.3.1].

Engel, T., Bornemann, P., and Bauer, E. (1979). *Surface Science*, **81**, 252–60. [6.2.4].

English, C. A. and Venables, J. A. (1974). *Proceedings of the Royal Society (London) A*, **340**, 57–80. [6.1.7, A.3].

Esbjerg, M. and Nørskov, J. K. (1980). *Physical Review Letters*, **45**, 807–10. [2.1.2.2].

Estermann, I. and Stern, O. (1930). *Zeitschrift für Physik*, **61**, 95–125. [2.4.2].

Estermann, I., Frisch, R., and Stern, O. (1932). *Zeitschrift für Physik*, **73**, 348–65. [2.4.2].

Estrup, P., Greene, E. F., Cardillo, M. J., and Tully, J. C. (1986). *Journal of Physical Chemistry*, **90**, 4099–104. [4.1.2].

Etters, R. D. (1989). In *Simple molecular systems at very high density*, (ed. A. Polian, P. Loubeyre, and N. Boccara), pp.109–39. Plenum, New York. [6.1.7].

Etters, R. D. and Duparc, O. B. M. H. (1985). *Physical Review B*, **32**, 7600–3. [6.1.7.3].

Etters, R. D. and Kobashi, K. (1984). *Journal of Chemical Physics*, **81**, 6249–53. [6.1.7.3].

Etters, R. D., Pan, R.-P., and Chandrasekharan, V. (1980). *Physical Review Letters*, **45**, 645–8. [5.2.1.1].

Etters, R. D., Roth, M. W., and Kuchta, B. (1990). *Physical Review Letters*, **65**, 3140–3. [5.2.1.6, 6.1.7.3].

Etters, R. D., Kuchta, B., and Belak, J. (1993). *Physical Review Letters*, **70**, 826–9. [5.2.1.4, 6.1.7.3].

Evans, D. J., Hoover, W. G., Failor, B. H., Moran, B., and Ladd, A. J. C. (1983). *Physical Review A*, **28** 1016–21. [5.2.1.6].

Evans, R. (1990). In *Liquids at interfaces*, Les Houches session XLVIII, (ed. J. Charvolin, J. F. Joanny, and J. Zinn-Justin), pp.1–98, North-Holland, Amsterdam. [5.2.1.5].

Evans, R., Marconi, U. M. B., and Tarazona, P. (1986). *Journal of Chemical Physics*, **84**, 2376–99. [5.2.1.5].

Fain, S. C. Jr., (1982). In *Chemistry and physics of solid surfaces IV*, (ed. R. Vanselow and R. Howe), pp.203–18. Springer, New York. [1.6.2.1, 1.7.4.3, 6.1.0].

Fain, S. C. Jr., Chinn, M. D., and Diehl, R. D. (1980). *Physical Review B*, **21**, 4170–2. [6.1.5].

Fain, S. C. Jr., Xia, B., and Peidle, J. (1994). *Physical Review B*, **50**, 14565–75. [6.1.2].

Fairobent, D. K., Saam, W. F., and Sander, L. M. (1982). *Physical Review B*, **26**, 179–83. [5.2.1.5].

Faisal, A. Q. D., Hamichi, M., Raynerd, G., and Venables, J. A. (1986). *Physical Review B*, **34**, 7440–3. [5.2.1.2, 6.1.6].

Falk, H. (1970). *American Journal of Physics*, **38**, 858–69. [5.1.1.2].

Fano, U. and Cooper, J. W. (1968). *Reviews of Modern Physics*, **40**, 441–507. [1.2].

Fehder, P. L. (1969). *Journal of Chemical Physics*, **50**, 2617–29. [5.2.2.1].

Feibelman, P. J. (1982). *Progress in Surface Science*, **12**, 287–402. [1.6.4, 1.7.4.1, 2.2.1.3].

Feibelman, P. J. (1990). *Physical Review Letters*, **65**, 729–32. [5.2.1.4].

Feng, Y. P. and Chan, M. H. W. (1993). *Physical Review Letters*, **71**, 3822–5. [6.1.7.3].

Ferrenberg, A. H. and Swendsen, R. H. (1989). *Physical Review Letters*, **63**, 1195–8. [5.2.1.6].

Ferry, D. and Suzanne, J. (1996). *Surface Science*, **345**, L19–22. [6.3].

Feynman, R. P. (1937). *Physical Review*, **56**, 340–3. [2.3.2.3].

Fisher, D. and Diehl, R. D. (1992). *Physical Review B*, **46**, 2512–22. [4.3.1].

Flurchick, K. and Etters, R. D. (1986). *Journal of Chemical Physics*, **84**, 4657–63. [5.2.1.1].

Fowler, R. H. (1966). *Statistical mechanics*. University Press, Cambridge. [5.1.1.2].

Fowler, R. H. and Guggenheim, E. A. (1939). *Statistical thermodynamics*. University Press, Cambridge. [5.1.1.3].

Fowler, P. W. and Hutson, J. M. (1986). *Physical Review B*, **33**, 3724–35. [2.3.1.3].

Fowler, P. W. and Tole, P. (1988). *Surface Science*, **197**, 457–73. [2.3.1.3, 6.3].

Frank, F. C. and Van der Merwe, J. H. (1949a). *Proceedings of the Royal Society (London) A*, **198**, 205–16. [5.2.2.2].

Frank, F. C. and Van der Merwe, J. H. (1949b). *Proceedings of the Royal Society (London) A*, **198**, 216–25. [5.2.2.2].

Frank, V. L. P., Lauter, H. J., and Leiderer, P. (1988). *Physical Review Letters*, **61**, 436–9. [6.1.2].

Frank, V. L. P., Lauter, H. J., Godfrin, H., and Leiderer, P. (1990). In *Phonons 89*, (ed. S. Hunklinger, W. Ludwig, and G. Weiss) pp.1001–3. World Scientific, Singapore. [6.1.1].

Frankl, D. R. (1983). *Progress in Surface Science*, **13**, 285–356. [1.7.4.2].

Freeman, D. L. (1975a). *Journal of Chemical Physics*, **62**, 941–9. [2.1.2.2].

Freeman, D. L. (1975b). *Journal of Chemical Physics*, **62**, 4300–7. [2.1.2.2].

Freimuth, H. and Wiechert, H. (1985). *Surface Science*, **162**, 432–8. [6.1.1].

Freimuth, H., Wiechert, H., Schildberg, H. P., and Lauter, H. J. (1990). *Physical Review B*, **42**, 587–603. [1.6.2,2, 5.2.2.2, 6.1.2, 6.2.2].

Frenkel, D. and McTague, J. P. (1979). *Physical Review Letters*, **42**, 1632–5. [4.3].

Frenkel, D. and McTague, J. P. (1980). *Annual Review of Physical Chemistry*, **31**, 491–521. [1.7.4.4].

Frenkel, J. and Kontorova, T. (1938). *Physikalische Zeitschrift der Sowjetunion*, **13**, 1–10. [5.2.2.2].

Frenken, J. W. M. and Hinch, B. J. (1992). In *Helium atom scattering from surfaces*, (ed. E. Hulpke), pp.287–313. Springer, Berlin. [1.6.3].

Frenken, J. W. M., Hinch, B. J., Toennies, J. P., and Wöll, Ch. (1990). *Physical Review B*, **41**, 938–46. [4.4].

Frigo, A., Toigo, F., Cole, M. W., and Goodman, F. O. (1986). *Physical Review B*, **33**, 4184–8. [2.3.3.1].

Frisch, R. and Stern, O. (1933). *Zeitschrift für Physik*, **84**, 430–42. [2.4.2].

Fuselier, C. R., Raich, J. C., and Gillis, N. S. (1980). *Surface Science*, **92**, 667–80. [3.1, 3.6].

Gallet, F., Deville, G., Valdes, A., and Williams, F. I. B. (1982). *Physical Review Letters*, **49**, 212–5. [4.3.1, 5.2.2.1].

Gangwar, R. and Suter, R. M. (1990). *Physical Review B*, **42**, 2711–4. [6.1.5].

Gangwar, R., Colella, N. J., and Suter, R. M. (1989). *Physical Review B*, **39**, 2459–71. [6.1.6, B].

Gay, J. M., Dutheil, A., Krim, J., and Suzanne, J. (1986). *Surface Science*, **177**, 25–35. [6.1.7.1].

Gay, J. M., Suzanne, J., Dash, J. G., and Lauter, H. J. (1991). *Journal de Physique France (I)*, **1**, 1279–90. [6.1.4].

Gay, J. M., Stocker, P., Degenhardt, D., and Lauter, H. J. (1992). *Physical Review B*, **46**, 1195–7. [6.3].

Gear, W. C. (1971). *Numerical initial value problems in ordinary differential equations*. Prentice Hall, Englewood Cliffs. [5.2.1.6].

Gerlach, R., Glebov, A., Lange, G., Toennies, J. P., and Weiss, H. (1995). *Surface Science*, **331–333**, 1490–5. [6.3].

Gibson, K. D. and Sibener, S. J. (1985). *Faraday Discussions of the Chemical Society (London)*, **80**, 203–15. [4.4, 5.2.1.2, 6.2.1].

Gibson, K. D. and Sibener, S. J. (1988a). *Journal of Chemical Physics*, **88**, 7862–92. [5.1.2.2, 6.2.0, 6.2.1].

Gibson, K. D. and Sibener, S. J. (1988b). *Journal of Chemical Physics*, **88**, 7893–910. [1.6.3, 6.2.1].

Gibson, K. D., Sibener, S. J., Hall, B. M., Mills, D. L., and Black, J. E. (1985). *Journal of Chemical Physics*, **83**, 4256–70. [1.3, 6.2.1].

Ginzburg, V. L. and Pitaevskii, L. P. (1958). *ZETF*, **34**, 1240–5. *Sov. Phys. JETP*, **7**, 858–61. [4.3.2].

Ginzburg, V. L. and Sobyanin, A. A. (1982). *Journal of Low Temperature Physics*, **49**, 507–43. [4.3.2].

Girard, C. and Girardet, C. (1987). *Chemical Physics Letters*, **138**, 83–9. [2.3.1, 6.3].

Girardet, C. and Hoang, P. N. M. (1993). *Surface Science*, **282**, 288–305. [6.3].

Girardet, C., Picaud, S., and Hoang, P. N. M. (1994). *Europhysics Letters*, **25**, 131–6. [6.0, 6.3].

Girardet, C., Ramseyer, C., Hoang, P. N. M., and Picaud, S. (1995). *Physical Review B*, **52**, 2144–53. [6.3].

Gittes, F. T. and Schick, M. (1984). *Physical Review B*, **30**, 209–14. [5.0].

Glachant, A. and Bardi, U. (1979). *Surface Science*, **87**, 187–202. [6.2.4].

Glachant, A., Jaubert, M., Bienfait, M., and Boato, G. (1981). *Surface Science*, **115**, 219–35. [6.2.0, 6.2.4].

Glachant, A., Bienfait, M., and Jaubert, M. (1984). *Surface Science*, **148**, L665–70. [6.2.4, B].

Glandt, E. D. (1978). *Journal of Chemical Physics*, **68**, 2952–8. [5.2.2.1].

Glandt, E. D. (1981). *Journal of Chemical Physics*, **74**, 1321–5. [5.2.1].

Glandt, E. D., Myers, A. L., and Fitts, D. D. (1979). *Journal of Chemical Physics*, **70**, 4243–7. [5.2.1.4].

Glaser, M. L. and Clarke, N. A. (1993). *Advances in Chemical Physics*, **83**, 543–709. [4.3].

Glattli, D. C., Andrei, E. Y., and Williams, F. I. B. (1988a). *Physical Review Letters*, **60**, 420–3. [4.3.1, 5.2.2.1].

Glattli, D. C., Andrei, E. Y., and Williams, F. I. B. (1988b). *Surface Science*, **196**, 17–23. [4.3.1, 5.2.2.1].

Glebov, A., Toennies, J. P. T., and Weiss, H. (1996). *Surface Science*, **351**, 200–8. [1.6.3].

Glebov, A., Miller, R. E., and Toennies, J. P. (in press). [6.3].

Glyde, H. R. (1995). *Excitations in Liquid and Solid Helium*. Oxford University Press. [5.2.1.2].

Glyde, H. R. and Goldman, V. V. (1976). *Journal of Low Temperature Physics*, **25**, 601–17. [5.2.1.2].

Glyde, H. R. and Khanna, F. C. (1971). *Canadian Journal of Physics*, **49**, 2997–3009. [5.2.1.2].

Godfrin, H. and Lauter, H. J. (1995). In *Progress in Low Temperature Physics*, Vol. XIV (ed. W. P. Halperin), Chap. 4, pp. 213–320. North-Holland, Amsterdam. [1.6.2.2, 1.7.4.3, 6.1.0, 6.1.1].

Goellner, C. J., Daunt, J. G., and Lerner, E. (1975). *Journal of Low Temperature Physics*, **21**, 347–58. [5.2.1.4].

Gomer, R. (1990). *Reports on Progress in Physics*, **53**, 917–1002. [1.6.5, 4.1.2].

Gooding, R. J., Joos, B., and Bergersen, B. (1983). *Physical Review B*, **27**, 7669–75. [2.3.3.2, 3.4, 5.2.1.1, 5.2.2.2, 6.1.5].

Goodman, F. O. and Wachman, H. Y. (1976). *Dynamics of gas–surface scattering*. Academic, New York. [1.7.3].

Goodstein, D. L. (1969). *Physical Review*, **183**, 327–34. [1.6.3].

Goodstein, D. L., Hamilton, J. J., Lysek, M. J., and Vidali, G. (1984). *Surface Science*, **148**, 187–99. [6.1.7.1].

Gordon, M. B. (1987). *Physical Review B*, **35**, 2052–4. [3.4, 5.2.2.2].

Gordon, M. B. and Lançon, F. (1985). *Journal of Physics C*, **18**, 3929–42. [5.2.1.1, 5.2.2.2, 6.1.5].

Gordon, M. B. and Villain, J. (1985). *Journal of Physics C*, **18**, 3919–28. [6.1.5, 6.2.2].

Gordon, R. G. and Kim, Y. S. (1972). *Journal of Chemical Physics*, **56**, 3122–33. [2.1.2.2].

Gottlieb, J. M. (1990a). *Physical Review B*, **42**, 5377–80. [2.3.3.2, 5.2.1.2, 6.2.2].

Gottlieb, J. M. (1990b). Unpublished Ph.D. thesis, University of Wisconsin–Madison. [6.1.2].

Gottlieb, J. M. and Bruch, L. W. (1989). *Physical Review B*, **40**, 148–57. [3.3, 6.1.2, B].

Gottlieb, J. M. and Bruch, L. W. (1991a). *Physical Review B*, **44**, 5750–8. [5.2.2.2, B].

Gottlieb, J. M. and Bruch, L. W. (1991b). *Physical Review B*, **44**, 5759–65. [5.2.2.2, 6.2.2, B].

Gottlieb, J. M. and Bruch, L. W. (1993). *Physical Review B*, **48**, 3943–8. [2.3.2.4, 2.4.4, 4.2.2.2, 6.1.1].

Graham, A. P., Bertino, M. F., Hofmann, F., Toennies, J. P., and Wöll, Ch. (in press). [6.2.4].

Gready, J. E., Bacskay, G. B., and Hush, N. S. (1978). *Journal of the Chemical Society. Faraday Transactions II*, **74**, 1430–40. [2.3.1.3, 6.3].

Greer, S. C., Meyer, L., and Barrett, C. S. (1969). *Journal of Chemical Physics*, **50**, 4299–304. [A.3].

Greif, J. M. (1982). Unpublished Ph.D. thesis, California Institute of Technology. [B].

Greif, J. M. and Goodstein, D. L. (1981). *Journal of Low Temperature Physics*, **44**, 347–66. [4.3.1, 5.2.1.2].

Greiser, N., Held, G. A., Frahm, R., Greene, R. L., Horn, P. M., and Suter, R. M. (1987). *Physical Review Letters*, **59**, 1706–9. [6.1.6, 6.2.1].

Grey, F. and Bohr, J. (1991). In *Phase Transitions in Surface Films 2*, (ed. H. Taub, G. Torzo, H. J. Lauter, and S. C. Fain, Jr.) pp.83–96. Plenum, New York. [3.6, 6.1.2].

Greywall, D. S. (1990). *Physical Review B*, **41**, 1842–62. [6.1.1].

Greywall, D. S. (1993). *Physical Review B*, **47**, 309–18. [2.4.4, 4.3.2, 6.1.1].

Greywall, D. S. (1994). *Physica B*, **197**, 1–12. [4.2.2.1].

Greywall, D. S. and Busch, P. A. (1990a). *Physical Review Letters*, **65**, 64–7. [6.1.1].

Greywall, D. S. and Busch, P. A. (1990b). *Physical Review Letters* **65**, 2788–91. [6.1.1].

Griffiths, R. B. (1980). In *Phase transitions in surface films*. (ed. J. G. Dash and J. Ruvalds), pp.1–27. Plenum, New York. [4.2.1, B].

Griffiths, R. B. (1990). In *Fundamental problems in statistical mechanics VII*. (ed. H. van Beijeren), pp.69–110. Elsevier, Amsterdam. [5.2.2.2].

Grigor, A. F. and Steele, W. A. (1968). *Journal of Chemical Physics*, **48**, 1032–37. [A.3].

Grillet, Y. J. and Rouquerol, F. (1979). *Journal of Colloid and Interface Science*, **70**, 239–44. [1.6.1.1].

Grimes, C. C. and Adams, G. (1979). *Physical Review Letters*, **42**, 795–8. [4.3.1].

Gross, E. K. U. and Dreizler, R. M. (ed.) (1995). *Density functional theory*. Plenum, New York. [2.1.2.2].

Grübel, G., Gibbs, D., Zehner, D. M., Abernathy, D. L., Sandy, A. R., and Mochrie, S. G. J. (1993). *Surface Science*, **287/288**, 842–6. [6.2.0].

Gunnarsson, O. and Lundqvist, B. I. (1976). *Physical Review B*, **13**, 4274–98. [2.1.2.2].

Guo, Z.-C. and Bruch, L. W. (1982). *Journal of Chemical Physics*, **77**, 1417–24. [4.2.2.2, 5.2.2.1].

Guo, Z.-C. and Bruch, L. W. (1992). *Journal of Chemical Physics*, **97**, 7748–56. [6.3].

Guryan, C. A., Lee, K. B., Stephens, P. W., Goldman, A. I., Larese, J. Z., Heiney, P. A., and Fontes, E. (1988). *Physical Review B*, **37**, 3461–6. [6.1.5].

Hakim, T. M. and Glyde, H. R. (1988). *Physical Review B*, **37**, 984–92. [5.2.1.2, 6.1.3].

Hakim, T. M., Glyde, H. R., and Chui, S. T. (1988). *Physical Review B*, **37**, 974–83. [5.2.1.2, 6.1.7.1].

Hall, B., Mills, D. L., and Black, J. E. (1985). *Physical Review B*, **32**, 4932–45. [2.3.2.4, 4.4, 5.2.1.2].

Hall, B., Mills, D. L., Zeppenfeld, P., Kern, K., Becher, U., and Comsa, G. (1989). *Physical Review B*, **40**, 6326–38. [2.3.2.4, 4.4, 5.2.1.2, 6.2.2].

Hallock, R. B. (1995). *Journal of Low Temperature Physics*, **101**, 31–40. [2.4.3, 6.4.4].

Halperin, B. I. and Nelson, D. R. (1978). *Physical Review Letters*, **41**, 121–4. [4.2.3, 4.3].

Halpin-Healy, T. M. and Kardar, M. (1985). *Physical Review B*, **31**, 1664–7. [5.1.3.1].

Halpin-Healy, T. M. and Kardar, M. (1986). *Physical Review B*, **34**, 318–24. [5.1.3.1].

Hamaker, H. C. (1937). *Physica*, **4**, 1058–72. [2.1.2.1].

Hamichi, M., Faisal, A. Q. D., Venables, J. A., and Kariotis, R. (1989). *Physical Review B*, **39**, 415–25. [6.1.6].

Hamichi, M., Kariotis, R., and Venables, J. A. (1991). *Physical Review B*, **43**, 3208–14. [1.6.2.1].

Hammonds, E. M., Heiney, P., Stephens, P. W., Birgeneau, R. J., and Horn, P. M. (1980). *Journal of Physics C*, **13**, L301–6. [6.1.6].

Hammonds, K. D., McDonald, I. R., and Tildesley, D. J. (1990). *Molecular Physics*, **70**, 175–95. [6.1.7.2].

Hanono, F., Gatts, C. E. N., and Lerner, E. (1985). *Journal of Low Temperature Physics*, **60**, 73–84. [6.1.3].

Hansen, F. Y. and Bruch, L. W. (1995). *Physical Review B*, **51**, 2515–36. [2.3.3.2, 2.4.4, 5.2.1.2, 5.2.1.4, 6.1.7.3].

Hansen, F. Y. and Taub, H. (1987). *Journal of Chemical Physics*, **87**, 3232–45. [6.1.7.4].

Hansen, F. Y. and Taub, H. (1992). *Physical Review Letters*, **69**, 652–5. [6.1.7.4].

Hansen, F. Y., Alldredge, G. P., Bruch, L. W., and Taub, H. (1985). *Journal of Chemical Physics*, **83**, 348–52. [2.3.1.2].

Hansen, F. Y., Frank, V. L. P., Taub, H., Bruch, L. W., Lauter, H. J., and Dennison, J. R. (1990). *Physical Review Letters*, **64**, 764–7. [6.1.7.3].

Hansen, F. Y., Bruch, L. W., and Roosevelt, S. E. (1992). *Physical Review B*, **45**, 11238–48. [2.3.3.2, 6.1.7.3].

Hansen, F. Y., Newton, J. C., and Taub, H. (1993). *Journal of Chemical Physics*, **98**, 4128–41. [6.1.7.4].

Hansen, F. Y., Bruch, L. W., and Taub, H. (1995). *Physical Review B*, **52**, 8515–27. [6.1.7.3].

Hansen, J.-P. and Verlet, L. (1969). *Physical Review*, **184**, 151–61. [5.2.2.1].

Hanson, F. and McTague, J. P. (1980). *Journal of Chemical Physics*, **72**, 6363–7. [6.1.5].

Hanson, F. E., Mandell, M. J., and McTague, J. P. (1977). *Journal de Physique (Paris)*, **38**, C4-76 to C4-81. [6.1.4].

Harris, A. B. and Berlinsky, A. J. (1979). *Canadian Journal of Physics*, **57**, 1852–69. [6.1.2, 6.1.7.3].

Harris, J. (1984). *Physical Review A*, **29**, 1648–59. [2.1.2.2].

Harris, J. and Jones, R. O. (1974). *Journal of Physics F (Metal Physics)*, **4**, 1170–86. [2.1.2.2].

Harris, J. and Liebsch, A. (1982). *Journal of Physics C*, **15**, 2275–91. [2.1.2.2, 2.2.2, 2.2.2.1, 2.2.2.2, 2.2.2.3].

Harris, J. and Zaremba, E. (1985). *Physical Review Letters*, **55**, 1940C. [2.3.3.1].

Harten, U., Lahee, A. M., Toennies, J. P., and Wöll, Ch. (1985). *Physical Review Letters*, **54**, 2619–22. [6.2.0].

Hegde, S. G. and Daunt, J. G. (1978). *Journal of Low Temperature Physics*, **32**, 765–88. [4.2.2.2].

Heidberg, J., Kandel, M., Meine, D., and Wildt, U. (1995a). *Surface Science*, **331-333**, 1467–72. [1.6.3, 6.3].

Heidberg, J., Schönekäs, O., Weiss, H., Lange, G., and Toennies, J. P. (1995b). *Berichte der Bunsen-Gesellschaft für Physikalische Chemie*, **99**, 1370-5. (1995b) [6.3].

Heiney, P. A., Stephens, P. W., Birgeneau, R. J., Horn, P. M., and Moncton, D. E. (1983a). *Physical Review B*, **28**, 6416–34. [6.1.6].

Heiney, P. A., Stephens, P. W., Mochrie, S. G. J., Akimitsu, J., Birgeneau, R. J., and Horn, P. M. (1983b). *Surface Science*, **125**, 539–64. [6.1.7.3].

Henk, J. and Feder, R. (1994). *Journal of Physics: Condensed Matter*, **6**, 1913–26. [6.2.2].

Henderson, D. (1977). *Molecular Physics*, **34**, 301–15. [5.2.2.1].

Henderson, D. R. (ed). (1992). *Fundamentals of inhomogeneous fluids*. M. Dekker, Amsterdam. [5.2.1.5].

Hering, S. V., van Sciver, S. W., and Vilches, O. E. (1976). *Journal of Low Temperature Physics*, **25**, 789–805. [5.1.3.1, 6.1.1].

Herminghaus, S., Sigel, U., Paatzsch, T., Musil, H., and Leiderer, P. (1994). *Berichte der Bunsen-Gesellschaft für Physikalische Chemie*, **98**, 443–6. [1.6.1.1].

Herwig, K. W., Newton, J. C., and Taub, H. (1994). *Physical Review B*, **50**, 15287–97. [6.1.7.4].

Hess, G. B. (1991). In *Phase transitions in surface films 2*, (ed. H. Taub, G. Torzo, H. J. Lauter, and S. C. Fain, Jr.), pp.357–89. Plenum, New York. [1.6.1.1, 6.1.0].

Hess, G. B., Sabatini, M. J., and Chan, M. H. W. (in press). [6.4.4].

Hijmans, T. W., Walraven, J. T. M., and Shlyapnikov, G. V. (1992). *Physical Review B*, **45**, 2561–4. [4.1.2].

Hilgers, G., Potthoff, M., Wirth, S., Müller, N., Heinzmann, U., Haunert, L., Braun, J., and Borstel, G. (1991). *Surface Science*, **251/252**, 612–5. [6.2.2, 6.2.3].

Hilgers, G., Potthoff, M., Müller, N., and Heinzmann, U. (1993). *Surface Science*, **287/288**, 414–7. [6.2.3].

Hilgers, G., Potthoff, M., Müller, N., and Heinzmann, U. (1995). *Surface Science*, **322**, 207–20. [6.2.3].

Hilliker, P. R. and Hakim, T. M. (1990). *Journal of Chemical Physics*, **93**, 2080–1. [5.2.1.2].

Hirschfelder, J. O., Curtiss, C. F., and Bird, R. B. (1964). *Molecular theory of gases and liquids*. Wiley, New York. [2.0, 2.1.1.1, 2.1.1.2, 2.3.1. 2.3.1.3, 2.4.1, 5.2.1, 5.2.1.3, 5.2.1.4, 5.2.2.1].

Hirshfeld, F. L. and Mirsky, K. (1979). *Acta Crystallographica*, **A35**, 366–70. [6.1.7.2].

Hoang, P. N. M. and Girardet, C. (1991). *Surface Science*, **243**, 361–72. [5.2.1.2].

Hoang, P. N. M., Girardet, C., Sidoumou, M., and Suzanne, J. (1993). *Physical Review B*, **48**, 12183–92. [6.3].

Hofmann, F. and Toennies, J. P. (1996). *Chemical Reviews*, **96**, 1307–26. [6.3].

Hoge, H. J. and Lassiter, J. W. (1951). *Journal of Research of the National Bureau of Standards (U. S.)*, **47**, 75–79. [A.1].

Höhler, G. and Niekisch, E. A. (ed.) (1982). *Structural studies of surfaces with atomic and molecular beam diffraction*. Springer, Berlin. [1.7.3].

Hohenberg, P. C. (1967). *Physical Review*, **158**, 383–6. [4.3.2].

Hohenberg, P. and Kohn, W. (1964). *Physical Review*, **136**, B864–71. [2.1.2.2].

Hohenberg, P. C., Kohn, W., and Sham, L. J. (1990). *Advances in Quantum Chemistry*, **21**, 7–26. [2.1.2.2].

Hoinkes, H. (1980). *Reviews of Modern Physics*, **52**, 933–70. [1.6.2.1, 1.7.4.2, 2.4.1, 2.4.2].

Holian, B. L. (1980). *Physical Review B*, **22**, 1394–404. [5.2.1.3].

Holloway, S. and Nørskov, J. K. (1991). *Bonding at surfaces*. Liverpool University Press, Liverpool. [1.7.3].

Holt, A. C., Hoover, W. G., Gray, S. G., and Shortle, D. R. (1970). *Physica*, **49**, 61–76. [5.2.1.3].

Hong, H. and Birgeneau, R. J. (1989). *Zeitschrift für Physik B*, **77**, 413–9. [6.1.6].

Hong, H., Birgeneau, R. J., and Sutton, M. (1986). *Physical Review B*, **33**, 3344–8. [3.5].

Hong, H., Peters, C. J., Mak, A., Birgeneau, R. J., Horn, P. M., and Suematsu, H. (1987). *Physical Review B*, **36**, 7311–4. [6.1.6].

Hong, H., Peters, C. J., Mak, A., Birgeneau, R. J., Horn, P. M., and Suematsu, H. (1989). *Physical Review B*, **40**, 4797–807. [5.2.1.2, 6.1.6].

Hooton, D. J. (1955). *Philosophical Magazine*, **46**, 422–32. [4.2.3].

Hoover, W. G. (1986). *Molecular dynamics*. Springer, New York. [1.7.4.4, 5.2.1.6].

Hoover, W. G. (1991). *Computational statistical mechanics*. Elsevier, Amsterdam. [5.2.1.6].

Hoover, W. C. and Alder, B. J. (1967). *Journal of Chemical Physics*, **46**, 686–91. [4.3.1, 5.2.2.1].

Horch, S., Zeppenfeld, P., and Comsa, G. (1995a). *Surface Science*, **331–333**, 908–12. [6.2.2].

Horch, S., Zeppenfeld, P., and Comsa, G. (1995b). *Applied Physics A*, **60**, 147–53. [6.2.2].

Horn, K., Scheffler, M., and Bradshaw, A. M. (1978). *Physical Review Letters*, **41**, 822–5. [1.6.4, 6.2.0, 6.2.3].

Horn, K., Mariani, C., and Cramer, L. (1982). *Surface Science*, **117**, 376–86. [6.2.0, 6.2.4].

Horne, J. M., Yerkes, S. C., and Miller, D. R. (1980). *Surface Science*, **93**, 47–63. [2.3.3.1, 6.2.1].

Houlrik, J. M. and Landau, D. P. (1991). *Physical Review B*, **44**, 8962–71. [3.4, 5.2.2.2].

Hozhabri, N., Sharma, S. C., and Wang, S. J. (1989). *Physical Review B*, **39**, 3990–5. [1.6.1.1].

Huff, G. B. and Dash, J. G. (1976). *Journal of Low Temperature Physics*, **24**, 155–74. [6.1.3].

Hulpke, E. (ed.) (1992). *Helium atom scattering from surfaces*. Springer, Berlin. [1.7.3].

Humes, R. P., Smalley, M. V., Rayment, T., and Thomas, R. K. (1988). *Canadian Journal of Chemistry*, **66**, 557–61. [6.1.7.1].

Hunzicker, K. A. and Phillips, J. M. (1986). *Physical Review B*, **34**, 8843–9. [5.2.1.3, 6.1.7.1].

Hurley, A. C. (1976). *Introduction to the electron theory of small molecules*. Academic, New York. [2.1.1.1].

Hurst, R. P. and Levelt, J. M. H. (1961). *Journal of Chemical Physics*, **34**, 54–63. [5.2.1.3].

Hutson, J. M. and Fowler, P. W. (1986). *Surface Science*, **173**, 337–50. [2.3.1.3].

Hutson, J. M., Fowler, P. W., and Zaremba, E. (1986). *Surface Science*, **175**, L775–81. [2.2.1.2, 2.2.1.3].

Ibach, H. (1991). *Electron energy loss spectrometers, their technology of high performance.* Springer, Berlin. [1.6.3].

Ibach, H. and Mills, D. L. (1982). *Electron energy loss spectroscopy and surface vibrations.* Academic, New York. [1.6.3, 2.2.1.2].

Ignatiev, A. and Rhodin, T. N. (1973). *Physical Review B*, **8**, 893–906. [3.5].

Ihm, G. and Cole, M. W. (1989). *Langmuir*, **5**, 550–7. [2.1.2.2].

Ihm, G., Cole, M. W., Toigo, F., and Scoles, G. (1987). *Journal of Chemical Physics*, **87**, 3995–9. [2.4.1].

Ihm, G., Cole, M. W., Toigo, F., and Klein, J. R. (1990). *Physical Review A*, **42**, 5244–52. [2.3.1].

Imry, Y. (1978). *Critical Reviews in Solid State and Materials Science*, **8**, 157–74. [3.5, 4.2.3].

Imry, Y. and Gunther, L. (1971). *Physical Review B*, **3**, 3939–45. [3.5, 4.2.3].

Inaba, A. and Morrison, J. (1986). *Physical Review B*, **34** 3238–42. [6.1.7.4].

Inaba, A., Shirakami, T., and Chihara, H. (1988). *Chemical Physics Letters*, **146**, 63–6. [6.1.7.3].

Inaba, A., Shirikami, T., and Chihara, H. (1991). *Journal of Chemical Thermodynamics*, **23**, 461–74. [6.1.7.3].

Ishi, S. and Viswanathan, B. (1991). *Thin Solid Films*, **201**, 373–402. [2.3.3.2, 6.2.0].

Israelachvili, J. N. (1992). *Intermolecular and surface forces.* Academic, London. [1.7.3, 2.1.2.1].

Jancovici, B. (1967). *Physical Review Letters*, **19**, 20–2. [4.2.3].

Janssen, W. B. J. M., Michiels, J., and van der Avoird, A. (1991). *Journal of Chemical Physics*, **94**, 8402–12. [5.2.1.2, A.3].

Janzen, A. R. and Aziz, R. A. (1995). *Journal of Chemical Physics*, **103**, 9626–30. [2.1.1.1].

Jaroniec, M. and Madey, R. (1988). *Physical adsorption on heterogeneous solids.* Academic, New York. [1.7.1].

Jhanwar, B. L. and Meath, W. J. (1980). *Molecular Physics*, **41**, 1061–70. [E.1].

Jhanwar, B. L. and Meath, W. J. (1982). *Chemical Physics*, **67**, 185–99. [E.1].

Jhanwar, B. L., Meath, W. J., and MacDonald, J. C. F. (1983). *Canadian Journal of Physics*, **61**, 1027–34. [E.1].

Jiang, S., Gubbins, K. E., and Zollweg, J. A. (1993). *Molecular Physics*, **80**, 103–16. [6.1.7.1].

Jiang, X.-P., Toigo, F., and Cole, M. W. (1984). *Surface Science*, **145**, 281–93. [2.2.1.3].

Jin, A. J., Bjurstrom, M. R., and Chan, M. H. W. (1989). *Physical Review Letters*, **62**, 1372–5. [6.1.6].

Jónsson, H., Weare, J. H., Ellis, T. H., and Scoles, G. (1987). *Surface Science*, **180**, 353–70. [2.3.1].

Joos, B. (1982). *Solid State Communications*, **42**, 709–13. [5.2.2.2].

Joos, B., Bergersen, B., and Klein, M. L. (1983). *Physical Review B*, **28**, 7219–24. [5.2.2.2, 6.1.6].

Joos, B., Ren, Q., and Duesberry, M. S. (1994). *Surface Science*, **302**, 385–94. [4.3.1].

Joshi, Y.P. and Tildesley, D. J. (1985). *Molecular Physics*, **55**, 999–1016. [2.3.3.2, 2.4.4, 6.1.7.3].

Juretschke, H. J. (1953). *Physical Review*, **92**, 1140–4. [2.2.2.3].

Kaindl, G., Chiang, T.-C., Eastman, D. E., and Himpsel, F. J. (1980). *Physical Review Letters*, **45**, 1808–11. [1.6.4, 2.3.2.1, 6.2.3].

Kaindl, G. and Mandel, T. (1987). *Physical Review Letters*, **59**, 2238(C). [2.3.2.1].

Kalia, R. K. and Vashishta, P. (1981). *Journal of Physics C*, **14**, L643–8. [4.3.1].

Kalia, R. K. and Vashishta, P. (ed.) (1982). *Melting, localization and chaos*. North Holland, New York. [4.3.1].

Kalos, M. H. and Whitlock, P. A. (1986). *Monte Carlo methods*. Wiley, New York. [1.7.4.4, 5.2.1.6].

Kappus, W. (1978). *Zeitschrift für Physik B*, **29**, 239–44. [2.3.2.4].

Kardar, M. and Berker, A. N. (1982). *Physical Review Letters*, **48**, 1552–5. [3.4].

Karimi, M. and Vidali, G. (1986). *Physical Review B*, **34**, 2794–8. [E.2].

Kellay, H., Bonn, D., and Meunier, J. (1993). *Physical Review Letters*, **71**, 2607–10. [6.4.4].

Kellogg, G. L. (1994). *Surface Science Reports*, **21**, 1–88. [5.2.1.4].

Kellogg, G. L. and Feibelman, P. J. (1990). *Physical Review Letters*, **64**, 3143–6. [5.2.1.4].

Kenny, T. W. and Richards, P. L. (1990). *Physical Review Letters*, **64**, 2386–9. [1.6.1.2, 6.2.0].

Kern, K. (1987). *Physical Review B*, **35**, 8265–8. [2.3.3.2, 6.2.2].

Kern, K. and Comsa, G. (1989). *Advances in Chemical Physics*, **76**, 211–80. [1.6.3, 1.7.4.2, 6.2.2].

Kern, K. and Comsa, G. (1991). In *Phase transitions in surface films 2*, (ed. H. Taub, G. Torzo, H. J. Lauter, and S. C. Fain, Jr.), pp.41–65. Plenum, New York. [6.2.0].

Kern, K., David, R., Palmer, R. L., and Comsa, G. (1986a). *Physical Review Letters*, **56**, 620–3. [2.3.3.1, 6.2.2].

Kern, K., David, R., Palmer, R. L., and Comsa, G. (1986b). *Physical Review Letters*, **56**, 2823–6. [6.2.2].

Kern, K., Zeppenfeld, P., David, R., Palmer, R. L., and Comsa, G. (1986c). *Physical Review Letters*, **57**, 3187–90. [6.1.4, 6.2.1, 6.2.2].

Kern, K., David, R., Palmer, R. L., and Comsa, G. (1986d). *Applied Physics A*, **41**, 91–3. [5.2.1.2, 6.2.2].

Kern, K., David, R., Palmer, R. L., and Comsa, G. (1986e). *Surface Science*, **175**, L669–74. [6.2.1].

Kern, K., Zeppenfeld, P., David, R., and Comsa, G. (1987a). *Physical Review Letters*, **59**, 79–82. [5.2.1.2, 5.2.2.2, 6.2.0, 6.2.2].

Kern, K., Zeppenfeld, P., David, R., and Comsa, G. (1987b). *Physical Review B*, **35**, 886–9. [4.4, 5.2.1.2, 6.2.2].

Kern, K., David, R., Zeppenfeld, P., Palmer, R. and Comsa, G. (1987c). *Solid State Communications*, **62**, 391–4. [6.2.2].

Kern, K., David, R., Zeppenfeld, P., and Comsa, G. (1988). *Surface Science*, **195**, 353–70. [1.6, 5.2.2.2, 6.1.4, 6.2.0, 6.2.2].

Kern, R. and Krohn, M. (1989). *Physica Status Solidi A*, **116**, 23–38. [2.3.2.4].

Kihara, T. and Honda, N. (1965). *Journal of the Physical Society of Japan*, **20**, 15–9. [2.3.2.1].

Kihara, T., Midzuno, Y., and Shizume, T. (1954). *Journal of the Physical Society of Japan*, **9**, 681–7. [5.1.1.2].

Kikuchi, K. (1951). *Physical Review*, **81**, 988–1003. [5.1.1.3].

Kim, H. K. and Chan, M. H. W. (1984). *Physical Review Letters*, **53**, 170–3. [5.1.2.1].

Kim, H. K., Zhang, Q. M., and Chan, M. H. W. (1986a). *Physical Review B*, **34**, 4699–709. [5.1.2.1, 6.1.7.1].

Kim, H. K., Zhang, Q. M., and Chan, M. H. W. (1986b). *Physical Review Letters*, **56**, 1579–82. [6.1.7.4].

Kim, H. K., Feng, Y. P., Zhang, Q. M., and Chan, M. H. W. (1988). *Physical Review B*, **37**, 3511–23. [6.1.7.4].

Kim, H.-Y., Cole, M. W., Toigo, F., and Nicholson, D. (1988). *Surface Science*, **198**, 555–70. [2.3.2.2].

Kim, H.-Y. and Steele, W. A. (1992). *Physical Review B*, **45**, 6226–33. [5.2.1.6, 6.1.7.1].

Kim, I. M. and Landau, D. P. (1981). *Surface Science*, **110**, 415–22. [5.1.1.2].

Kingsbury, D. L. (1991). Unpublished Ph. D. thesis, University of Washington–Seattle. [6.3].

Kinoshita, M. and Lado, F. (1994). *Molecular Physics*, **83**, 351–9. [5.2.1.4].

Kiselev, A. V. and Kovaleva, N. N. (1956). *Zhur. Fiz. Khim.*, **30**, 2775–86. [6.1.7].

Kiselev, A. V. and Kovaleva, N. N. (1959). *Izvest. Akad. Nauk. SSSR Otdel. Khim. Nauk.* p. 955 [*Bull. Acad. Sci. USSR Chem. Div.* pp.989–98] [6.1.7].

Kjaer, K., Nielsen, M., Bohr, J., Lauter, H. J., and McTague, J. P. (1982). *Physical Review B*, **26**, 5168–74. [3.3].

Kjems, J. K., Passell, L., Taub, H., and Dash, J. G. (1974). *Physical Review Letters*, **32**, 724–7. [1.6.2.2].

Kleiman, G. G. and Landman, U. (1973a). *Physical Review Letters*, **31**, 707–10. [2.1.2.2].

Kleiman, G. G. and Landman, U. (1973b). *Physical Review B*, **8**, 5484–95. [2.1.2.2].

Kleiman, G. G. and Landman, U. (1976). *Solid State Communications*, **18**, 819–22. [2.1.2.2].

Klein, M. L. and Venables, J. A. (ed.) (1976). *Rare gas solids*, Vol. I. Academic, New York. [5.0].

Klein, M. L. and Venables, J. A. (ed.) (1977). *Rare gas solids*, Vol. II. Academic, New York. [5.0].

Klein, M. L. and Koehler, T. R. (1976). Chap. 6 In *Rare gas solids*, Vol. I, (ed. M. L. Klein and J. A. Venables), pp.301–81. Academic, New York. [5.2.1.2].

Klein, M. L., O'Shea, S., and Ozaki, Y. (1984). *Journal of Physical Chemistry*, **88**, 1420–5. [2.3.1].

Kleinekathöfer, U., Tang, K. T., Toennies, J. P., and Yiu, C. L. (1996). *Chemical Physics Letters*, **249**, 257–63. [2.1.1.1].

Knapp, H., Teller, M., and Langhorst, R. (1987). *Solid–liquid equilibrium data collection. Binary systems*, Chemistry Data Series Vol. VIII, Part 1. DECHEMA, Frankfurt/Main. [A.3].

Knight, J. F. and Monson, P. A. (1986). *Journal of Chemical Physics*, **84**, 1909–15. [5.2.2.1].

Knorr, G. (1992). *Physics Reports*, **214**, 113–57. [1.6.1.1, 1.7.4.3, 6.1.7].

Kobashi, K. and Etters, R. D. (1985a). *Surface Science*, **150**, 252–62. [B].

Kobashi, K. and Etters, R. D. (1985b). *Journal of Chemical Physics*, **82**, 4341–57. [B].

Koch, S. W. and Abraham, F. F. (1983). *Physical Review B*, **27**, 2964–79. [6.1.6].

Koch, S. W. and Abraham, F. F. (1986). *Physical Review B*, **33**, 5884–5. [5.2.2.2, 6.1.5].

Koch, S. W., Rudge, W. E., and Abraham, F. F. (1984). *Surface Science*, **145**, 329–44. [5.2.2.2, 6.1.5].

Kohn, W. (1994). *Surface Reviews and Letters*, **1**, 129–32. [4.1.2].

Kohn, W. and Lau, K. H. (1976). *Solid State Communications*, **18**, 553–5. [2.3.2.3, 6.2.3].

Kohn, W. and Sham, L. J. (1965). *Physical Review*, **140**, A1133–8. [2.1.2.2].

Kołos, W. and Roothaan, C. C. J. (1960). *Reviews of Modern Physics*, **32**, 219–32. [2.1.1.1].

Kołos, W. and Wolniewicz, L. (1965). *Journal of Chemical Physics*, **43**, 2429–41. [2.1.1.1].

Korpiun, P. and Lüscher, E. (1977). Chap. 12 In *Rare Gas Solids*, Vol. II., (ed. M. L. Klein and J. A. Venables), pp.729–822. Academic, New York. [6.1.4, 6.1.6, A.3].

Kosterlitz, J. M. (1980). In *Phase transitions in surface films*, (ed. J. G. Dash and J. Ruvalds), pp.193–231. Plenum, New York. [4.3].

Kosterlitz, J. M. and Thouless, D. J. (1973). *Journal of Physics C*, **6**, 1181–203. [4.2.3, 4.3, 4.3.1, 4.3.2].

Kosterlitz, J. M. and Thouless, D. J. (1978). *Progress in Low Temperature Physics*, Vol. 7B, pp.371–433. North Holland, New York. [4.3].

Kreimer, B. C., Oh, B. K., and Kim, S. K. (1973). *Molecular Physics*, **26**, 297–309. [5.2.2.1].

Kreuzer, H. J. and Gortel, Z. W. (1986). *Physisorption kinetics*. Springer, Berlin. [1.7.1, 4.1.2].

Krim, J. (1991). In *Phase transitions in surface films 2*, (ed. H. Taub, G. Torzo, H. J. Lauter, and S. C. Fain, Jr.), pp.169–82. Plenum, New York. [1.6.1.1, 6.2.0].

Krim, J., Suzanne, J., Shechter, H., Wang, R., and Taub, H. (1985). *Surface Science*, **162**, 446–51. [6.1.7.4].

Krim, J., Solina, D. H., and Chiarello, R. (1991). *Physical Review Letters*, **66**, 181–4. [6.2.1].

Kromhout, R. A. and Linder, B. (1979). *Chemical Physics Letters*, **61**, 283–7. [2.3.2.3].

Kubik, P. R., Hardy, W. N., and Glattli, H. (1985). *Canadian Journal of Physics*, **63**, 605–20. [6.1.2].

Kuchta, B. and Etters, R. D. (1987a). *Physical Review B*, **36**, 3400–6. [2.3.1, 5.2.1.1, 6.1.7.3].

Kuchta, B. and Etters, R. D. (1987b). *Physical Review B*, **36**, 3407–12. [5.2.1.1, 6.1.7.3].

Kuchta, B. and Etters, R. D. (1993). *Journal of Computational Physics*, **108**, 353–6. [5.2.1.6].

Kumar, A. and Meath, W. J. (1992). *Molecular Physics*, **75**, 311–24. [E.1].

Kumar, S. and Etters, R. D. (1991). *Journal of Chemical Physics*, **94**, 5190–7. [2.3.1].

Kusner, R. E., Mann, J. A., and Dahm, A. J. (1995). *Physical Review B*, **51**, 5746–91. [4.3.1].

Lakhlifi, A. and Girardet, C. (1991). *Surface Science*, **241**, 400–15. [2.3.1, 6.3].

Landau, D. P., Mon, K. K., and Schüttler, H.-B. (ed.) (1993). *Computer simulation in condensed matter physics V*. Springer, Berlin. [1.7.4.4, 5.2.1.6].

Lander, J. J. and Morrison, J. (1967). *Surface Science*, **6**, 1–32. [1.6.2.1].

Landolt-Börnstein (1971). III/5a Crystal Structure Data for tables. Springer, Berlin. [A.3].

Lane, J. E. and Spurling, T. H. (1976). *Australian Journal of Chemistry*, **29**, 2103–21. [5.2.1.6].

Lang, N. D. (1973). In *Solid State Physics*, **28** (ed. H. Ehrenreich, F. Seitz, and D. Turnbull), pp.225–300. Academic, New York. [2.1.2.2].

Lang, N. D. (1981). *Physical Review Letters*, **46**, 842–5. [2.1.2.2, 2.2.2, 2.3.2.3].

Lang, N. D. (1994). *Surface Science*, **299/300**, 284–97. [2.1.2.2].

Lang, N. D. and Kohn, W. (1970). *Physical Review B*, **1**, 4555–68. [2.1.2.2, 2.2.1.2, 2.2.2.3].

Lang, N. D. and Kohn, W. (1971). *Physical Review B*, **3**, 1215–23. [2.2.1.2, 2.2.2.3].

Lang, N. D. and Kohn, W. (1973). *Physical Review B*, **7**, 3541–50. [2.2.1.3, 2.3.2.1, 2.4.3].

Lang, N. D. and Nørskov, J. K. (1983). *Physical Review B*, **27**, 4612–6. [2.1.2.2, 2.2.2].

Lang, N. D. and Williams, A. R. (1982). *Physical Review B*, **25**, 2940–2. [2.1.2.2, 2.2.2, 2.3.2.3].

Lang, N. D., Williams, A. R., Himpsel, F. J., Reihl, B., and Eastman, D. E. (1982). *Physical Review B*, **26**, 1728–37. [2.3.2.1, 2.3.2.3].

Lang, N. D., Holloway, S., and Nørskov, J. K. (1985). *Surface Science*, **150**, 24–38. [2.3.2.1].

Langbein, D. (1974). *Van der Waals attraction*, Springer Tracts in Modern Physics, Vol.**72**. Springer, Berlin. [1.7.3].

Lange, G., Toennies, J. P., Vollmer, R., and Weiss, H. (1993). *Journal of Chemical Physics*, **98**, 10096–9. [6.3].

Lange, G., Schmicker, D., Toennies, J. P., Vollmer, R., and Weiss, H. (1995). *Journal of Chemical Physics*, **103**, 2308–19. [6.3].

Langmuir, I. (1918). *Journal of the American Chemical Society*, **40**, 1361–403. [1.3, 4.1.1, 4.1.2, 5.1].

Larese, J. Z. (1993). *Accounts of Chemical Research*, **26**, 353–60. [6.1.4].

Larese, J. Z. and Rollefson, R. J. (1983). *Surface Science*, **127**, L172–8. [6.1.7.4].

Larese, J. Z. and Rollefson, R. J. (1985). *Physical Review B*, **31**, 3048–50. [6.1.7.1].

Larese, J. Z. and Zhang, Q. M. (1991). *Physical Review B*, **43**, 938–46. [6.1.4].

Larese, J. Z., Passell, L., Heidemann, A. D., Richter, D., and Wicksted, J. P. (1988a). *Physical Review Letters*, **61**, 432–5. [6.1.7.4].

Larese, J., Passell, L., and Ravel, B. (1988b). *Canadian Journal of Chemistry*, **66**, 633–6. [6.1.7.4].

Larese, J. Z., Harada, M., Passell, L., Krim, J., and Satija, S. (1988c). *Physical Review B*, **37**, 4735–42. [3.5, 6.1.7.1].

Larese, J. Z., Zhang, Q. M., Passell, L., Hastings, J. M., Dennison, J. R., and Taub, H. (1989). *Physical Review B*, **40**, 4271–5. [3.5].

Larese, J. Z., Hastings, J. M., Passell, L., Smith, D., and Richter, D. (1991). *Journal of Chemical Physics*, **95**, 6997–7000. [6.3].

Larher, Y. (1971). *Journal of Colloid and Interface Science*, **37**, 836–48. [B].

Larher, Y. (1978). *Journal of Chemical Physics*, **68**, 2257–63. [6.1.4].

Larher, Y. (1983). *Surface Science*, **134**, 469–75. [6.1.4].

Larher, Y. (1992). In *Surface properties of layered structures*, (ed. G. Benedek), pp.261–315. Kluwer, Dordrecht. [1.6.1.1].

Larher, Y. and Gilquin, B. (1979). *Physical Review A*, **20**, 1599–1602. [6.1.4].

Larher, Y. and Terlain, A. (1980). *Journal of Chemical Physics*, **72**, 1052–4. [6.1.5].

Larsen, S. Y., Kilpatrick, J. E., Lieb, E. H., and Jordan, H. F. (1965). *Physical Review*, **140**, A129–30. [4.2.1].

Lau, K. H. (1978). *Solid State Communications*, **28**, 757–62. [2.3.2.4].

Lau, K. H. and Kohn, W. (1977). *Surface Science*, **65**, 607–18. [2.3.2.4].

Lau, K. H. and Kohn, W. (1978). *Surface Science*, **75**, 69–85. [2.3.2.2].

Lauter, H. J. (1990). *Phonons 89*, (ed. S. Hunklinger, W. Ludwig, and G. Weiss), pp.871–9. World Scientific, Singapore. [1.6.3, 6.1.2].

Lauter, H. J., Schildberg, H. P., Godfrin, H., Wiechert, H., and Haensel, R. (1987). *Canadian Journal of Physics*, **65**, 1435–9. [6.1.1].

Lauter, H. J., Frank, V. L. P., Taub, H., and Leiderer, P. (1990). In *Proceedings of the 19th International Conference on Low Temperature Physics*, (ed. D. S. Betts) *Physica B*, **165/166**, 611–2. [6.1.2, 6.1.7.3].

Lauter, H. J., Godfrin, H., and Leiderer, P. (1992). *Journal of Low Temperature Physics*, **87**, 425–43. [6.1.1].

Layet, J. M., Bienfait, M., Ramseyer, C., Hoang, P. N. M., Girardet, C., and Coddens, G. (1993). *Physical Review B*, **48**, 9045–53. [6.3].

Lee, J. and Kosterlitz, J. M. (1991). *Physical Review B*, **43**, 1268–71. [5.2.1.6].

Lee, J. and Strandburg, K. J. (1992). *Physical Review B*, **46**, 11190–3. [4.3.1, 5.2.1.6, 5.2.2.1].

Lee, S. A. and Lindsay, S. M. (1990). *Physica Status Solidi. (b)*, **157**, K83–6. [6.1].

Lefèvre-Seguin, V., Nacher, P. J., Brossel, J., Hardy, W. N., and Laloë, F. (1985). *Journal de Physique (Paris)*, **46**, 1145–72. [6.4.2].

Leptoukh, G., Strickland, B., and Roland, C. (1995). *Physical Review Letters*, **74**, 3636–9. [6.4.1].

Lennard-Jones, J. E. (1932a). *The adsorption of gases by solid surfaces*. Faraday Society, London. [1.2, 1.7.3].

Lennard-Jones, J. E. (1932b). *Transactions of the Faraday Society*, **28**, 333–59. [1.2, 2.1.2.1].

Lennard-Jones, J. E. and Dent, B. M. (1928). *Transactions of the Faraday Society*, **24**, 92–108. [2.3.1.3, 6.3].

Lennard–Jones, J. E. and Devonshire, A. F. (1936). *Proceedings of the Royal Society (London) A*, **156**, 6–28. [4.1.2].

Lerner, E., Hedge, S. G., and Daunt, J. G. (1972). *Physics Letters*, **41A**, 239–40. [6.1.3].

Lerner, E. and Hanono, F. (1979). *Journal of Low Temperature Physics*, **35**, 363–70. [6.1.3].

Lerner, E., Hanono, F., and Gatts, C. E. N. (1985). *Surface Science*, **160**, L524–8. [6.1.3].

Le Roy, R. J. (1976). *Surface Science*, **59**, 541–53. [2.4.2].

Leung, P. W. and Chester, G. V. (1991). *Physical Review B*, **43**, 735–51. [5.2.2.1].

Levelt, J. M. H. and Hurst, R. P. (1960). *Journal of Chemical Physics*, **32**, 96–104. [5.2.1.3].

Levesque, D. and Verlet, L. (1969). *Physical Review*, **182**, 307–16. [5.2.2.1].

Li, W., Shrestha, P., Migone, A. D., Marmier, A., and Girardet, C. (1996). *Physical Review B*, **54**, 8833–43. [6.4.3].

Liebsch, A. (1986a). *Physical Review B*, **33**, 7249–51. [2.2.1.2, 2.2.1.3].

Liebsch, A. (1986b). *Journal of Physics C*, **19**, 5025–47. [2.2.1.2, 2.2.1.3].

Liebsch, A. (1987). *Physical Review B*, **35**, 9030–36. [2.2.1.2].

Liebsch, A., Harris, J., and Weinert, M. (1984). *Surface Science*, **145**, 207–22. [2.4.3].

Lifshitz, E. M. (1956). *Soviet Physics JETP*, **2**, 73–93. *Zh. Eksp. Teor. Fiz.* **29**, 94–110 (1955). [2.1.2.1, 2.2.1.3].

Lim, T. K. (1985). *Journal of Chemical Physics*, **82**, 1616–7. [5.2.2.1].

Linder, B. and Kromhout, R. A. (1976). *Physical Review B*, **13**, 1532–5. [2.3.2.3].

Litzinger, J. A. and Stewart, G. A. (1980). In *Ordering in two dimensions*, (ed. S. K. Sinha), pp.267–9. North Holland, New York. [6.1.6].

Liu, F.-C., Liu, Y.-M., and Vilches, O. E. (1995). *Physical Review B*, **51**, 2848–56. [6.1.2].

Liu, G.-Y., Robinson, G. N., Scoles, G., and Heiney, P. (1992). *Surface Science*, **262**, 409–21. [6.3].

Liu, K. S., Kalos, M. H., and Chester, G. V. (1976). *Physical Review B*, **13**, 1971–4. **17**, 4479 (1978) (E). [5.2.2.1, 6.1.1].

Liu, W. and Fain, S. C. Jr. (1992). *Journal of Vacuum Science and Technology A*, **10**, 2231–6. [6.1.2].

Liu, W. and Fain, Jr. S. C. (1993). *Physical Review B*, **47**, 15965–8. [6.1.2].

Liu, W.-K. (1985). *Physical Review B*, **32**, 868–73. [2.3.2.2].

Liu, Y.-M., Ebey, P. S., Vilches, O. E., Dash, J. G., Bienfait, M., Gay, J.-M., and Coddens, G. (1996). *Physical Review B*, **54**, 6307–14. [6.1.2].

London, F. (1930a). *Zeitschrift für Physik*, **63**, 245–79. [1.2, 2.1.1.1, 2.1.2.1].

London, F. (1930b). *Zeitschrift für Physikalische Chemie (B)* **11**, 222–51. [1.2, 2.1.1.1, 2.1.2.1].

Lundqvist, S. and March, N. H. (ed.) (1983). *Theory of the inhomogeneous electron gas*. Plenum, New York. [2.1.2.2].

Luo, F., McBane, G. C., Kim, G., Giese, C. F., and Gentry, W. R. (1993). *Journal of Chemical Physics*, **98**, 3564–7. [2.1.1.3].

Luo, F., Giese, C. F., and Gentry, W. R. (1996). *Journal of Chemical Physics*, **204**, 1151–4. [2.1.1.3].

Luty, T., van der Avoird, A., and Berns, R. M. (1980). *Journal of Chemical Physics*, **73**, 5305–9. [5.2.1.2].

Lynden-Bell, R. M. (1990). *Surface Science*, **230**, 311–22. [6.3].

Lynden-Bell, R. M., Talbot, J., Tildesley, D. J., and Steele, W. A. (1985). *Molecular Physics*, **54**, 183–95. [6.1.7.3].

Lyuksyutov, I., Naumovets, A. G., and Pokrovsky, V. (1992). *Two-dimensional crystals*. Academic, Boston. [5.2.2.2].

Ma, J., Kingsbury, D. L., Liu, F., and Vilches, O. E. (1988). *Physical Review Letters*, **61**, 2348–51. [1.6.1.2, 6.3].

MacMillen, D. B. and Landman, U. (1984). *Journal of Chemical Physics*, **80**, 1691–702. [2.1.2].

MacRury, T. B. and Linder, B. (1971). *Journal of Chemical Physics*, **54**, 2056–66. [2.2.1.3, 2.3.2.2].

MacRury, T. B. and Linder, B. (1972). *Journal of Chemical Physics*, **56**, 4368–77. [2.2.1.3, 2.3.2.2].

Mahan, G. D. (1981). *Many-particle physics*. Plenum, New York. [2.1.2.2].

Mahanty, J. and Ninham, B. W. (1976). *Dispersion forces*. Academic, London. [1.7.3, 2.0, 2.2.1].

Mandel, T., Domke, M., Kaindl, G., Laubschat, C., Prietsch, M., Middelmann, U., and Horn, K. (1985). *Surface Science*, **162**, 453–60. [2.3.2.3].

Mann, K., Celli, V., and Toennies, J. P. (1987). *Surface Science*, **185**, 269–82. [F].

Manninen, M., Nørskov, J. K., Puska, M. J., and Umrigar, C. (1984). *Physical Review B*, **29**, 2314–6. [2.1.2.2].

Maradudin, A. A. and Wallis, R. F. (1980). *Surface Science*, **91**, 423–39. [2.3.2.4].

March, N. H. (1986). *Chemical bonds outside solid surfaces*. Plenum, New York. [1.7.3].

Marchese, M., Jacucci, G., and Klein, M. L. (1984). *Surface Science*, **145**, 364–70. [5.2.1.2].

Margenau, H. and Pollard, W. G. (1941). *Physical Review*, **60**, 128–34. [2.1.2.1].

Margeneau, H. and Kestner, N. R. (1971). *Theory of Intermolecular Forces*, (2nd edn). Pergamon, Oxford. [2.0, 2.1, 2.1.1.1].

Martin, A. J. and Bilz, H. (1979). *Physical Review B*, **19**, 6593–600. [6.3]

Martin, P. C. (1968). In *Many-body physics*, (ed. C. DeWitt and R. Balian), pp.39–136. Gordon and Breach, New York. [2.2.1].

Marty, D. and Poitrenaud, J. (1984). *Journal de Physique (Paris)*, **45**, 1243–55. [4.3.1].

Marvin, A. M. and Toigo, F. (1982). *Physical Review A*, **25**, 782–802. [2.1.2.1].

Marx, R. (1985). *Physics Reports*, **125**, 1–67. [1.6.1.2, 1.7.4.3].

Marx, D. and Wiechert, H. (1996). *Advances in Chemical Physics*, **95**, 213–394. (ed. I. Prigogine and S. A. Rice) [6.1.7, 6.1.7.3, 6.4.3].

Marx, D., Sengupta, S., Nielaba, P., and Binder, K. (1994a). *Physical Review Letters*, **72**, 262–5. [6.1.7.3].

Marx, D., Sengupta, S., Opitz, O., Nielaba, P. and Binder, K. (1994b). *Molecular Physics*, **83**, 31–62. [6.1.7.3].

Marx, D., Sengupta, S., Nielaba, P., and Binder, K. (1994c). *Surface Science*, **321**, 195–216. [6.1.7.3].

Maschhoff, B. L. and Cowin, J. P. (1994). *Journal of Chemical Physics*, **101**, 8138–51. [2.3.2.3].

Mason, B. F. and Williams, B. R. (1983). *Surface Science*, **130**, 295–312. [6.2.4].

Mason, B. F. and Williams, B. R. (1984). *Surface Science*, **139**, 173–84. [6.2.1].

Masuda, T., Barnes, C. J., Hu, P., and King, D. A. (1992). *Surface Science*, **276**, 122–38. [3.4].

Mattera, L., Rosatelli, F., Salvo, C., Tommasini, F., Valbusa, U., and Vidali, G. (1980). *Surface Science*, **93**, 515–25. [2.4.2].

Mavroyannis, C. (1963). *Molecular Physics*, **6**, 593–600. [2.1.2.1].

McConville, G. T. (1974). *Journal of Chemical Physics*, **60**, 4093. [A.3].

McCoy, B. M. and Wu, T. T. (1973). *The two-dimensional Ising model*. Harvard, Cambridge. [5.1.1].

McIntosh, R. L. (1992). *Dielectric behavior of physically adsorbed gases*. M. Dekker, New York. [1.6.1.1].

McLachlan, A. D. (1963). *Proceedings of the Royal Society (London) A*, **271**, 387–401. [2.1.2.1].

McLachlan, A. D. (1964). *Molecular Physics*, **7**, 381–8. [1.2, 2.1.2.1, 2.2.1.3, 2.3.2.1, 2.3.2.2].

McMillan, W. L. (1977). *Physical Review B*, **16**, 4655–8. [5.2.2.2].

McTague, J. P. and Novaco, A. D. (1979). *Physical Review B*, **19**, 5299–306. [5.2.1.2, 6.2.1].

McTague, J. P., Frenkel, D., and Allen, M. P. (1980). In *Ordering in two dimensions*, (ed. S. K. Sinha), pp.147–53. North Holland, New York. [4.3].

McTague, J. P., Nielsen, M. and Passell, L. (1978). *Critical Reviews in Solid State and Materials Science*, **8**, 135–55. [1.7.4.1].

Meath, W. J. and Koulis, M. (1991). *Journal of Molecular Structure (THEO-CHEM 72)*, **226**, 1–37. [2.1.1.2].

Meichel, T., Suzanne, J., Girard, C., and Girardet, C. (1988). *Physical Review B*, **38**, 3781–97. [6.3].

Meijer, E. J., Frenkel, D., LeSar, R., and Ladd, A. J. C. (1990). *Journal of Chemical Physics*, **92**, 7570–5. [5.2.1.6].

Meixner, D. L. and George, S. M. (1993a). *Surface Science*, **297**, 27–39. [6.2.2].

Meixner, D. L. and George, S. M. (1993b). *Journal of Chemical Physics*, **98**, 9115–25. [6.2.2].

Meixner, D. L., Arthur, D. A., and George, S. M. (1992). *Surface Science*, **261**, 141–54. [6.3].

Meixner, W. C. and Antoniewicz, P. R. (1976). *Physical Review B*, **13**, 3276–83. [2.3.2.3].

Meldrim, J. M. and Migone, A. D. (1995). *Physical Review B*, **51**, 4435–40. [6.4.3].

Menaucourt, J., Thomy, A., and Duval, X. (1980). *Journal de Chimie Physique*, **77**, 959–65. [6.1.7.4].

Menzel, D. (1982). In *Chemistry and Physics of Solid Surfaces IV*, Springer Series in Chemical Physics #20, (ed. R. Vanselow and R. Howe), pp.389–406. Springer, Berlin. [1.6.3].

Mermin, N. D. (1965). *Physical Review*, **137**, A1441–3. [5.2.1.5].

Mermin, N. D. (1968). *Physical Review*, **176**, 250–4. *B*, **20**, 4762 (1979) (E). [4.2.3].

Metropolis, N., Rosenbluth, A. W., Rosenbluth, M. N., Teller, A. H., and Teller, E. (1953). *Journal of Chemical Physics*, **21**, 1087–92. [5.2.1.6].

Meyer, E. S., Mester, J. C., and Silvera, I. F. (1994). *Journal of Chemical Physics*, **100**, 4021–2. [2.1.1.3].

Migone, A. D., Chan, M. H. W., Niskanen, K. J., and Griffiths, R. B. (1983a). *Journal of Physics C*, **16**, L1115–20. [2.3.3.2, 6.1.0, 6.1.5, 6.1.7.3].

Migone, A. D., Kim, H. K., Chan, M. H. W., Talbot, J., Tildesley, D. J., and Steele, W. A. (1983b). *Physical Review Letters*, **51**, 192–5. [6.1.7.3].

Migone, A. D., Li, Z. R., and Chan, M. H. W. (1984). *Physical Review Letters*, **53**, 810–3. [6.1.4].

Migone, A. D., Alkhafaji, M. T., Vidali, G., and Karimi, M. (1993). *Physical Review B*, **47**, 6685–96. [6.4.3].

Miller, M. D. and Nosanow, L. H. (1978). *Journal of Low Temperature Physics*, **32**, 145–57. [5.2.2.1, 6.1.1].

Miller, M. D., Woo, C.-W., and Campbell, C. E. (1972). *Physical Review A*, **6**, 1942–7. [6.1.1].

Miller, M. D., Nosanow, L. H., and Parish, L. J. (1977). *Physical Review B*, **15**, 214–29. [5.2.2.1].

Millot, F. (1979). *Journal de Physique Lettres*, **40**, L9–10. [6.1.4].

Millot, F., Larher, Y., and Tessier, C. (1982). *Journal of Chemical Physics*, **76**, 3327–35. [6.0].

Mills, D. L. and Burstein, E. (ed.) (1994). *Scattering from surfaces, In Surface Review and Letters*, **1**, 47–210. World Scientific, Singapore. [1.7.2].

Minot, C., Van Hove, M. A., and Biberian, J.-P. (1996). *Surface Science*, **346**, 283–93. [6.3].

Miranda, R., Albano, E. V., Daiser, S., Ertl, G., and Wandelt, K. (1983a). *Physical Review Letters*, **51**, 782–5. [6.2.3].

Miranda, R., Daiser, S., Wandelt, K., and Ertl, G. (1983b). *Surface Science*, **131**, 61–91. [6.2.3].

Mistura, G., Lee, H. C., and Chan, M. H. W. (1994). *Journal of Low Temperature Physics*, **96**, 221–44. [6.4.4].

Miura, K. and Morimoto, T. (1991). *Langmuir*, **7**, 374–9. [6.1.7].

Mochel, J. M. and Chen, M.-T. (1994). *Physica B*, **197**, 278–82. [4.3.2, 6.4.2].

Mochrie, S. G. J., Sutton, M., Birgeneau, R. J., Moncton, D. E., and Horn, P. M. (1984). *Physical Review B*, **30**, 263–73. [1.6.2.2, 6.1.7.4, A.3].

Moiseiwitsch, B. L. (1953). *Proceedings of the Royal Society (London) A*, **219**, 102–9. [2.1.2.2].

Moleko, L. K., Joos, B., Hakim, T. M., Glyde, H. R., and Chui, S. T. (1986). *Physical Review B*, **34**, 2815–22. [5.2.1.2].

Moller, M. A. and Klein, M. L. (1988). *Canadian Journal of Chemistry*, **66**, 774–8. [6.1.7.4].

Moller, M. A. and Klein, M. L. (1989). *Journal of Chemical Physics*, **90**, 1960–7. [2.3.1].

Moncton, D. E., Stephens, P. W., Birgeneau, R. J., Horn, P. M., and Brown, G. S. (1981). *Physical Review Letters*, **46**, 1533–6; **49**, 1679 (C)(1982). [5.2.2.2].

Monson, P. A., Steele, W. A., and Henderson, D. (1981). *Journal of Chemical Physics*, **74**, 6431–9. [5.2.1.4].

Moog, E. R. and Webb, M. B. (1984). *Surface Science*, **148**, 338–64. [2.3.2.3, 6.2.0, 6.2.3].

Moog, E. R., Unguris, J., and Webb, M. B. (1983). *Surface Science*, **134**, 849–64. [1.6.2.1, 6.2.0, 6.2.1].

Morales, J. J. (1994). *Physical Review E*, **49**, 5127–30. [5.2.1.6, 5.2.2.1].

Morf, R. H. (1979). *Physical Review Letters*, **43**, 931–5. [4.3.1].

Morf, R. (1984). *Helvetica Physica Acta*, **56**, 743–54. [4.3, 4.3.1].

Morishige, K. (1993). *Molecular Physics*, **78**, 1203–9. [6.1.7.2].

Morishige, K., Mowforth, C., and Thomas, R. K. (1985). *Surface Science*, **151**, 289–300. [6.1.7.3].

Morrison, I. D. and Ross, S. (1973). *Surface Science*, **39**, 21–36. [5.2.2.1].

Motteler, F. C. (1986). Unpublished Ph.D. thesis, University of Washington–Seattle. [5.1.3.1, 6.1.1].

Motteler, F. C. and Dash, J. G. (1985). *Physical Review B*, **31**, 346–9. [6.1.2].

Motteler, F. C. and Dash, J. G. (1986 unpublished) [5.1.3.1].

Mouritsen, O. G. (1984). *Computer studies of phase transitions and critical phenomena*. Springer, Berlin. [1.7.4.4].

Mouritsen, O. G. and Berlinsky, A. J. (1982). *Physical Review Letters*, **48**, 181–4. [6.1.7.3].

Müller, J. E. (1990). *Physical Review Letters*, **65**, 3021–4. [2.3.3.2, 6.2.0, 6.2.2].

Müller, J. E. (1993). In *The chemical physics of solid surfaces*, Vol.6, (ed. D. A. King and D. P. Woodruff), pp.29–49. Elsevier, Amsterdam. [2.3.3.2, 6.2.0].

Mugele, F., Albrecht, U., and Leiderer, P. (1994). *Journal of Low Temperature Physics*, **96**, 177–84. [1.6.3, 6.4.2].

Muirhead, R. J., Dash, J. G., and Krim, J. (1984). *Physical Review B*, **29**, 5074–80. [5.0].

Mulliken, R. S. (1932). *Reviews of Modern Physics*, **4**, 1–86. [2.1.1.1].

Mutaftschiev, B. (ed.) (1982). *Interfacial aspects of phase transformations*. Reidel, Dordrecht. [1.7.2].

Muto, Y. (1943). *Journal of the Physico-Mathematical Society of Japan*, **17**, 629–31. [6.1.4].

Mysels, K. J. and Scholten, P. C. (1991). *Langmuir*, **7**, 209–11. [2.1.2.1].

Nagler, S. E., Horn, P. M., Rosenbaum, T. F., Birgeneau, R. J., Sutton, M., Mochrie, S. G. J., Moncton, D. E., and Clarke, R. (1985). *Physical Review B*, **32**, 7373–83. [6.1.6].

Naidoo, K. J., Schnitker, J., and Weeks, J. D. (1993). *Molecular Physics*, **80**, 1–24. [4.3].

Narasimhan, S. and Vanderbilt, D. (1992). *Physical Review Letters*, **69**, 1564–7. [2.3.2.4, 5.2.2.2].

Nelson, D. R. (1980). In *Fundamental problems in statistical mechanics V*, (ed. E. G. D. Cohen), pp.53–108. North Holland, New York. [1.7.3, 4.3].

Nelson, D. R. (1983). In *Phase transitions and critical phenomena*, Vol. 7, (ed. C. Domb and J. L. Lebowitz), pp.1–99. Academic, New York. [1.7.3, 4.3, 4.3.1, 5.2.2.1].

Nelson, D. R. and Halperin, B. I. (1979). *Physical Review B*, **19**, 2457–84. [3.5, 4.3, 4.3.1].

Newton, J. C. and Taub, H. (1996). *Surface Science*, **364**, 273–78. [6.1.7.4].

Ni, X.-Z. and Bruch, L. W. (1986). *Physical Review B*, **33**, 4584–95. [2.4.4, 5.2.1.3, 5.2.2.1, 6.1.1, 6.1.2, B],

Nicholson, D. and Parsonage, N. G. (1982). *Computer simulation and the statistical mechanics of adsorption*. Academic, London. [1.7.1, 2.0, 5.2.1.6].

Nicholson, D. and Parsonage, N. G. (1986). *Journal of the Chemical Society. Faraday Transactions II*, **82**, 1657–67. [6.1.4].

Niebel, K. F. and Venables, J. A. (1976). Chap. 9 In *Rare gas solids*, Vol. I, (ed. M. L. Klein and J. A. Venables), pp.558–89. Academic, London. [6.2.1].

Nielsen, M. and McTague, J. P. (1979). *Physical Review B*, **19**, 3096–106. [6.1.7.3].

Nielsen, M., McTague, J. P., and Ellenson, W. (1977). *Journal de Physique (Paris)*, **38**, C4-10 to C4-18. [1.6.3, 6.1.0, 6.1.2].

Nielsen, M., McTague, J. P., and Passell, L. (1980a). In *Phase Transitions in Surface Films*, (ed. J. G. Dash and J. Ruvalds) pp.127–64. Plenum, New York. [1.6.3, 6.1.0].

Nielsen, M., Als-Nielsen, J., and McTague, J. P. (1980b). In *Ordering in Two Dimensions*, (ed. S. K. Sinha) pp.135–41. North Holland, New York. [6.1.2].

Nielsen, M., Als-Nielsen, J., Bohr, J., McTague, J. P., Moncton, D. E., and Stephens, P. W. (1987). *Physical Review B*, **35**, 1419–25. [6.1.4].

Niemeijer, Th. and van Leeuwen, J. M. J. (1976). In *Phase transitions and critical phenomena*, Vol. 6, (ed. C. Domb and M. S. Green), pp.425–505. Academic, New York. [5.1.3].

Nienhuis, B., Berker, A. N., Riedel, E., and Schick, M. (1979). *Physical Review Letters*, **43**, 737–40. [5.1.1.1, 5.1.3.1].

Nieuwenhuys, B. E., Meijer, D. Th., and Sachtler, W. M. H. (1974). *Physica Status Solidi A*, **24**, 115–22. [6.2.2].

Nijboer, B. R. A. (1984). *Physica*, **125A**, 275–9. [6.3].

Nijboer, B. R. A. and Renne, M. J. (1968). *Chemical Physics Letters*, **2**, 35–8. [2.1.2.1, 2.2.1.3].

Nijboer, B. R. A. and Renne, M. J. (1971). *Physica Norvegica*, **5**, 243–51. [2.1.2.1, 2.2.1.3].

Nijboer, B. R. A. and Ruijgrok, Th. W. (1985). *Journal of Physics C*, **18**, 2043–54. [3.1].

Nilsson, A., Björneholm, O., Hernnäs, B., Sandell, A., and Mårtensson, N. (1993). *Surface Science*, **293**, L835–40. [6.2.1].

Niskanen, K. J. and Griffiths, R. B. (1985). *Physical Review B*, **32**, 5858–73. [6.1.5].

Nordlander, P. and Harris, J. (1984). *Journal of Physics C*, **17**, 1141–52. [2.1.1.2, 2.2.0, 2.2.2, 2.2.2.2, 2.2.2.3, 2.4.3].

Nordlander, P., Holmberg, C., and Harris, J. (1986). *Surface Science*, **175**, L753–8. [2.1.2, 2.1.2.1].

Nørskov, J. and Lang, N. D. (1980). *Physical Review B*, **21**, 2131–6. [2.1.2.2].

Nosanow, L. H. (1977). *Journal of Low Temperature Physics*, **26**, 613–26. [5.2.2.1, 6.1.3].

Nosanow, L. H. and Shaw, G. L. (1962). *Physical Review*, **128**, 546–50. [5.2.1.3].

Nosanow, L. H., Parish, L. J., and Pinski, F. J. (1975). *Physical Review B*, **11**, 191–204. [5.2.2.1].

Nosé, S. and Klein, M. L. (1983). *Journal of Chemical Physics*, **78**, 6928–39. [A.3].

Novaco, A. D. (1973a). *Physical Review A*, **7**, 678–84. [6.1.1].

Novaco, A. D. (1973b). *Physical Review A*, **7**, 1653–9. [6.1.1].

Novaco, A. D. (1973c). *Physical Review A*, **8**, 3065–70. [6.1.1].

Novaco, A. D. (1975). *Journal of Low Temperature Physics*, **21**, 359–68. [5.2.1.4].

Novaco, A. D. (1979). *Physical Review B*, **19**, 6493–501. [5.2.1.2, 6.1.3, 6.1.4].

Novaco, A. D. (1992). *Physical Review B*, **46**, 8178–94. [5.2.1.2, 6.1.2].

Novaco, A. D. (1994). (private communication). [5.2.1.2, 5.2.2.2, 6.1.2].

Novaco, A. D. and Campbell, C. E. (1975). *Physical Review B*, **11**, 2525–34. [4.2.2.2, 6.1.1].

Novaco, A. D. and McTague, J. P. (1977a). *Physical Review Letters*, **38**, 1286–9. [3.0, 3.6, 5.2.1.2, 6.1.3].

Novaco, A. D. and McTague, J. P. (1977b). *Journal de Physique (Paris)*, **38**, C4-116 to C4-120. [3.0, 3.6, 5.2.1.2, 6.1.3].

Novaco, A. D. and Milford, F. J. (1972). *Physical Review A*, **5**, 783–9. [4.2.2.2].

Novaco, A. D. and Shea, P. A. (1982). *Physical Review B*, **26**, 284–94. [4.3].

Novaco, A. D. and Wroblewski, J. P. (1989). *Physical Review B*, **39**, 11364–71. [6.1.0].

Nuttall, W. J., Fahey, K. P., Young, M. J., Keimer, B., Birgeneau, R. J., and Suematsu, H. (1993). *Journal of Physics: Condensed Matter*, **5**, 8159–76. [3.5, 6.1.6].

Nuttall, W. J., Noh, D. Y., Wells, B. O., and Birgeneau, R. J. (1994). *Surface Science*, **307-309**, 768–74. [1.3, 1.6.2.2, 4.3.1, 6.0, 6.1.6].

Nuttall, W. J., Noh, D. Y., Wells, B. O., and Birgeneau, R. J. (1995). *Journal of Physics: Condensed Matter*, **7**, 4337–50. [6.1.6].

O'Connor, D. J., Sexton, B. A., and Smart, R. St. C. (ed). (1992). *Surface analysis methods in materials science*, Springer Series in *Surface Science*, Vol.**23**. Springer, New York. [1.7.3].

Ohtani, H., Kao, C.-T., Van Hove, M. A., and Somorjai, G. A. (1986). *Progress in Surface Science*, **23**, 155–316. [1.7.4.3, 5.1, 5.1.2].

Okano, K. (1965). *Journal of the Physical Society of Japan*, **20**, 2085. [2.3.2.1].

Okwamoto, Y. and Bennemann, K. H. (1987). *Surface Science*, **186**, 511–22. [5.2.2.2].

Onsager, L. (1944). *Physical Review*, **65**, 117–49. [5.1.1.3, 5.1.2.1].

Opitz, O., Marx, D., Sengupta, S., Nielaba, P., and Binder, K. (1993). *Surface Science*, **297**, L122–6. [6.1.7.3].

Osen, J. W. and Fain, S. C. Jr. (1987). *Physical Review B*, **36**, 4074–7. [6.1.7.4].

Osório, R. and Koiller, B. (1985). *Physica*, **131A**, 263–77. [5.1.3.2].

Owers-Bradley, J. R., Cowan, B. P., Richards, M. G., and Thomson, A. L. (1978). *Physics Letters A*, **65**, 424–6. [4.2.2.2].

Packard, W. E. and Webb, M. B. (1988). *Surface Science*, **195**, 371–91. [4.2.1].

Palmberg, P. W. (1971). *Surface Science*, **25**, 598–608. [1.6.4, 6.2.3].

Pan, R. P., Etters, R. D., Kobashi, K., and Chandrasekharan, V. (1982). *Journal of Chemical Physics*, **77**, 1035–47. [5.2.1.1, 6.1.7.3].

Panagiotopoulos, A. Z. (1994). *International Journal of Thermophysics*, **15**, 1057–72. [5.2.2.1].

Panagiotopoulos, A. Z. (1995). In *Observations, prediction and simulation of phase transitions in complex fluids*, (ed. M. Baus, L. F. Rall, and J.-P. Ryckaert), pp.463–501. Kluwer Academic, Dordrecht. [5.2.1.6, 5.2.2.1].

Pandit, R., Schick, M., and Wortis, M. (1982). *Physical Review B*, **26**, 5112–40. [5.0, 5.1.1.1, B].

Panella, V., Suzanne, J., Hoang, P. N. M., and Girardet, C. (1994). *Journal de Physique I (France)*, **4**, 905–20. [6.3].

Parr, R. G. and Yang, W. (1989). *Density-functional theory of atoms and molecules*. Oxford. [2.1.2.2].

Parrinello, M. and Rahman, A. (1980). *Physical Review Letters*, **45**, 1196–9. [5.2.1.6].

Parsonage, N. G. and Nicholson, D. (1978). *Critical Reviews in Solid State and Materials Science*, **8**, 175–97. [5.2.1.6].

Patrykiejew, A. and Sokolowski, S. (1989). *Advances in Colloid and Interface Science*, **30**, 203–334. [4.2.2, 5.2.1.4, 6.1.7].

Patrykiejew, A., Sokolowski, S., Zientarski, T., and Binder, K. (1995). *Journal of Chemical Physics*, **102**, 8221–34. [5.2.2.1].

Peierls, R. E. (1934). *Helvetica Physica Acta*, **7**, Supp. II, 81–3. [4.2.3].

Pereyra, V., Nielaba, P., and Binder, K. (1993). *Journal of Physics: Condensed Matter*, **5**, 6631–46. [6.1.7.3].

Persson, B. N. J. (1992a). *Surface Science Reports*, **15**, 1–135. [5.1, 5.2.1.6].

Persson, B. N. J. (1992b). *Surface Science*, **269/270**, 103–12. [5.2.1.4].

Persson, B. N. J. (1993). *Physical Review B*, **48**, 18140–58. [5.2.1.4, 6.2.1].

Persson, B. N. J. and Apell, P. (1983). *Physical Review B*, **27**, 6058–65. [2.2.1.3].

Persson, B. N. J. and Zaremba, E. (1984). *Physical Review B*, **30**, 5669–79; **32**, 6916 (E) (1985). [2.2.1.2, 2.2.1.3, 2.3.2.2].

Pestak, M. W., Goldstein, R. E., Chan, M. H. W., de Bruyn, J., Balzarini, D. A., and Ashcroft, N. W. (1987). *Physical Review B*, **36**, 599–614. [5.1.1.1].

Peters, C. and Klein, M. L. (1983). *Physical Review B*, **32**, 6077–9. [6.1.7.3].

Peters, C. and Klein, M. L. (1985). *Molecular Physics*, **54**, 895–909. [6.1.7.3].

Peters, C., Morrison, J. A., and Klein, M. L. (1986). *Surface Science*, **165**, 355–74. [6.1.7.4].

Peters, G. H. and Tildesley, D. J. (1996). *Langmuir*, **12**, 1557–65. [6.1.7.4].

Pettersen, M. S., Lysek, M. J., and Goodstein, D. L. (1986). *Surface Science*, **175**, 141–56. [6.1.7.1].

Phelps, R. B., Birmingham, J. T., and Richards, P. L. (1993). *Journal of Low Temperature Physics*, **92**, 107–25. [6.2.0]

Phillips, J. M. (1984). *Physical Review B*, **29**, 5865–71. [6.1.7.1].

Phillips, J. M. (1986). *Physical Review B*, **34**, 2823–33. [5.2.1.3, 6.1.7.1].

Phillips, J. M. (1989). *Langmuir*, **5**, 571–5. [6.1.7.1].

Phillips, J. M. (1995). *Physical Review B*, **51**, 7186–91. [5.2.1.1].

Phillips, J. M. and Bruch, L. W. (1979). *Surface Science*, **81**, 109–24. [5.2.1, 5.2.1.3].

Phillips, J. M. and Bruch, L. W. (1983). *Journal of Chemical Physics*, **79**, 6282–6. [5.2.1.3].

Phillips, J. M. and Bruch, L. W. (1985). *Journal of Chemical Physics*, **83**, 3660–7. [5.2.2.1, 6.2.1].

Phillips, J. M. and Bruch, L. W. (1988). *Physical Review Letters*, **60**, 1681 (C). [6.2.1].

Phillips, J. M. and Story, T. R. (1990). *Physical Review B*, **42**, 6944–53. [6.1.7.1].

Phillips, J. M., Bruch, L. W., and Murphy, R. D. (1981). *Journal of Chemical Physics*, **75**, 5097–109. [4.3.1, 5.2.1.3, 5.2.2.1].

Phillips, J. M., Zhang, Q. M., and Larese, J. Z. (1993). *Physical Review Letters*, **71**, 2971–4. [6.1.4].

Picaud, S. and Girardet, C. (1993). *Chemical Physics Letters*, **209**, 340–6. [2.3.1.2, 6.3].

Picaud, S., Hoang, P. N. M., and Girardet, C. (1992). *Surface Science*, **278**, 339–52. [6.3].

Picaud, S., Hoang, P. N. M., Girardet, C., Meredith, A., and Stone, A. J. (1993a). *Surface Science*, **294**, 149–60. [6.3].

Picaud, S., Lakhlifi, A., and Girardet, C. (1993b). *Journal of Chemical Physics*, **98**, 3488–96. [6.3].

Picaud, S., Hoang, P. N. M., and Girardet, C. (1995). *Surface Science*, **322**, 381–90. [6.3].

Pierre, L., Guignes, H., and Lhuillier, C. (1985). *Journal of Chemical Physics*, **82**, 496–507. [6.4.2].

Pines, D. and Nozières, P. (1966). *The theory of quantum liquids*, Vol. I. Benjamin, New York. [5.2.1.2].

Poelsema, B. and Comsa, G. (1989). *Scattering of thermal energy atoms from disordered surfaces*. Springer, Berlin. [4.1.2].

Poelsema, B., Verheij, L. K., and Comsa, G. (1983). *Physical Review Letters*, **51**, 2410–3. [6.2.2, B].

Poelsema, B., Verheij, L. K., and Comsa, G. (1985). *Surface Science*, **152/153**, 851–8. [6.1.6, 6.2.2].

Pokrovskii, V. L. (1981). *Journal de Physique (Paris)*, **42**, 761–6. [5.2.2.2].

Pokrovsky, V. L. and Talapov, A. L. (1978). *Sov. Phys. JETP*, **48**, 579–82. [5.2.2.2].

Pokrovskii, V. L. and Talapov, A. L. (1980). *Sov. Phys. JETP*, **51**, 134–47. [3.4].

Pokrovsky, V. L. and Talapov, A. L. (1984). *Theory of incommensurate crystals*. Harwood Academic Publishers, London. [5.2.2.2].

Polanco, S. E. and Bretz, M. (1978). *Physical Review B*, **17**, 151–8. [6.1.1]

Pollack, G. L. (1964). *Reviews of Modern Physics*, **36**, 748–91. [5.0, 6.1.4].

Pomeau, Y. and Résibois, P. (1975). *Physics Reports*, **19**, 63–139. [4.4].

Potthoff, M., Hilgers, G., Müller, N., Heinzmann, U., Haunert, L., Braun, J., and Borstel, G. (1995). *Surface Science*, **322**, 193–206. [1.6.2.1, 2.4.1, 2.4.4, 6.2.2].

Prenzlow, C. F. and Halsey, G. D. Jr. (1957). *Journal of Physical Chemistry*, **61**, 1158–65. [6.1.4].

Press, W. (1972). *Journal of Chemical Physics*, **56**, 2597–609. [6.1.7.1].

Press, W., Hüller, A., Stiller, H., Stirling, W., and Currat, R. (1970). *Physical Review Letters*, **32**, 1354–6. [6.1.7.1].

Price, G. L. and Venables, J. A. (1976). *Surface Science*, **59**, 509–32. [1.6.2.1, 5.2.1.3, 6.1.5, 6.1.6, B].

Pullman, B., Jortner, J., Nitzan, A., and Gerber, R. B. (ed.) (1984). *Dynamics on surfaces*. Reidel, Dordrecht. [1.7.2].

Puska, M. J., Nieminen, R. M., and Manninen, M. (1981). *Physical Review B*, **24**, 3037–47. [2.1.2.2].

Qian, X. and Bretz, M. (1988). *Physical Review Letters*, **61**, 1498–500. [1.6.1.1, 6.2.1].

Quateman, J. H. and Bretz, M. (1984). *Physical Review B*, **29**, 1159–75. [6.1.7.1].

Quattrocci, L. M. and Ewing, G. E. (1992). *Chemical Physics Letters*, **197**, 308–13. [1.6.3, 6.3].

Quentel, G., Rickard, J. M., and Kern, R. (1975). *Surface Science*, **50**, 343–59. [1.6.1.1, 6.1.0].

Rabedeau, T. A. (1989). *Physical Review B*, **39**, 9643–5. [5.1.3.1, 6.1.1].

Radin, C. (1984). *Journal of Statistical Physics*, **35**, 109–17. [3.1].

Radin, C. (1986). *Communications in Mathematical Physics*, **105**, 385–90. [3.1].

Raether, H. (1988). *Surface plasmons on smooth and rough surfaces and on gratings. Springer Tracts in Modern Physics*, **111**. Springer, Heidelberg. [1.6.1.1].

Ramakrishnan, T. V. (1982). *Physical Review Letters*, **48**, 541–5. [4.3.1].

Ramakrishnan, T. V. and Yussouff, M. (1979). *Physical Review B*, **19**, 2775–94. [5.2.1.5].

Ramseyer, C. and Girardet, C. (1995). *Journal of Chemical Physics*, **103**, 5767–75. [6.3].

Ramseyer, C., Hoang, P. N. M., and Girardet, C. (1992). *Surface Science*, **265**, 293–304. [6.3].

Ramseyer, C., Hoang, P. N. M., and Girardet, C. (1994a). *Physical Review B*, **49**, 2861–8. [6.2.2].

Ramseyer, C., Girardet, C., Zeppenfeld, P., Goerge, J., Büchel, M., and Comsa, G. (1994b). *Surface Science*, **313**, 251–65. [6.2.0, 6.2.4].

Ranganathan, S. (1994). *Journal of Physics: Condensed Matter*, **6**, 1299–308. [5.2.2.1].

Ranganathan, S. and Dubey, G. S. (1993). *Journal of Physics: Condensed Matter*, **5**, 387–96. [5.2.2.1].

Rapp, R. E., de Souza, E. P., and Lerner, E. (1981). *Physical Review B*, **24**, 2196–204. [6.1.3].

Rauber, S., Klein, J. R., Cole, M. W., and Bruch, L. W. (1982). *Surface Science*, **123**, 173–8. [2.0, 2.1.2.3, 2.3.2.2, E].

Ravelo, R. and El-Batanouny, M. (1989). *Physical Review B*, **40**, 9574–89. [5.2.2.2].

Ravelo, R. and El-Batanouny, M. (1993). *Physical Review B*, **47**, 12771–84. [5.2.2.2].

Reddy, M. R. and O'Shea, S. F. (1986). *Canadian Journal of Physics*, **64**, 677–84. [5.2.2.1].

Redhead, P. A. (1962). *Vacuum*, **12**, 203–11. [1.6.3].

Rehr, J. J. and Tejwani, M. (1979). *Physical Review B*, **20**, 345–8. [4.2.2.2].

Rejto, P. A. and Andersen, H. C. (1993). *Journal of Chemical Physics*, **98**, 7636–47. [2.4.4, 5.2.2.2, 6.2.2].

Renne, M. J. (1971). *Physica*, **53**, 193–209. [2.1.2.1, 2.2.1].

Ricca, F. (ed.) (1972). *Adsorption–desorption phenomena*. Academic, New York. [1.7.2].

Richards, M. G. (1980). In *Phase transitions in surface films*, (ed. J. G. Dash and J. Ruvalds), pp.165–92. Plenum, New York. [1.6.3, 6.1.0, 6.1.1].

Roberts, R. H. and Pritchard, J. (1976). *Surface Science*, **54**, 687–91. [6.2.1].

Rollefson, R. J. (1972). *Physical Review Letters*, **29**, 410–2. [6.1.1].

Roosevelt, S. E. and Bruch, L. W. (1990). *Physical Review B*, **41**, 12236–49. [3.1, 4.2.3, 5.2.1.2, 6.1.7.3].

Ross, S. and Olivier, J. P. (1964). *On physical adsorption*. Wiley Interscience, New York. [1.4, 1.7.1].

Roth, M. and Etters, R. D. (1991). *Physical Review B*, **44**, 6581–9. [5.2.1.6].

Rouquerol, J., Partyka, J., and Rouquerol, F. (1977). *Journal of the Chemical Society. Faraday Transactions I*, **73**, 306–14. [1.6.1.2, 6.1.4].

Rovere, M., Nielaba, P., and Binder, K. (1993). *Zeitschrift für Physik B*, **90**, 215–28. [5.2.2.1].

Rowlinson, J. S. and Widom, B. (1989). *Molecular theory of capillarity*. Clarendon, Oxford. [B].

Rudzinski, W. and Everett, D. H. (1992). *Adsorption of gases on heterogeneous surfaces*. Academic, New York. [1.7.1].

Ruiz, J. C., Scoles, G., and Jónsson, H. (1986). *Chemical Physics Letters*, **129**, 139–43. [2.4.2].

Rutledge, J. E. and Taborek, P. (1992). *Physical Review Letters*, **69**, 937–40. [6.4.4].

Ryckaert, J. P., Ciccotti, G., and Berendsen, H. J. C. (1977). *Journal of Computational Physics*, **23**, 327–41. [5.2.1.6].

Saam, W. F. and Ebner, C. (1977). *Physical Review A*, **15**, 2566–8. [5.2.1.5].

Safran, S. A. (1994). *Statistical thermodynamics of surfaces, interfaces and membranes*. Addison-Wesley, New York. [1.7.3].

Sahni, V. and Levy, M. (1986). *Physical Review B*, **33**, 3869–72. [2.1.2.2].

Sahni, V., Gruenebaum, J., and Perdew, J. (1982). *Physical Review B*, **26**, 4371–7. [2.1.2.2].

Sams, J. R. Jr., Constabaris, G., and Halsey, G. D. Jr. (1960). *Journal of Physical Chemistry*, **64**, 1689–96. [1.4, 2.3.1, 4.2.2].

Sander, L. M. and Hautman, J. (1984). *Physical Review B*, **29**, 2171–4. [5.1.3.2, 5.2.1.4, 5.2.1.5].

Sandy, A. R., Mochrie, S. G. J., Zehner, D. M., Huang, K. G., and Gibbs, D. (1991). *Physical Review B*, **43**, 4667–87. [6.2.0].

Sandy, A. R., Mochrie, S. G. J., Zehner, D. M., Grübell, G., Huang, K. G., and Gibbs, D. (1992). *Physical Review Letters*, **68**, 2192–5. [6.2.0].

Schabes-Retchkiman, P. S. and Venables, J. A. (1981). *Surface Science*, **105**, 536–64. [6.1.6].

Scheffler, M., Horn, K., Bradshaw, A. M., and Kambe, K. (1979). *Surface Science*, **80**, 69–77. [6.2.3].

Schick, M. (1980). In *Phase transitions in surface films*, (ed. J. G. Dash and J. Ruvalds), pp.65–113. Plenum, New York. [5.1.1.1, 5.1.3.1, 5.2.1.4, 6.1.0, 6.1.1].

Schick, M. (1981). *Progress in Surface Science*, **11**, 245–92. [1.7.4.3, 5.1, 5.1.2, 5.1.2.2].

Schick, M. (1990). In *Liquids at interfaces* Les Houches session XLVIII, (ed. J. Charvolin, J. F. Joanny, and J. Zinn-Justin), pp.415–97, North-Holland, Amsterdam. [5.0].

Schick, M., Walker, J. S., and Wortis, M. (1977). *Physical Review B*, **16**, 2205–19. [5.1.3.1].

Schlijper, A. G. (1985). *Journal of Statistical Physics*, **40**, 1–27. [5.1.1.3].

Schmeits, M. and Lucas, A. A. (1983). *Progress in Surface Science*, **14**, 1–52. [1.7.4.2].

Schöbinger, M. and Abraham, F. F. (1985). *Physical Review B*, **31**, 4590–6. [5.2.2.2].

Schöllkopf, W. and Toennies, J. P. (1994). *Science*, **266**, 1345–8. [2.1.1.3].

Schöllkopf, W. and Toennies, J. P. (1996). *Journal of Chemical Physics*, **104**, 1155–8. [2.1.1.3].

Schönhense, G. (1986). *Applied Physics A*, **41**, 39–60. [1.6.2.1].

Schwartz, C., Cole, M. W., and Pliva, J. (1978). *Surface Science*, **75**, 1–16. [2.4.2].

Schwennicke, C., Shimmelpfennig, J., and Pfnür, H. (1993). *Physical Review B*, **48**, 8928–37. [6.3].

Scoles, G. (1980). *Annual Review of Physical Chemistry*, **31**, 81–96. [2.4.1, 2.4.4].

Scoles, G. (ed.) (1988). *Atomic and molecular beam methods*. Oxford, New York. [1.7.4.1, 6.1.0].

Scoles, G. (1990). *International Journal of Quantum Chemistry. Symposium*, **24**, 475–9. [2.3.1].

Scott, T. A. (1976). *Physics Reports C*, **27**, 89–157. [6.1.5, 6.1.7.3, A.3].

Seguin, J. L., Suzanne, J., Bienfait, M., Dash, J. G., and Venables, J. A. (1983). *Physical Review Letters*, **51**, 122–5. [6.1.4].

Selke, W. (1992). In *Phase transitions and critical phenomena*, Vol. 15, (ed. C. Domb and J. L. Lebowitz), pp.1–72. Academic, London. [5.2.2.2].

Ser, F., Larher, Y., and Gilquin, B. (1988). *Molecular Physics*, **67**, 1077–84. [6.1.4].

Severin, E. S. and Tildesley, D. J. (1980). *Molecular Physics*, **41**, 1401–18. [2.3.1].

Shaw, C. G. and Fain, S. C. Jr. (1979). *Surface Science*, **83**, 1–10. [6.1.4].

Shaw, C. G. and Fain, S. C. Jr. (1980). *Surface Science*, **91**, L1–6. [6.1.4, 6.2.1, B].

Shaw, C. G., Fain, S. C. Jr., and Chinn, M. D. (1978). *Physical Review Letters*, **41**, 955–7. [6.1.4].

Shaw, C. G., Fain, S. C. Jr., and Chinn, M. D. (1980). *Surface Science*, **97**, 128–36. [3.5]

Shechter, H., Dash, J. G., Mor, M., Ingalls, R., and Bukhspan, S. (1976). *Physical Review B*, **14**, 1876–86. [1.6.3].

Shechter, H., Brener, R., and Suzanne, J. (1982). *Surface Science*, **114**, L10–4. [1.6.3].

Shiba, H. (1979). *Journal of the Physical Society of Japan*, **46**, 1852–60. [5.2.1.2, 6.1.5].

Shiba, H. (1980). *Journal of the Physical Society of Japan*, **48**, 211–8. [3.6, 5.2.1.2, 5.2.2.2, 6.1.5].

Shirakami, T., Inaba, A., and Chihara, H. (1990). *Thermochimica Acta*, **163**, 233–40. [6.1.5].

Sholl, D. S. and Skodje, R. T. (1994). *Physica D*, **71**, 168–84. [5.2.1.4].

Shrestha, P., Alkhafaji, M. T., Lukowitz, M. K., Yang, G., and Migone, A. D. (1994). *Langmuir*, **10**, 3244–9. [1.6.1.1, 6.4.3].

Shrimpton, N. D. and Joos, B. (1990). *Canadian Journal of Physics*, **68**, 587–98. [5.2.1.2].

Shrimpton, N. D. and Steele, W. A. (1991). *Physical Review B*, **44**, 3297–303. [2.3.3.2, 5.2.1.2, 6.1.5].

Shrimpton, N. D., Bergersen, B., and Joos, B. (1984). *Physical Review B*, **29**, 6999–7002. [5.2.2.2].

Shrimpton, N. D., Bergersen, B., and Joos, B. (1986). *Physical Review B*, **34**, 7334–41. [5.2.1.2, 5.2.2.2].

Shrimpton, N. D., Joos, B., and Bergersen, B. (1988). *Physical Review B*, **38**, 2124–39. [5.2.1.2, 5.2.2.2, 6.1.5].

Shrimpton, N. D., Cole, M. W., Steele, W. A., and Chan, M. H. W. (1992). In *Surface properties of layered structures*, (ed. G. Benedek), pp.219–60. Kluwer, Dordrecht. [1.7.3].

Siddon, R. L. and Schick, M. (1974a). *Physical Review A*, **9**, 907–24. [4.2.1, 4.2.2.2, 5.2.2.1].

Siddon, R. L. and Schick, M. (1974b). *Physical Review A*, **9**, 1753–6. [4.2.2.2].

Siddon, R. L. and Schick, M. (1974c). *Journal of Low Temperature Physics*, **17**, 489–95. [4.2.2.2].

Sidoumou, M., Panella, V., and Suzanne, J. (1994). *Journal of Chemical Physics*, **101**, 6338–43. [6.3].

Sikkink, J. H., Hilhorst, H. J., and Bakker, A. F. (1985). *Physica A*, **131**, 587–98. [5.2.2.1].

Silva-Moreira, A. F., Condona, F. J., and Goodstein, D. L. (1980). *Physics Letters A*, **76**, 324–6. [2.3.3.2, 4.2.2.2, 6.1.1].

Silvera, I. F. (1980). *Reviews of Modern Physics*, **52**, 393–452. [6.1.0, 6.1.2, A.3].

Silvera, I. F. (1995). *Journal of Low Temperature Physics*, **101**, 49–58. [6.4.2].

Simmrock, K. H., Janowsky, R., and Ohnsorge, A. (1986). *Critical data of pure substances*, Chemistry Data Series, Vol. II, parts 1 and 2. DECHEMA, Frankfurt/Main. [A.1].

Sinanoğlu, O. and Pitzer, K. S. (1960). *Journal of Chemical Physics*, **32**, 1279–88. [2.3.2.2].

Sinha, S. K. (ed.) (1980). *Ordering in two dimensions*. North Holland, New York. [1.7.2].

Sinha, S. K. (1987). In *Methods of experimental physics*, Vol.**23**-part B, (ed. D. L. Price and K. Sköld), pp.1–84. Academic, San Diego. [1.6.2.2, 1.7.4.1].

Skofronick, J. G. and Toennies, J. P. (1992). In *Surface properties of layered structures*, (ed. G. Benedek), pp.151–218. Kluwer, Dordrecht. [1.6.2.1, 1.6.3].

Slater, J. C. (1963). *Quantum Theory of Molecules and Solids*, Vol. 1. McGraw-Hill, New York. [2.1, 2.1.1.1].

Smalley, M. V., Hüller, A., Thomas, R. K., and White, J. W. (1981). *Molecular Physics*, **44**, 533–55. [3.5, 6.1.7.1].

Smit, B. (1992). Journal of Chemical Physics, **96**, 8639-40. [5.2.2.1].

Smit, B. and Frenkel, D. (1991). *Journal of Chemical Physics*, **94**, 5663–8. [5.2.2.1].

Sokolowski, S. and Stecki, J. (1981). *Journal of the Chemical Society, Faraday Transactions II*, **77**, 405–17. [4.2.2].

Sokolowski, S. and Steele, W. A. (1985a). *Journal of Chemical Physics*, **82**, 3413–9. [5.2.1.5].

Sokolowski, S. and Steele, W. A. (1985b). *Journal of Chemical Physics*, **82**, 2499–506. [5.2.1.5].

Somorjai, G. A. (1981). *Chemistry in two dimensions: surfaces*. Cornell, Ithaca. [1.7.3].

Specht, E. D., Birgeneau, R. J., D'Amico, K. L., Moncton, D. E., Nagler, S. E., and Horn, P. M. (1985). *Journal de Physique (Paris)*, **46**, L561–7. [6.0].

Specht, E. D., Mak, A., Peters, C., Sutton, M., Birgeneau, R. J., D'Amico, K. L., Moncton, D. E., Nagler, S. E., and Horn, P. M. (1987). *Zeitschrift für Physik B*, **69**, 346–77. [1.6.2.2, 5.2.2.2, 6.1.0, 6.1.5, 6.2.2].

Standard, J. M. and Certain, P. R. (1985). *Journal of Chemical Physics*, **83**, 3002–8. [2.2.1.1, A.2, E].

Stanley, H. E. (1971). *Introduction to phase transitions and critical phenomena*. Clarendon Press, Oxford. [5.1.1, 5.1.1.1, 5.1.1.3, 5.1.2.1, 5.1.2.2].

Statiris, P., Lu, H. C., and Gustafsson, T. (1994). *Physical Review Letters*, **72**, 3574–7. [6.2.0].

Steele, W. A. (1973). *Surface Science*, **36**, 317–52. [2.3, 2.3.1, 3.2].

Steele, W. A. (1974). *The interaction of gases with solid surfaces*. Pergamon, Oxford. [1.7.1, 2.3, 2.3.1, 3.2, 4.2.2, 4.2.2.1, 6.3, B].

Steele, W. A. (1977). In *Chemistry and physics of solid surfaces*, (ed. R. Vanselow and S. Y. Tong), pp.139–53. CRC Press, Cleveland. [1.7.3].

Steele, W. A. (1977). *Journal de Physique (Paris)*, *Colloq.*, **38**, C4-61 to C4-68. [2.3.1].

Steele, W. A. (1978). *Journal of Physical Chemistry*, **82**, 817–21. [2.3.1].

Steele, W. A. (1993). *Chemical Reviews*, **93**, 2355–78. [1.6.3, 1.7.4.2, 6.1.7].

Steele, W. A. (1996). *Langmuir*, **12**, 145–54. [1.7.4.3, 4.2.2.1, 6.1.7].

Stephens, P. W., Heiney, P., Birgeneau, R. J., and Horn, P. M. (1979). *Physical Review Letters*, **43**, 47–51. [1.6.2.2, 5.2.1.2, 5.2.2.2, 6.0].

Stephens, P. W., Birgeneau, R. J., Majkrjak, C. F., and Shirane, G. (1983). *Physical Review B*, **28**, 452–4. [A.3].

Stephens, P. W., Heiney, P. A., Birgeneau, R. J., Horn, P. M., Moncton, D. E., and Brown, G. S. (1984). *Physical Review B*, **29**, 3512–32. [5.2.1.2, 5.2.2.2, 6.1.5].

Stewart, G. A. (1974). *Physical Review A*, **10**, 671–81. [4.2.3, 4.3.1, 5.2.2.2, G].

Stewart, G. A. (1977). *Journal de Physique (Paris)*, **38**, C4-207 to C4-213. [4.2.3, 4.3.1, 5.2.2.2].

Stone, A. J. (1996). *The theory of intermolecular forces*. Oxford University Press. [2.0].

Stoneham, A. M. (1977). *Solid State Communications*, **24**, 425–8. [2.3.2.4].

Stoner, N., Van Hove, M. A., Tong, S. Y., and Webb, M. B. (1978). *Physical Review Letters*, **40**, 243–6. [2.4.1, 2.4.4, 3.5, 6.2.2].

Stott, M. J. and Zaremba, E. (1980). *Physical Review B*, **22**, 1564–83. [2.1.2.2].

Stott, M. J. and Zaremba, E. (1982). *Canadian Journal of Physics*, **60**, 1145–51. [2.1.2.2].

Strandburg, K. J. (1988). *Reviews of Modern Physics*, **60**, 161–207. [1.7.4.3, 4.3, 4.3.1, 5.2.2.1].

Strandburg, K. J. (ed.) (1992). *Bond orientational order in condensed matter systems*. Springer, Berlin. [1.7.2, 5.2.1.6].

Strandburg, K. J., Zollweg, J. A., and Chester, G. V. (1984). *Physical Review B*, **30**, 2755–9. [5.2.2.1].

Stringari, S. and Treiner, J. (1987). *Physical Review B*, **36**, 8369–75. [5.2.1.5].

Stroscio, J. A. and Eigler, D. M. (1991). *Science*, **254**, 1319–26. [1.6.5].

Suh, S.-H., Lermer, N., and O'Shea, S. F. (1989). *Chemical Physics*, **129**, 273–84. [6.1.6].

Sullivan, N. S. and Vaissiere, J. M. (1983). *Physical Review Letters*, **51**, 658–61. [6.1.7.3].

Sullivan, T. S., Migone, A. D., and Vilches, O. E. (1985). *Surface Science*, **162**, 461–9. [6.3].

Suter, R. M., Colella, N. J., and Gangwar, R. (1985). *Physical Review B*, **31**, 627–30. [6.1.5].

Sutherland, B. (1973). *Physical Review A*, **8**, 2514–6. [5.2.2.2].

Suzanne, J. and Gay. J. M. (1997). In *Physical structure of solid surfaces*, (ed. W. N. Unertl). [1.7.4.3].

Suzanne, J., Coulomb, J. P., and Bienfait, M. (1974). *Surface Science*, **44**, 141–56. [1.6.1.1, 1.6.4, 6.1.0].

Suzanne, J., Coulomb, J. P., and Bienfait, M. (1975). *Surface Science*, **47**, 204–6. [6.1.0].

Suzanne, J., Seguin, J., Taub, H., and Biberian, J. P. (1983). *Surface Science*, **125**, 153–70. [6.1.7.4].

Suzanne, J., Panella, V., Ferry, D., and Sidoumou, M. (1993). *Surface Science*, **293**, L912–6. [6.3].

Tabony, J. (1980). *Progress in Nuclear Magnetic Resonance Spectroscopy*, **14**, 1–26. [1.7.4.1].

Taborek, P. (1990). *Physical Review Letters*, **65**, 2612(C). [6.2.0].

Taborek, P. and Goodstein, D. (1979). *Review of Scientific Instruments*, **50**, 227–30. [4.0, B].

Takada, Y. and Kohn, W. (1985). *Physical Review Letters*, **54**, 470–2. [2.1.2.2, 2.3.3.1].

Takada, Y. and Kohn, W. (1988). *Physical Review B*, **37**, 826–37. [2.1.2.2, 2.2.2.2, 2.2.2.4].

Tang, K. T. and Toennies, J. P. (1984). *Journal of Chemical Physics*, **80**, 3726–41. [2.1.1.2].

Tang, K. T. and Toennies, J. P. (1986). *Zeitschrift für Physik D*, **1**, 91–101. [2.1.1.2, A.2].

Tang, K. T. and Toennies, J. P. (1992). *Surface Science*, **279**, L203–6. [2.1.1.2, 2.1.2].

Tang, K. T., Norbeck, J. M., and Certain, P. R. (1976). *Journal of Chemical Physics*, **64**, 3063–74. [E].

Tang, K. T., Toennies, J. P., and Yiu, C. L. (1995). *Physical Review Letters*, **74**, 1546–9. [2.1.1.2, 2.1.1.3].

Taub, H. (1980). In *Vibrational spectroscopies of adsorbed species*, (ed. A. T. Bell and M. L. Hair). American Chemical Society, Washington, DC. [1.6.2.2, 6.1.7.4].

Taub, H. (1988). In *The time domain in surface and structural dynamics*, (ed. G. J. Long and F. Grandjean), pp.467–97. Reidel, Dordrecht. [1.6.2.2, 6.1.7.4].

Taub, H., Carneiro, K., Kjems, J. K., Passell, L., and McTague, J. P. (1977). *Physical Review B*, **16**, 4551–68. [1.6.2.2, 1.6.3, 6.1.0, 6.1.4].

Taub, H., Trott, G. J., Hansen, F. Y., Danner, H. R., Coulomb, J. P. , Biberian, J. P., Suzanne, J., and Thomy, A. (1980). In *Ordering in two dimensions*, (ed. S. K. Sinha), pp.91–7. North Holland, New York. [6.1.7.4].

Taub, H., Torzo, G., Lauter, H. J., and Fain, S. C. Jr. (ed.) (1991). *Phase transitions in surface films 2*. Plenum, New York. [1.7.2].

Tejwani, M. J., Ferreira, O., and Vilches, O. E. (1980). *Physical Review Letters*, **44**, 152–5. [5.1.2.2].

Terlain, A. and Larher, Y. (1980). *Surface Science*, **93**, 64–70. [6.1.7.3].

Terlain, A. and Larher, Y. (1983). *Surface Science*, **125**, 304–11. [6.1.7.2].

Tessier, C. (1984). Unpublished These d'Etat–University of Nancy. Paris. Rapport CEA-R-5250. Centres d'Etudes Nucléaires de Saclay, France. [6.1.6].

Tessier, C. and Larher, Y. (1980). In *Ordering in two dimensions*, (ed. S. K. Sinha), pp.163–8. North Holland, New York. [6.1.6].

Tessier, C., Terlain, A., and Larher, Y. (1982). *Physica*, **113A**, 286–92. [A.3].

Theodorou, G. and Rice, T. M. (1978). *Physical Review B*, **18**, 2840–56. [3.3, 5.2.1.2].

Thiel, P. A., Hoffmann, F. M., and Weinberg, W. H. (1981). *Journal of Chemical Physics*, **75**, 5556–72. [1.6.3].

Thomas, R. K. (1982). *Progress in Solid State Chemistry*, **14**, 1–93. [1.7.4.1].

Thomy, A. and Duval, X. (1970). *Journal de Chimie Physique*, **67**, 1101–10. [1.3, 6.1].

Thomy, A., Duval, X., and Régnier, J. (1969). *Comptes Rendus de l'Academie des Sciences (Paris)*, **C268**, 1416–8. [6.1.3].

Thomy, A., Duval, X., and Régnier, J. (1981). *Surface Science Reports*, 1, 1–38. [1.3, 1.6.1.1, 4.2.0, 6.1.0].

Thorel, P., Coulomb, J. P., and Bienfait, M. (1982). *Surface Science*, 114, L43–7. [6.1.7.1].

Tiby, C., Wiechert, H., and Lauter, H. J. (1982). *Surface Science*, 119, 21–34. [3.3, 6.1.3].

Tiersten, S. C., Reinecke, T. L., and Ying, S. C. (1989). *Physical Review B*, **39**, 12575–84. [2.3.2.4].

Tillborg, H., Nilsson, A., Hernnäs, B., Mårtensson, N., and Palmer, R. E. (1993). *Surface Science*, 295, 1–12. [6.1.7].

Toennies, J. P. and Vollmer, R. (1989). *Physical Review B*, 40, 3495–8. [5.2.1.2].

Toney, M. F. and Fain, S. C. Jr. (1984). *Physical Review B*, 30, 1115–8. [6.1.7.3].

Toney, M. F. and Fain, S. C. Jr. (1987). *Physical Review B*, 36, 1248–58. [6.1.7.3].

Toney, M. F., Diehl, R. D., and Fain, S. C. Jr. (1983). *Physical Review B*, 27, 6413–7. [6.1.7.3].

Toxvaerd, S. (1978). *Journal of Chemical Physics*, 69, 4750–2. [4.3.1, 5.2.2.1].

Toxvaerd, S. (1979). *Physical Review Letters*, 43, 529–31. [4.4].

Toxvaerd, S. (1980). *Physical Review Letters*, 44, 1002–4. [4.3, 4.3.1].

Toxvaerd, S. (1983). *Physical Review Letters*, 51, 1971–4. [5.2.1.6, 5.2.2.1].

Trabelsi, M. and Coulomb, J. P. (1992). *Surface Science*, 272, 352–7. [6.3].

Trabelsi, M. and Larher, Y. (1992). *Surface Science*, 275, L631–5. [6.1.7.4].

Tsang, T. (1979). *Physical Review B*, 20, 3497–501. [5.2.1.3, 6.1.4].

Tsien, F. and Valleau, J. P. (1974). *Molecular Physics*, 27, 177–83. [5.2.2.1].

Tsong, T. T. (1983). *Field ion microscopy*. Cambridge University Press. [1.6.5].

Tsong, T. T. (1990). *Atom-probe field ion microscopy*. Cambridge University Press. [1.6.5].

Tully, J. C. (1981). *Surface Science*, 111, 461–78. [1.6.3, 4.1.2].

Tully, J. C. (1994). *Surface Science*, **299/300**, 667–77. [1.6.3, 4.1.2].

Udink, C. and van der Elsken, J. (1987). *Physical Review B*, 35, 279–83. [5.2.1.6].

Unertl, W. N. (1997). *Physical structure of solid surfaces*, Vol.1 in the Handbook of Surface Science, Series editors S. Holloway and N. V. Richardson. [1.7.3].

Unguris, J., Bruch, L. W., Moog, E. R., and Webb, M. B. (1979). *Surface Science*, 87, 415–36. [2.4.4, 4.2.1, 6.1.4, 6.2.0, 6.2.1, B].

Unguris, J., Bruch, L. W., Moog, E. R., and Webb, M. B. (1981). *Surface Science*, 109, 522–56. [2.4.4, 5.1.1.2, 6.2.0, 6.2.1, B].

Unguris, J., Bruch, L. W., Webb, M. B., and Phillips, J. M. (1982). *Surface Science*, 114, 219–39. [5.2.2.1, 6.2.1].

Van der Hoef, M. A. and Frenkel, D. (1991). *Physical Review Letters*, 66, 1591–4. [4.4].

Van der Merwe, J. H. (1982a). *Philosophical Magazine A*, 45, 127–43. [3.6].

Van der Merwe, J. H. (1982b). *Philosophical Magazine A*, 45, 145–57. [3.6].

Van der Merwe, J. H. (1982c). *Philosophical Magazine A*, **45**, 159–70. [3.6].

Van Gunsteren, W. F. and Berendsen, H. J. C. (1977). *Molecular Physics*, **34**, 1311–27. [5.2.1.6].

Van Himbergen, J. E. and Silbey, R. (1977). *Solid State Communications*, **23**, 623–7. [2.1.2.2].

Van Kampen, N. G. and Lodder, J. J. (1984). *American Journal of Physics*, **52**, 419–24. [4.2.3].

Van Kampen, N. G., Nijboer, B. R. A., and Schram, K. (1968). *Physics Letters A*, **26**, 307–8. [2.1.2.1, 2.2.1].

Van Rensburg, E. J. J. (1993). *Journal of Physics A*, **26**, 4805–18. [5.2.1.4].

Van Swol, F., Woodcock, L. V., and Cape, J. N. (1980). *Journal of Chemical Physics*, **73**, 913–22. [4.3.1, 5.2.2.1].

Venables, J. A. and Schabes-Retchkiman, P. S. (1978). *Surface Science*, **71**, 27–41. [6.1.5].

Venables, J. A., Spiller, G. D. T., and Hanbücken, M. (1984a). *Reports on Progress in Physics*, **47**, 399–459. [6.1.4, 6.1.7.3].

Venables, J. A., Seguin, J. L., Suzanne, J., and Bienfait, M. (1984b). *Surface Science*, **145**, 345–63. [6.1.4, 6.1.7.3].

Ventevogel, W. J. (1978). *Physica*, **92A**, 343–61. [3.1]

Verlet, L. (1967). *Physical Review*, **159**, 98–103. [5.2.2.1].

Vernov, A. and Steele, W. A. (1992). *Langmuir*, **8**, 155–9. [2.3.3.2, 6.1.7.3].

Vidali, G. and Cole, M. W. (1980). *Physical Review B*, **22**, 4661–5; **23**, (1981) 5649(E). [4.2.2.2].

Vidali, G. and Cole, M. W. (1981). *Surface Science*, **110**, 10–8. [2.0, E].

Vidali, G., Cole, M. W., Rauber, S., and Klein, J. R. (1983a). *Chemical Physics Letters*, **95**, 213–6. [2.4.1].

Vidali, G., Cole, M. W., and Klein, J. R. (1983b). *Physical Review B*, **28**, 3064–73. [2.4.1].

Vidali, G., Ihm, G., Kim, H.-Y., and Cole, M. W. (1991). *Surface Science Reports*, **12**, 133–81. [1.1, 1.7.4.2, 2.0, 2.1.2.3, 2.3, 2.4.1, 2.4.2, 2.4.3].

Vilches, O. E. (1980). *Annual Review of Physical Chemistry*, **31**, 463–90. [1.7.4.3, 5.1].

Vilches, O. E. (1992). *Journal of Low Temperature Physics*, **89**, 267–76. [6.1.2, 6.3].

Vilches, O. E., Liu, Y.-M., Ebey, P. S., and Liu, F.-C. (1994). *Physica B*, **194-196**, 665–6. [6.1.2].

Villain, J. (1980a). In *Ordering in strongly fluctuating condensed matter systems*, (ed. T. Riste), pp.221–60. Plenum, New York. [5.2.2.2].

Villain, J. (1980b). *Surface Science*, **97**, 219–42. [3.4, 5.2.2.2].

Villain, J. (1980c). In *Ordering in two dimensions*, (ed. S. K. Sinha), pp.123–9. North Holland, New York. [3.4, 5.2.2.2].

Villain, J. and Gordon, M. B. (1983). *Surface Science*, **125**, 1–50. [5.2.2.2].

Vives, E. and Lindgård, P. A. (1993). *Physical Review B*, **47**, 7431–45. [5.2.1.2].

Vora, P., Sinha, S. K., and Crawford, R. K. (1979). *Physical Review Letters*, **43**, 704–8. [6.1.7.1].

Vosko, S. H., Wilk, L., and Nusair, M. (1980). *Canadian Journal of Physics*, **58**, 1200–11. [2.1.2.2].

Wagner, M. and Ceperley, D. M. (1992). *Journal of Low Temperature Physics*, **89**, 581–4. [4.3.2, 6.4.2].

Wagner, M. and Ceperley, D. M. (1994). *Journal of Low Temperature Physics*, **94**, 185–217. [4.3.2, 6.4.2, B].

Wallace, D. C. (1972). *Thermodynamics of crystals*. Wiley, New York. [4.3.1, 5.2.2.2].

Walraven, J. T. M. (1992). In *Fundamental systems in quantum optics*, Les Houches Session LIII, (ed. J. Dalibard, J.-M. Raimond, and J. Zinn-Justin), pp.485–544. Elsevier, Amsterdam. [1.1, 6.4.2].

Walraven, J. T. M. and Hijmans, T. W. (1994). *Physica B*, **197**, 417–25. [6.4.2].

Wandelt, K. and Hulse, J. E. (1984). *Journal of Chemical Physics*, **80**, 1340–52. [1.6.3, 6.2.0, 6.2.3].

Wang, C. and Gomer, R. (1980). *Surface Science*, **91**, 533–50. [1.6.3, 6.2.4].

Wang, S. C., Senbetu, L., and Woo, C.-W. (1980). *Journal of Low Temperature Physics*, **41**, 611–28. [2.3.1].

Wang, R., Taub, H., Shechter, H., Brener, R., Suzanne, J., and Hansen, F. Y. (1983). *Physical Review B*, **27**, 5864–7. [1.6.3].

Wang, R., Taub, H., Lauter, H. J., Biberian, J. P., and Suzanne, J. (1985). *Journal of Chemical Physics*, **82**, 3465–9. [6.1.7.4].

Wang, S.-K., Newton, J. C., Wang, R., Taub, H., Dennison, J. R., and Shechter, H. (1989). *Physical Review B*, **39**, 10331–41. [6.1.7.3].

Watanabe, F. and Ehrlich, G. (1992). *Journal of Chemical Physics*, **96**, 3191–9. [2.3.2.4].

Weaver, J. H., Krafka, C., Lynch, D. W., and Koch, E. E. (1981). *Optical Properties of Metals*, Physics Data No. 18, (Fachinformationszentrum Karlsuhe). [E].

Webb, M. B. and Bruch, L. W. (1982). In *Interfacial aspects of phase transformations*, (ed. B. Mutaftschiev), pp.365–409. Reidel, Dordrecht. [B].

Webb, M. B. and Lagally, M. G. (1973). In *Solid State Physics*, **28** (ed. H. Ehrenreich, F. Seitz, and D. Turnbull), pp.301–405. Academic, New York. [1.6.2.1, 1.7.4.3, 3.5].

Wei, M. S. and Bruch, L. W. (1981). *Journal of Chemical Physics*, **75**, 4130–41. [5.2.1.3, 5.2.2.1, B].

Weinberg, W. H. (1983). *Annual Review of Physical Chemistry*, **34**, 217–43. [1.7.4.3, 5.1, 5.1.2].

Weiss, H. (1995). *Surface Science*, **331-333**, 1453–9. [6.3].

Weiss, P. S. and Eigler, D. M. (1992). *Physical Review Letters*, **69**, 2240–3. [1.6, 6.2.2].

Weling, F. and Griffin, A. (1981). *Physical Review Letters*, **46**, 353–6. [4.2.3].

Weling, F. and Griffin, A. (1982). *Physical Review B*, **25**, 2450–62. [4.2.3].

Werthamer, N. R. (1976). Chap.5 In *Rare gas solids*, Vol. I, (ed. M. L. Klein and J. A. Venables), pp.265–300. Academic, New York. [5.1.2].

Wheeler, J. C. and Gordon, R. G. (1969). *Journal of Chemical Physics*, **51**, 5566–83. [5.2.1.2].

Whitehouse, D. B. and Buckingham, A. D. (1993). *Journal of the Chemical Society. Faraday Transactions II*, **89**, 1909–13. [2.3.2.2, 6.1.0].

Whitlock, P. A., Chester, G. V., and Kalos, M. H. (1988). *Physical Review B*, **38**, 2418–25. [4.2.2.2, 5.2.1, 5.2.2.1, 6.1.1].

Wiechert, H. (1991a). *Physica B*, **169**, 144–52. [4.2.2.1, 6.1.0, 6.1.1, 6.1.2].

Wiechert, H. (1991b). In *Excitations in 2-D and 3-D quantum fluids*, (ed. A. F. G. Wyatt and H. J. Lauter), pp.499–510. Plenum, New York. [6.1.2].

Wiechert, H. and Freimuth, H. (1984). In *Proceedings of the 17th international conference on low temperature physics*, (ed. U. Eckern, A. Schmid, W. Weber, and H. Wühl). North–Holland, Amsterdam. pp. 1015-6. [6.1.1].

Wiechert, H. and Arlt, St.-A. (1993). *Physical Review Letters*, **71**, 2090–3. [6.1.7.3].

Wilks, J. (1967). *The properties of liquid and solid helium*. Oxford University Press. [6.1.1, A.3].

Williams, F. I. B. (1993). *Journal de Physique (Paris) IV*, **3**, C2, pp. 3 to 8. [4.3.1].

Wilson, K. G. (1983). *Reviews of Modern Physics*, **55**, 583–600. [5.1, 5.1.3].

Wolf, D. E. and Griffiths, R. B. (1985). *Physical Review B*, **32**, 3194–202. [B].

Woo, C-.W. (1976). In *The physics of liquid and solid helium*, Part I, (ed. K. H. Bennemann and J. B. Ketterson), pp.349–501. Wiley, New York. [6.1.1].

Wood, E. A. (1964). *Journal of Applied Physics*, **35**, 1306–12. [3.1].

Woodruff, D. P. and Delchar, T. A. (1986). *Modern techniques of surface science*. Cambridge, New York. [1.7.3].

Woolley, H. W., Scott, R. B., and Brickwedde, F. G. (1948). *Journal of Research of the National Bureau of Standards (U. S.)*, **41**, 379–475. [A.3].

Wyatt, A. F. G., Klier, J., and Stefanyi P. (1995). *Physical Review Letters*, **74**, 1151–4. [4.3.2].

Xia, B. and Fain, S. C. Jr. (1994). *Physical Review B*, **50**, 14576–9. [6.1.2].

Yang, C. N. (1952). *Physical Review*, **85**, 808–16. [5.1.2.1].

Yanuka, M., Yinnon, A. T., Gerber, R. B., Zeppenfeld, P., Kern, K., Becher, U., and Comsa, G. (1993). *Journal of Chemical Physics*, **99**, 8280–9. [6.4.1].

Yeomans, J. M. (1992). *Statistical mechanics of phase transitions*. Oxford. [5.1.1.1, 5.1.2.2].

Ying, S. C. (1971). *Physical Review B*, **3**, 4160–71. [5.2.2.2].

You, H. and Fain, S. C. Jr. (1985a). *Faraday Discussions of the Chemical Scociety (London)*, **80**, 159–70. [6.1.7.3].

You, H. and Fain, S. C. Jr. (1985b). *Surface Science*, **151**, 361–73. [6.1.7.3].

You, H. and Fain, S. C. Jr. (1986). *Physical Review B*, **34**, 2840–51. [6.1.7.3].

You, H., Fain, S. C. Jr., Satija, S., and Passell, L. (1986). *Physical Review Letters*, **56**, 244–7. [6.1.7.3].

Youn, H. S. and Hess, G. B. (1990). *Physical Review Letters*, **64**, 918–21. [6.1.4].

Youn, H. S., Meng, X. F., and Hess, G. B. (1993). *Physical Review B*, **48**, 14556–76. [6.1.4].

Young, A. P. (1979). *Physical Review B*, **19**, 1855–66. [4.3].

Young, D. M. and Crowell, A. D. (1962). *Physical adsorption of gases*. Butterworths, London. [1.7.1].

Yu, I. A., Doyle, J. M., Sandberg, J. C., Cesar, C. L., Kleppner, D., and Greytak, T. J. (1993). *Physical Review Letters*, **71**, 1589–92. [1.1, 4.1.2].

Zabel, H. and Robinson, I. K. (ed.) (1992). *Surface X-ray and neutron scattering*. Springer, Berlin. [1.6.2.2, 1.7.2].

Zangwill, A. (1988). *Physics at surfaces*. Cambridge, New York. [1.7.3].

Zaremba, E. (1976). *Physics Letters A*, **57**, 156–8. [2.3.2.3].

Zaremba, E. and Kohn, W. (1976). *Physical Review B*, **13**, 2270–85. [2.1.2.1, 2.2.1, 2.2.1.2, 2.2.1.3].

Zaremba, E. and Kohn, W. (1977). *Physical Review B*, **15**, 1769–81. [2.0, 2.2.2, 2.2.2.1, 2.2.2.2, 2.2.2.3, 2.2.2.4, D].

Zeiss, G. D. and Meath, W. J. (1977). *Molecular Physics*, **33**, 1155–76. [E.1].

Zeppenfeld, P., Becher, U., Kern, K., David, R., and Comsa, G. (1992a). *Physical Review B*, **45**, 5179–86. [5.2.2.2, 6.2.0, 6.2.2].

Zeppenfeld, P., Comsa, G., and Barker, J. A. (1992b). *Physical Review B*, **46**, 8806–10. [2.3.3.2, 6.2.2].

Zeppenfeld, P., Kern, K., David, R., Becher, U., and Comsa, G. (1993a). *Surface Science*, **285**, L461–7. [6.4.1].

Zeppenfeld, P., Becher, U., Kern, K., David, R., and Comsa, G. (1993b). *Surface Science*, **297**, L141–7. [6.4.1].

Zeppenfeld, P., Horch, S., and Comsa, G. (1994a). *Physical Review Letters*, **73**, 1259–62. [1.6.5, 6.2.2].

Zeppenfeld, P., Goerge, J., Büchel, M., David, R., and Comsa, G. (1994b). *Surface Science*, **318**, L1187–92. [6.2.4].

Zeppenfeld, P., Büchel, M., David, R., Comsa, G., Ramseyer, C., and Girardet, C. (1994c). *Physical Review B*, **50**, 14667–70. [6.2.4].

Zerrouk, T. E. A., Hamichi, M., Pilkington, J. D. H., and Venables, J. A. (1994). *Physical Review B*, **50**, 8946–9. [4.3.1, 6.1.6].

Zhang, Q. M., Kim, H. K., and Chan, M. H. W. (1986). *Physical Review B*, **34**, 8050–63. [5.1.2.2].

Zhang, S. and Migone, A. D. (1989). *Surface Science*, **222**, 31–7. [6.1.7.4].

Zhou, J. B., Lu, H. C., Gustafsson, T., and Häberle, P. (1994). *Surface Science*, **302**, 350–62. [6.3].

Zimmerli, G., Mistura, G., and Chan, M. H. W. (1992). *Physical Review Letters*, **68**, 60–3. [6.1.1].

Zollweg, J. A. and Chester, G. V. (1992). *Physical Review B*, **46**, 11186–9. [4.3.1, 5.2.1.6, 5.2.2.1].

Zollweg, J. A., Chester, G. C., and Leung, P. W. (1989). *Physical Review B*, **39**, 9518–30. [5.2.1.6].

Zucker, I. J. (1974). *Journal of Mathematical Physics*, **15**, 187. [5.2.2.1].

AUTHOR INDEX

Numbers in italic indicate entries in the list of references rather than in the text.

SUBJECT INDEX

In the following: n means footnote, f means figure.